SYMBOLS, COMPUTATION, AND INTENTIONALITY

Symbols, Computation, and Intentionality

A Critique of the Computational Theory of Mind

Steven W. Horst

Also by Steven Horst:

Laws, Mind, and Free Will. MIT Press, 2011

Beyond Reduction: Philosophy of Mind and Post-Reductionist Philosophy of Science. Oxford University Press, 2007

Please address all inquiries about rights and permissions to:
Steven Horst
Department of Philosophy
Wesleyan University
Middletown, CT 06459
      shorst@wesleyan.edu

Printed by CreateSpace, Charleston, SC, USA
First edition published in hardcover by University of
California Press, Berkeley and Los Angeles, 1996

Library of Congress Control Number: 95037156

ISBN-13: 978-1466348134
ISBN-10: 1466348135

For my parents, William and Erin Horst

From this circumstance alone, that a controversy has been long kept on foot, and remains still undecided, we may presume that there is some ambiguity in the expression, and that the disputants affix different ideas to the terms employed in the controversy.

—*David Hume*, An Enquiry Concerning Human Understanding, *Section VIII, Part 1*

Contents

PREFACE

This book was a long time in the making. There are parts of it that date back to about 1987 and other parts that are very recent indeed. It started out as an attempt to bring my own peculiar philosophical background (which is an unusual one in cognitive science circles) into contact with what was at that time (and to some extent still is) the mainline view of the mind in analytic philosophy of mind: the Computational Theory of Mind (CTM for short). I came to the study of cognitive science with three kinds of relevant background, each of which is at least a little bit off center with respect to the contemporary scene in the philosophy of mind. Perhaps the most prosaic of these is that I worked much of my way through graduate school teaching courses in computer programming, software design, and artificial intelligence. It would be a most heinous exaggeration if I were to describe myself as either a computer scientist or a hacker, but I learned some theory, did some programming, and made my way through most of the three-volume History of Artificial Intelligence with a class of undergraduate students. I knew computers in theory and in practice before I began to think about them as a philosopher.

This, however, was not my first exposure to computer models of the mind. I had studied as an undergraduate with Stephen Grossberg of Boston University, one of the few people doing continuous work in neural network models from the 1960s up until the present, even through the two decades when it was not a particularly popular thing to be doing. So whereas most people in the philosophy of cognitive science came to cognitive science by way of the symbol-processing paradigm embracing

Turing, Minsky, Newell, Colby, Winograd, and Marr (to name but a few), I cut my teeth on neural network models. In a way, I have been in exactly the opposite position of most philosophers doing cognitive science in recent years: whereas they have had to learn about the "new" neural network paradigm (which is in fact as old as the symbol-processing paradigm), I had to do exactly the opposite in the late 1980s.

The third element in my background, of course, was my philosophical training. My first philosophical love was speech act theory, on which I wrote an undergraduate thesis in 1981 with Bruce Fraser. The major philosophical writers on the subject at that juncture were Grice, Austin, Strawson, Vendler, Searle, Bach, and Harnish. With the possible exception of the last two, these philosophers practiced their trade in a style markedly different from that commonly found in cognitive science today. At the same time, I was beginning to read some of the works of Edmund Husserl, which was to have a profound influence on how I came to view philosophy. In spite of praise from people like Chisholm, Sellars, Drey-fus, Føllesdal, Haugeland, and (more recently) Putnam, Husserl is not adequately appreciated among American analytic philosophers. Much of what has transpired since his day in philosophy of language is already present in the first "Logical Investigation," and no one was more keenly aware than he of the difficulties and pitfalls of coming to a philosophical understanding of the mind. Husserl's focus on the centrality of intentionality made what seems to have been a permanent impression on me. Somewhere in 1982 I became convinced that the study of speech acts could not progress further without a study of intentionality. (It was gratifying to see a year later that John Searle had come to a similar conclusion.) That focus became central to my philosophical thinking for the next ten years, and it still occupies an important (though no longer central) place for me in the scheme of important philosophical problems.

As a graduate student, I worked with Kenneth Sayre at Notre Dame, one of the first philosophers to write about artificial intelligence in the early 1960s, and a longtime proponent of an alternative vision of the mind centering around the Mathematical Theory of Communication articulated by Shannon and Weaver. It was, in fact, only after I started working with Sayre that I began to read what most people consider "mainstream" artificial intelligence and philosophy of cognitive science, so the symbol-processing paradigm was actually the third paradigm I was exposed to in cognitive science. Along the way in my philosophical studies I felt some influence from the writings of Plato, Aristotle, Kant, and Wittgenstein.

In short, I came to the contemporary scene in cognitive science with a very different list of philosophical and scientific heroes from those of most of my colleagues in the field; and much of what I found in the "mainstream" initially struck me as eminently wrongheaded. But it is one thing to think that something is wrongheaded; it is of course quite another to understand why people would believe it and to identify just where you think the problem lies. This book is in large measure the product of a long process of trying to do two things: first, to understand the project from the inside, as it were, in terms that its own advocates would embrace; and, second, to articulate what seem to me its major flaws in a fashion that does not depend too much upon an alternative philosophical viewpoint and which might be accessible to someone who does not share my own philosophical leanings.

As a result, the first two chapters of this book attempt to lay out computationalism in its historical context and to explain to the reader why one might very sensibly think that it is offering some tempting philosophical fruit. At the same time, I have tried to emphasize elements in the historical context and connections with other strands of philosophical psychology that seem important yet are often passed over by those who consider themselves to be within the computationalist camp. I hope that these chapters will serve as a good introduction to the computational theory for a wide philosophical audience. I suspect that they may also prove useful for the initiate who wishes to read the critical sections of the book, as they attempt to lay out computationalism with more exactitude than is normally done and with a minimum of rhetoric. In a sense, the moral of the book is just this: if you are not extremely careful about how you use words like 'computer', 'symbol', 'syntax', and 'meaning', you are likely to stumble into some pernicious confusions about computation and the mind and to be tempted by some subtly fallacious arguments that seem to deliver philosophical results but in fact mislead.

The rest of the book grew gradually. It started as a purely critical project of debunking claims that CTM provides an account of the inten-tionality of the mental and a "vindication" of realism about mental states. Once I had satisfied myself that I had proven my case there to my own satisfaction, I began to be more interested in what could be said in a positive way about the importance of computational psychology as a way of understanding the mind, and how far away from the views ar-ticulated by writers like Jerry Fodor and Zenon Pylyshyn one would have to go in order to do so. Unlike some critics of CTM, I do not believe that empirical research in cognitive science stands or falls with the *philo-*

sophical claims of CTM. The several chapters of the book that explore alternative ways of looking at computation and the mind were first drafted at the National Endowment for the Humanities Summer Seminar on Mental Representation held at the University of Arizona in the Summer of 1991. Rob Cummins, who directed that institute, has gone far beyond the call of duty by reading three separate drafts of the book in the three years since that time. Chapter 6, which contains some of the most provocative material in the entire book, was written after the rest of the manuscript to respond to criticisms that Rob raised about an earlier draft. Rob has my undying gratitude for his responses along the way, and not least of all for occasionally admitting that I had convinced him about something. Thanks to him and to the NEH, which helped finance that extremely productive summer.

Major thanks also go to Ken Sayre, who forced me to take CTM seriously and on its own terms, and forced me to a higher standard of clarity and exactitude than I might otherwise have attained. Thanks to Ken also for the useful way he engaged in the process of helping me hammer out views that were strongly related to his own yet contrary to his own formulations.

Jay Garfield of Hampshire College read and commented upon the entire manuscript very late in the game, made some very supportive comments, and also made a number of very important suggestions that have ultimately made the final product a much better and more readable book than it otherwise might have been. The chapter on naturalization in particular is much expanded as a result of Jay's (deservedly) pitiless attack upon a former incarnation of the same. (I fear it still does not meet with his entire approval—the credit for its expansion is his, any residual faults are my own.)

Richard DeWitt of Fairfield University also went beyond the call of duty in reading multiple drafts of four or five chapters, and has been supportive of the project via numerous e-mail exchanges since we met at Cummins's NEH seminar. Likewise, my Wesleyan colleague Sanford Shieh made some very helpful suggestions on the chapters in Part II, and probably saved me from some grave embarrassments in my use of terms that had technical meanings in logic of which I was blessedly unaware.

Many other people read or commented on all of part of the manuscript along the way. All of the following people were at least so kind as to agree to read parts of the manuscript for me somewhere along the line. Many made important contributions to the present form of the work by their comments and criticisms: Michael Anderson, Robert Audi, Lynne

Rudder Baker, David Burrell, Hubert Dreyfus, Aaron Edidin, Brian Fay, Tim Fischer, Pat Franken, Bruce Fraser, Heather Gert, Ruth Ginzberg, Victor Gourevitch, Robert Losonsky, Vaughn McKim, Chris Menzel, Mark Moes, Hans Müller, Shelly Park, Bill Ramsey, Bill Robinson, and Joe Rouse.

Finally, I should like to thank the many people whose love and friendship over the years have made it possible for me to pursue something as demanding as a book in philosophy. In particular, I wish to thank my parents, who have provided ample support throughout my life, and who have been in my corner for many years while I worked on a project whose merits they could only take on faith. Plato somewhere describes intellectual creations as a kind of progeny. I hope that they will be pleased with their first grandchild.

Middletown, Connecticut
October 1994

Preface to the Paperback Edition (2011)

The original printing of this book by the University of California Press came in 1996. Like most academic monographs, or at least those by first-time authors, it was released initially in hardcover, and at a price that was a bit high for most individual readers. It was my expectation that a paperback edition would be released a year or two later. But I did not foresee that the Press would be discontinuing new releases in its philosophy line, and that, once the copies of the first printing were gone, it would go out of print. Eventually I realized that this presented me with the opportunity to ask to have the rights returned to me and to prepare a paperback edition to be sold at a reasonable cost, and the Press was very cooperative in facilitating this.

Unfortunately, the original proofs were from an age when books were formatted for linotype and not as PDF files. As a result, there were no print-ready files to be had. I have gone through the book, page by page, to assure that the pagination is the same in the paperback edition as in the hardcover, but producing a line-by-line replica I decided to be pointless. There are two exceptions to the parallel pagination. The first is this Preface to the Paperback Edition, which of course was not a part of the hardcover, which adds to the roman-numeraled frontmatter. The second is the Appendix and backmatter. The original Appendix was printed in a smaller font than the rest of the text, making its already rather technical content a difficult read. I have set it in the same font and size as the rest of the text, and so it is longer in this paperback edition, and of course the sections that follow, Notes, Bibliography, and Index, are numbered differently as well. In neither case should this interfere with consistent citations, unless they are citations of material in the Notes. (Perhaps readers will call these the A and B editions.)

The criticisms I have offered of the Computational Theory of Mind are ones I still stand by, though in subsequent publications about it I have placed less focus on the criticisms. I will note one matter on which it would be very easy for the reader to misunderstand my position. In the central chapters, I argue that the notion of symbolic representation cannot be made to do a particular sort of work in philosophy of mind – namely, to explain the intentional and semantic properties of mental states. In retrospect, I wish that I had said a bit more on the topic of what *constructive* roles *other* notions of representation have played or might play in the cognitive sciences. The other principal difference between my philosophical inclinations then and now is that, when I wrote this book, I was much more trusting of philosophical intuitions about necessity and possibility. Readers interested in how my mind has changed on this subject may wish to pick up a copy of my second book, *Beyond Reduction: Philosophy of Mind and Post-Reductionist Philosophy of Science* (Oxford University Press, 2007).

Middletown, CT
August 2011

INTRODUCTION

There are few things that hold a greater fascination for us human beings than the project of explaining ourselves to ourselves. On the one hand, the human mind is the part of us that makes us who we are as individuals. It is also our minds that set us as a species apart from brute matter and from other members of the animal kingdom. The centrality of self-knowledge in Western philosophy goes back at least to Socrates' adherence to the motto inscribed over the temple of Apollo at Delphi: "Know thyself." On the other hand, the mind has proved one of the most intractable mysteries for modern science. Indeed, modern science, conceived as a discipline concerned with the lawful causal interactions of material bodies, has been hard pressed to accommodate the world of thoughts and concepts and images that seem essential to any treatment of the mind. One might even go so far as to say that the *central problem of modern philosophy* has been one of somehow closing the gap between two apparently incommensurable discourses: a discourse about our minds that speaks of ideas and images, and a discourse about the world of nature that speaks of causal relations between bodies in motion.

Since Alan Turing's introduction of the notion of a computing machine in the late 1930s, there has been a growing interest in a new paradigm for understanding the mind: a paradigm that treats the mind as a digital computer. The arrival of machine computation upon our intellectual landscape has had a profound and widespread impact upon research in the many disciplines that are concerned with the study of the mind. In fields such as cognitive psychology, ethology, linguistics, the philosophy of mind, and cognitive neuroscience, the computational view of the mind has become a mainstream view—perhaps even the dominant view in recent years. Even though there is no monolithic consensus about how the computer paradigm is to be applied to the mind, and even though there are many researchers in all of the disciplines that study the mind who are working out of other traditions, it is by now generally agreed that the computational approach has emerged as a force to be reckoned with. And thus even writers who view the computer metaphor as essentially bankrupt have nonetheless felt moved to devote considerable ink to refuting it or establishing the merits of their own views against it.

I believe that there are two very different approaches that a philosopher may take to this very rich body of "computationalist" work in the study of cognition or "cognitive science." The first is that of the historian and philosopher of science. As a philosopher of science, one may look at the computationalist paradigm in psychology with an eye towards *issues that are internal to the various sciences of cognition:* What are the methodological assumptions of computational psychology? How do they differ from those of, say, behaviorism or associationism or neuroscience? What are psychological theorists really committed to in their use of theoretical terms such as 'representation' or 'syntax'? What are the issues that really stand between rival research programmes such as "good old-fashioned AI," which emphasizes rules and representations, and neural network approaches? What are the implicit assumptions of different theorists about the "good-making" qualities of scientific theories in a domain such as psychology?

On the other hand, the philosopher of *mind* may also look to the computational paradigm for answers to long-standing *philosophical* problems, such as the mind-body problem, issues about the metaphysical nature of the mind and the relationship between thought and matter, the relationship between psychology and the natural sciences, and the nature of intentionality. While there have been some welcome contributions of late to the history and philosophy of psychology that take a careful look at actual research in the sciences of cognition,[1] by far the greater portion of philosophical interest in the computer paradigm has been concentrated on the more distinctively philosophical enterprises of explaining intentionality and "naturalizing" psychology by rendering its commitment to mental states and processes compatible with materialism and the generality of physics.

This book is intended as a contribution towards such an understanding of the nature of the computer paradigm and its importance to the empirical study of cognition and to the philosophy of mind. It combines an extended examination of a "mainstream" approach to the importance of computation—the "Computational Theory of Mind" (CTM) championed by Jerry Fodor and Zenon Pylyshyn—with a preliminary articulation of an alternative approach to examining the importance of computational psychology. The thesis, in a sentence, is that CTM does *not* provide a solution to the philosophical problems that it is heralded as solving—indeed, it involves some deep confusions about computers, symbols, and meaning—but that this does not undercut the possibility that the computer paradigm may provide an important resource (for all we know, perhaps the key resource) for the development of a mature science of cognition. In short, we will be disappointed if we look to CTM for solutions to long-standing philosophical problems about the mind. But computational psychology is nonetheless a robust research programme that is deserving of philosophical study, and the final section of this book suggests an alternative approach to viewing computational psychology from the standpoint of the philosophy of science rather than that of metaphysics.

CTM claims that the mind literally is a computer. And what it is to be a computer, according to CTM, is to be a device that stores symbols and performs transformations upon those symbols in accordance with formal (or, more precisely, syntactic) rules. There are two distinct and important strands to this theory. The first strand is representational and consists in the claim that individual mental states, such as particular beliefs and desires, are relationships between an organism and mental *representations* . These mental representations are physically instantiated symbol tokens having both semantic and syntactic properties. This view, taken alone, Fodor sometimes calls the "Representational Theory of Mind." The Representational Theory of Mind (RTM) is a theory about the nature of individual mental states. The second thread of CTM is the claim that mental *processes,* such as forming and testing a hypothesis or reasoning to a conclusion, are *computational* processes that the mind performs upon these representations. That is, when the mind moves from one thought to another, it is generating new mental representations, and it does so by applying syntactically based rules to the representations already present in it, just as a digital computer generates new symbolic representations by applying syntactically based rules to existing representations.

CTM has generated a great deal of interest among philosophers because it goes beyond claims of (mere!) empirical utility for the computer paradigm and makes substantive philosophical claims as well. Two such claims are of particular importance. First, it is claimed that CTM—or, more specifically, the representational component of CTM, RTM—provides an account of how mental states have such properties as meaning, reference, and intentionality. According to Fodor, mental states "inherit" their semantic properties from those of the representations they involve. The second claim is at least as bold: namely, that CTM provides a "vindication" of "intentional psychology" (that is, of psychology that is committed to a realistic construal of explanations in the intentional idiom) by showing that intentional explanations can be tied to nomologically based causal explanations that are in no way incompatible with materialism or with the generality of physics.

These two claims are bold and ambitious, to say the least. A theory that could accomplish either of these goals in isolation would be of considerable importance. A theory that accomplished both, while also being closely linked to a burgeoning methodological approach to actual research in cognition, could hardly draw more attention than it deserved. Indeed, if CTM succeeds as its advocates claim, the emergence of the notion of computation will have provided the basis for a revolution in the study of mind as fundamental and important as the Copernican revolution in astronomy.

I shall argue, however, that while the computer may ultimately provide the basis for the extension into psychology of the Galilean project of the mathematization of science, CTM's attempts to explain intentionality and to vindicate intentional psychology are based upon subtle but fundamental confusions. At the heart of CTM is the claim that mental states are relations to mental representations—to meaningful symbols—and that this accounts for their semantic properties and their intentionality. The crucial questions one must ask of CTM, therefore, are these: (1) Just what does it mean to say that mental states are "relations to meaningful symbols"? And (2) just how is the postulation of meaningful symbols supposed to explain the semantic properties and the intentionality of mental states? The first question calls for an examination of just what we are saying of a thing when we call it a "meaningful symbol." The second calls for an application of the results of such an examination to the formulations of CTM offered by Fodor and Pylyshyn.

It is both curious and unfortunate that these questions have received so little attention from philosophers of mind: *curious* because the questions seem so crucial to the assessment of a theory that is generally acknowledged to be of great interest and importance; *unfortunate* because an examination of these questions uncovers significant ambiguities in such notions as "representation," "symbol," "meaning," and "intentionality." Until we have acknowledged these ambiguities, it is impossible either to assess CTM or even to determine exactly what it is that it is claiming.

This problem has, I think, been touched upon by some writers—notably by Kenneth Sayre and John Searle, both of whom urge upon us the conclusion that there is something about symbols, and particularly about symbols in computers, that renders them unsuitable for an explanation of the meaningfulness of mental states. My criticisms of CTM run in the same vein. Where I part ways with Searle and Sayre is that they look for the problem specifically in the use of symbols in computers. In my view, however, the fundamental issue turns out in the end to have curiously little to do with computers. The issue, rather, is whether the notion of symbolic representation provides the bedrock upon which a theory of the intentionality of mental states may be built. My answer to that is *no;* and insofar as the project of vindicating intentional psychology (at least as envisioned by advocates of CTM) can be shown to depend upon its ability provide a theory of intentionality, that vindication fails as well.

Why can't one establish a theory of intentionality for mental states upon a foundation of symbolic representations in the mind? I am afraid that I do not know how to give an answer to that question that satisfies my own standards of rigor in less space than the several chapters it occupies in this book, but I shall try to give a short answer here that may prove helpful and not too inaccurate. When one takes a close look at what one is saying when one calls something a "meaningful symbol" or a "symbolic representation," it turns out that one is tacitly saying things about the conventions and intentions of symbol users. This is just part of what we are saying of a thing in calling it a symbol, and of what we are saying of a symbol when we say that it has semantic properties. But conventions and intentions of symbol users are ultimately facts about people's mental states. And so any explanation of the intentionality of mental states that rests upon the meaningfulness of symbolic representations ends up explaining the intentionality of mental states in a way that refers to other meaningful mental states. Thus one important problem with CTM's account of intentionality is that it turns out to be circular and regressive: circular because it explains the meaningfulness of mental states by appealing to the meanings of symbols, while one must also explain the meaningfulness of symbols by appealing to the meanings of mental states; regressive because the explanation of any particular mental state will ultimately refer back to other mental states.

Moreover, it is not only the *semantic* properties of symbols that are conventional in nature; *syntactic* properties, and the very *symbol types themselves* are ultimately dependent upon conventions. (The fact that something is a letter *p* or an inscription of the English word 'dog' depends upon conventions that establish the existence of those symbol types.) In particular, the kinds of syntactically based rules that are necessary for *compositionality* are conventional in nature: in order to generate semantic properties for complex representations, it is not enough to have interpretations for the primitives and "syntax" in the weak sense of *rules for legal concatenation* or *equivalence classes of legal transformations*. Rather, one needs a stronger kind of syntax that involves rules for how syntactic patterns contribute to meanings of complex representations—for example, a rule to the effect that if *'A'* means "X" and *'B'* means "Y," then *'A-&-B'* will mean "X and Y." The only way we know of getting this kind of compositionality is by way of conventions. It is not clear that there is any other way of getting compositionality; at very least, CTM's advocates would have to show how semantic composition could be achieved without the aid of conventions.

Now of course this argument rests upon a particular construal of what it is to be a symbolic representation or to be a meaningful symbol. But, as far as I am aware, this is the only sense of 'symbolic representation' and 'meaningful symbol' that we have. One is, of course, inclined to wonder whether perhaps writers like Fodor really mean something different when they speak of the mind containing "meaningful symbols." But if they do, it is curious that they never inform the reader that they are using familiar expressions in novel ways. Indeed, in one place Fodor gives a brief glance at this possibility only to dismiss the issue as unlikely to prove important.

> It remains an open question whether internal representation, so construed, is sufficiently like natural language representation so that both can be called representation 'in the same sense'. But I find it hard to care much how this question should be answered. There is an analogy between the two kinds of representation. Since public languages are conventional and the language of thought is not, there is unlikely to be *more* than an analogy. If you are impressed by the analogy, you will want to say that the inner code is a language. If you are unimpressed by the analogy, you will want to say that the inner code is in some sense a representational system that is not a language. But in

neither case will what you say affect what I take to be the question that is seriously at issue: whether the methodological assumptions of computational psychology are coherent. (Fodor 1975: 78-79)

My contention, by contrast, is that it does indeed matter a great deal whether words like 'representation' are used in their usual sense within CTM, because I believe that the conventionality of linguistic symbols is not something that can be divorced from their symbolhood. It is not that we find a nonconventional property called "meaning" in linguistic symbols that additionally happens to be conventional in nature. Rather, the very notion of "meaning" that we apply to symbols is interwoven with conventionality through and through. And thus if we apply these familiar notions of "representation," "meaning," and "syntax" to CTM, we are led to circularity and regress. As Fodor says, this criticism does not undercut the methodological assumptions of computational *psychology* . But this is only so because computational psychology (that is, empirical science inspired by the computer paradigm) is not committed to the view that its "mental representations" literally are symbols in a language, as I shall argue in later chapters. A merely analogous usage of the word 'representation' is just fine for computational psychology. The plausibility of CTM's *philosophical* claims, by contrast, would seem to turn *precisely* upon the assumption that the "symbols" in question are "symbols" in precisely the same sense that we speak of "symbols" in a language. Formalization and computation show us how to tie meaning to causation for (convention-based) *linguistic symbols,* and not for anything else. If mental representations are something other than linguistic symbols, we need to see how the link from meaning to causation works for some new class of entities. The arguments Fodor and others give for their claims about intentionality and the vindication of intentional psychology simply do not go through as stated if words like 'representation' and 'symbol' are used in a merely analogous or metaphorical manner.

On the other hand, it is clear that one might try to develop CTM in a way that divorces the technical notion of mental representation from convention-based linguistic signs. The fact that CTM's advocates do not try to do this in any explicit detail does not mean that this avenue might not prove more fruitful in the end. One might, for example, say that the "semantic properties" of mental representations are not the same sort of "semantic properties" possessed by garden-variety symbols. That is, one might say that expressions such as 'semantic property' are homonymous, and have different senses when applied to garden-variety symbols and to mental representations, in which case results of conceptual analysis of semantic terminology as applied to discursive symbols cannot be used to create problems for a theory of mental representations.

Unfortunately, to the best of my knowledge, no advocates of CTM explicitly pursue this course. But since it does seem to be the only way of saving the theory from the results of my conceptual analysis, I develop two ways of pursuing this line of thought. The first is to take causal theories of content like that supplied in Fodor (1987) as supplying a causal *definition* of semantic terminology as applied to mental representations. The second is to treat semantic terminology as applied to mental representations as being theoretical and open-ended in character: that is, to treat terms such as 'meaningful' and 'referential' as applied to mental representations as terms whose meaning we do not presently know but might discover as the result of further investigation. I shall argue in Part III that neither of these strategies seems likely to be able to provide an account of intentionality or to vindicate intentional psychology.

These problems for CTM as a philosophical thesis, however, do not entail that the computer paradigm is of no use for the philosopher or the empirical researcher. For I think that there is a much better way to understand the nature and importance of the computer paradigm for the study of cognition. If one adopts this alternative view, the importance of providing an account of intentionality wanes significantly, while the need to justify intentional psychology disappears altogether. To arrive at this standpoint, however, we must cease looking to CTM as a source of solutions to old philosophical puzzles and begin to look at computational psychology as a research programme in psychology from the perspective of historians and philosophers of science.

The basis of the alternative approach is the premise that two of the traditional distinguishing marks of a mature science have been the mathematization of its explanations and the clarification of connections between the domain and the laws of that science and those of other areas of knowledge. So, to take a paradigm example, chemistry progressed towards mathematical maturity through the development of the periodic table, the notion of *valences,* and the discovery of rules governing reactions between different classes of molecules. It progressed towards connective maturity as the explanations given in chemical terms were able to provide explanations for phenomena described at a higher level (e.g., as described in the vocabularies of metallurgy or genetics) and as categories such as *valence* were in turn explained at a lower level in terms of such ideas as elementary particles and orbitals.

The interpretation of the importance of the computer paradigm that I wish to urge is the following: what the notion of computation *may* be able to provide for the empirical scientist is the right kind of technical machinery for the mathematization of the study of cognition— particularly, cognitive psychology. (I emphasize the word 'may' be- cause I seek only to illuminate what computational psychology *would provide if successful,* and not to make any predictions about its eventual successes or failures.) That is, what computer science gives us is an abstract vocabulary that might turn out to provide the resources for psychology to progress towards mathematical maturity. I think that it should be clear that this is of enormous interest, even without the philo- sophical benefits claimed for CTM. Surely a large part of what psy- chology is about is providing an inventory of cognitive processes, "mapping" the relations between these, "unlocking the black boxes" underlying high-level processes by specifying lower-level processes that would account for them, and showing how mental processes are connected to behavior. That is, *part* of what psychology is about is specifying the *form* of the mind by tracing out the functional relations mental states bear to one another and to behavior. Many researchers interested in cognition have staked their careers upon their belief that computational notions allow them to carry out this project in ways that were previously unavailable. Indeed, the strength of this belief is evi- denced by the emergence of "cognitive science" as an approach to the mind that is organized around the premise that cognitive processes can be described in computational terms. I think this research project is of great interest regardless of whether the notion of computation can con- tribute to the solution of any philosophical problems as well.

Moreover, viewed in this way, cognitive science as an empirical re- search programme is not imperiled by my criticisms of CTM. What I argue against CTM is that if you take it as central to the very notion of *computation* that computation consists in the manipulation of *meaning- ful symbols,* then there are serious problems involved in saying that cognition is computation. If, on the other hand, what is essential to the notion of computation is *functional specifiability*—in, say, the form of a machine table—these problems do not arise. If cognitive science is oriented towards the thesis that *cognitive processes are functionally specifiable,* then it can attempt to apply the technical resources of com- puter science to the domain of psychology without worrying about problems with the notions of *symbol or representation.* Indeed, one might even propose theories that depend upon the premise that there are mental states or brain states that play a role in thought, a role that is *formally analogous to* the roles played by symbols in the execution of particular computer programs, without threat of incoherence from misuse of such words as 'symbol' and 'meaningful'. (One might, with some risk, even use the word 'symbol' in describing such states, so long as one was careful that the illegitimate importation of the ordinary meaning of the word 'symbol' did not do any illicit work in one's explanations.)

Of course, what one loses in this alternative is the hope CTM excited of finding a level of explanation—a domain of meaningful mental representations over which mental computations are defined—at which there is a clear meeting of the ways between mentalistic description cast in the intentional idiom and one of the natural sciences. On this view, cognitive science does not "close the gap" between mind and nature. Here, however, there is a parting of the ways between the interests of the philosopher of mind and those of the empirical scientist. For the computer paradigm might help psychology progress to one or both types of scientific maturity without providing a philosophical account of intentionality in the process. First, it might provide the tools for the mathematization of psychology without providing for connective maturity as well. But it does seem likely that a good mathematization of cognitive explanation is just the sort of thing that would be helpful in correlating states specified in the intentional idiom with states specified in neurological terms: that is, it is arguable that the only way of finding out how cognitive states are instantiated is to find out what in the brain has the right (functional) "shape" to realize them. So if connection between cognitive explanation and other kinds of explanation is to take place, it may partially be *through* the mathematization of both levels of explanation. And for this it is a plausible hypothesis that computer science provides the appropriate resources.

But it is important to see that one might get the kind of connectivity that the scientist desires without thereby solving any philosophical problems. The researcher committed to intentional explanation *and* natural explanation wants to find out what neural processes are *specially associated with* what intentionally specified processes. And she is interested in this association just to the extent that intentional and naturalistic predictions will *track* one another. The metaphysical nature of this "special association" really does not matter as far as empirical science is concerned. Empirical science is largely blind to the differences between relationships stronger than empirical adequacy, and hence a good integrated psychological theory could be equally compatible with materialism and supervenience or with thoroughgoing parallelist dualism. And this, I think, should be viewed as a virtue rather than a vice: in my book, consistency with a wide range of ontological options counts as a significant virtue for an empirical research programme.

Now there is a sense in which such an integrated psychology would provide an "account" of intentionality and a sense in which it would not. If by "an account of intentionality" one means *(a)* a model of the relations between intentional states, stimuli, and behavior, and *(b)* a specification of the natural systems through which intentional states and processes are—to use an intentionally neutral term—*realized,* then an integrated psychology might well involve an "account of intentionality." But if "an account of intentionality" means something *stronger* — say, if it involves providing natural conditions upon which intentional properties would *have* to supervene, then an integrated psychology might well not provide *this* kind of "account of intentionality." I believe, however, that it is fundamentally misguided to seek such a naturalistic account of intentionality, for reasons that I shall develop in chapters 9 and 11. If I am right, then inability to provide an account of intentionality in this strong sense is not a fault.

I shall argue for a similar attitude towards the other goal CTM has sought to achieve: that of vindicating intentional psychology. To put it very briefly, I do not believe that intentional psychology is presently in need of vindication. The perceived need for a vindication turned upon some concerns about methodology and ontology that came to prominence in the writings of behaviorists and reductionists. One might do well to ask whether these concerns ought to have survived the theories that brought them to prominence. But even if one finds these concerns to be serious ones, they must at very least be put off for the present. By just about everyone's reckoning, any full-scale meeting of the ways that might take place between intentional explanation and neuroscience (much less physics) is a long ways away and depends upon a great deal of research, much of which almost has to be pursued through top-down strategies in cognitive psychology. So, in a sense, any real assessment of cognitivism's compatibility with the generality of physical explanation could only take place once we had a reasonably successful predictive cognitive psychology.

Of course we all look for occasions when our top-down strategies get us to a level where we can find some plausible candidate for a known neurological mechanism that has the right functional features to support the kind of cognitive process we have postulated. Such moments are landmarks that provide some of the best kinds of reasons to believe one's research is on the right track. That being said, it is nonetheless the case that (1) intentional explanation seems an indispensable starting point for cognitive psychology, regardless of whether such research would ultimately allow us to "throw away the ladder"; (2) it could, in principle, turn out that research in cognitive science could produce a good predictive psychology without ever hooking up with neuroscience in a comprehensive fashion (e.g., we would not throw out psychophysics if we could not produce neutral models to account for the data); (3) if this were to happen, it is not at all clear that we should, as a result, regard such psychological theories as flawed, much less metaphysically perverse; and (4) in the meantime, it is absolutely pointless to expect empirical researchers to care about whether their work meets such ideological tests as conformity with one's favorite ontological theory.

In short, I do not think that intentional psychology is in need of vindication at the present time. The pressing question for the philosophy of psychology is whether intentional explanation can be systematized and mapped out using something like the techniques afforded us by the notion of computation or by some alternative notions, and whether in the course of this project our ordinary mentalistic notions like "desire," "belief," and "judgment" will be retained, built into a larger framework, transformed, or abandoned altogether. Certain outcomes of this project might call for the reassessment of intentional psychology. (And of course there are already those who believe that it is a mistake to view it as an explanatory science in the first place.) There is a separate, and largely empirical question about how cognitive states are realized through specific physiological structures. The connections between the success or failure of this project and the status of intentional psychology are far more tenuous, but really need not be fretted over at this stage of the game.

To repeat, on my reading of the significance of the computer paradigm, what it offers is a project that might hasten the progress of psychology towards scientific maturity by providing the right technical resources for mathematizing the functional relationships that mental states bear to one another and to behaviors. Interpreted in this fashion, the computational approach to cognition is one that is distinguished principally by the conceptual tools it borrows from computer science. However, the computational approach is only one research programme among several that seek to provide the right formal tools for studying cognition. It is a research programme that has rivals that supply different tools for the mathematization of psychology. Notable among these are the information-theoretic approach favored by Sayre and the network-based mathematical models of various sorts offered over the past thirty years by Pitts and McCulloch, Grossberg, Anderson, and others. What formal techniques end up providing the best descriptions is a question to be answered by the fertility of these research projects.

What this book calls for, then, is a separation of two kinds of issues. The first set of issues involves questions about how to compare competing theories about the mind that emerge out of empirical science. For example, apart from their abilities to give some description of the phenomena in their own canonical vocabularies, just where do two approaches to cognition, such as CTM and connectionism, really differ? What are the "good-making" qualities that are relevant to the assessment of empirical theories in psychology, and which are possessed in greater abundance by whose theories? The second set of issues is made up of more purely philosophical questions about the mind-body problem, the exact metaphysical relationship between mental states and the physical states through which they are realized, and attempts to give a logically necessary and sufficient account of notions such as meaning and intentionality. It is the thesis of this book that, contrary to popular rumor, CTM does nothing to solve the latter problems. Nonetheless, it is quite possible to "bowdlerize" CTM in a fashion that avoids the problems of interpretive regress and to construe it as a special version of machine functionalism; and interpreted in this fashion, computational psychology can be seen as an interesting contender with respect to the first set of issues. With CTM's claims to solving philosophical problems out of the way, however, there is now a level playing field, and computational psychology may be compared with its competitors in terms of their purely scientific merits. And in a roundabout way, I think this counts as progress.

A BRIEF GUIDE TO THIS BOOK

This book is divided into four sections. Part I, comprising chapters 1 through 3, gives an exposition of CTM and of its claims to solve important philosophical problems. It also provides an initial statement of some potential problems for CTM arising from criticisms raised by Searle and Sayre. I make a case that their criticisms are not definitive, and call for a more careful analysis of the notions of "symbol," "syntax," and "symbolic meaning." This analysis is provided in Part II. Chapter 4 presents a conventionalist analysis of symbols, syntax, and symbolic meaning, which is then applied to symbols in computers in chapter 5 and defended against some likely objections in chapter 6. (The reader who comes out of chapter 4 with burning objections will lose nothing by reading chapter 6 before chapter 5.)

The results of this analysis are then applied in Part III (chapters 7 through 9) in a critique of the philosophical claims of CTM. Chapter 7 argues that, if you interpret CTM's talk of "symbols," "syntax," and "semantics" in the ordinary convention-laden way, you are left with an account that is circular and regressive. Chapter 8 argues that CTM fares no better if the semiotic vocabulary is reconstructed in a nonconventionalist way. In short, CTM maintains an illusion of explaining intentionality only by slipping back and forth between semiotic notions based on conventional symbols and talk of an alternative "pure semantics." Chapter 9 briefly makes the case that CTM is unlikely to be supplemented by an independent naturalization of content: some features of the mind do not seem susceptible to naturalization at all, while others seem likely to be naturalized (if at all) only in a fashion incompatible with the constraints laid down by CTM. The dialectical situation at the end of Part III is that CTM's claims to producing distinctively philosophical fruit have been undermined.

Part IV then presents an alternative view of the importance of the computer paradigm. Chapter 10 outlines how computation might provide psychology with important good-making qualities without naturalizing intentionality or vindicating intentional psychology. The book concludes, in chapter 11, with a philosophical examination of the assumptions that intentionality needs naturalizing and mental states need vindicating. I argue that, in the absence of strong aprioristic arguments for naturalism, we are better off letting the special sciences flourish as best they may and shaping our metatheoretic views about intertheoretic connections on the basis of the shape that real psychology takes rather than upon any preconceived notions of what it *should* look like.

PART I

Computationalism and its Critics

CHAPTER ONE

The Computational Theory of Mind

The past thirty years have witnessed the rapid emergence and swift ascendency of a truly novel paradigm for understanding the mind. The paradigm is that of machine computation, and its influence upon the study of mind has already been both deep and far-reaching. A significant number of philosophers, psychologists, linguists, neuroscientists, and other professionals engaged in the study of cognition now proceed upon the assumption that cognitive processes are in some sense computational processes; and those philosophers, psychologists, and other researchers who do *not* proceed upon this assumption nonetheless acknowledge that computational theories are now in the mainstream of their disciplines.

But if there is general agreement that the paradigm of machine computation may have significant implications for both the philosopher of mind and the empirical researcher interested in cognition, there is no such agreement about what these implications are. There is, perhaps, little doubt that computer modeling can be a powerful tool for the psychologist, much as it is for the physicist and the meteorologist. But not all researchers are agreed that the cognitive processes they may model on a computer are themselves computations, any more than the storms that the meteorologist models are computations.

Similarly, there is significant disagreement among philosophers about whether the paradigm of machine computation provides a literal characterization of the mind or merely an alluring metaphor. Three alternative ways of assessing the importance of the computer paradigm stand out. The most modest possibility is that the computer metaphor will prove an able catalyst for generating theories in psychology, in much the sort of way that numerous other metaphors have so often played a role in the development of other sciences, yet in such a fashion that little or nothing about computation per se will be of direct relevance to the explanatory value of the resulting theories. A second and slightly stronger possibility is that the conceptual machinery employed in computer science will provide the right sorts of tools for allowing psychology (or at least parts of psychology) to become a rigorous science, in much the fashion that conceptual tools such as Cartesian geometry and the calculus provided a basis for the emergence of Newtonian mechanics, and differential geometry made possible the relativistic physics which supplanted it. On this view, which will be discussed in the final chapter of this book, what the computer paradigm might contribute is the basis for the maturation of psychology by way of the mathematization of its explanations and the connections between intentional explanation and explanation cast at the level of some lower-order (e.g., neurological) processes through which intentional states and processes are realized. This view is committed to the thesis that the mind is a computer only in the very weak sense that the interrelations between mental states have formal properties for which the vocabulary associated with computation provides an apt characterization—that is, to the view that there is a description of the interrelations of mental states and processes that is isomorphic to a computer program. This thesis involves no commitment to the stronger view that terms like 'representation', 'symbol', and 'computation' play any stronger role in explaining why mental states and processes are *mental* states and processes, but only the weaker view that, given that we may posit such states and processes, their "form" may be described in computational terms. (You might say that, on this view, the mind is "computational" in the same sense that a relativistic universe is "differential.") The third and strongest view of the relevance of machine computation to psychology—one example of which will be the main focus of this book—is that notions such as "representation" and "computation" not only provide the psychologist with the formal tools she needs to do her science in a rigorous fashion, but also provide the philosopher with fundamental tools that allow for an analysis of *the essential nature of cognition* and for the solution of important and long-standing philosophical problems.

This book examines one particular application of the paradigm of machine computation to the study of mind: namely, the "Computational

Theory of Mind" (CTM) advocated in recent years by Jerry Fodor (1975, 1980a, 1981, 1987, 1990) and Zenon Pylyshyn (1980, 1984). Over the past two decades, CTM has emerged as the "mainstream" view of the significance of computation in philosophy. Its advocates have articulated a very strong position: namely, that cognition literally is computation and the mind literally is a digital computer. CTM is comprised of two theses. The first is a thesis about the nature of intentional states, such as individual beliefs and desires. According to CTM, intentional states are relational states involving an organism (or other cognizer) and mental representations. These mental representations, moreover, are to be understood on the model of representations in computer storage: in particular, they are symbol tokens that have both syntactic and semantic properties. These symbols include both semantic primitives and complex symbols whose semantic properties are a function of their syntactic structure and the semantic values of the primitives they contain. The second thesis comprising CTM is about the nature of cognitive processes—processes such as reasoning to a conclusion, or forming and testing a hypothesis, which involve chains of beliefs, desires, and other intentional states. According to CTM, cognitive processes are computations over mental representations. That is, they are causal sequences of tokenings of mental representations in which the relevant causal regularities are determined by the syntactic properties of the symbols and are describable in terms of formal (i.e., syntactic) rules. The remainder of this chapter will be devoted to clarifying the nature and status of these two claims.

As we shall see in chapter 2, CTM's advocates have also made a very persuasive case that viewing the mind as a computer allows for the solution of significant philosophical problems: notably, they have argued (1) that it provides an account of the intentionality of mental states, and (2) that it shows that psychology can employ explanations in the intentional idiom without involving itself in methodological or ontological difficulties. The claims made on behalf of CTM thus fall into the third and strongest category of attitudes towards the promise of the computer paradigm. The task undertaken in the subsequent chapters of this book is to evaluate these claims that have been made on behalf of CTM and to provide the beginnings of an alternative understanding of the importance of the computer paradigm for the study of cognition. In particular, we shall examine (1) whether CTM succeeds in solving these philosophical problems, and (2) whether the weaker possibility of its providing the basis for a rigorous psychology in any way depends upon either the understanding of cognition and computation endorsed by CTM or its ability to explain intentionality and vindicate intentional psychology.

1.1 INTENTIONAL STATES

CTM is a theory about the nature of intentional states and cognitive processes. To understand what this means, however, we must first become clear about the meanings of the expressions 'intentional state' and 'cognitive process'. The expression 'intentional state' is used as a generic term for mental states of a number of kinds recognized in ordinary language and commonsense psychology. Some paradigm examples of intentional states would be

—*believing* (judging, doubting) that such-and-such is the case,
—*desiring* that such-and-such should take place,
—*hoping* that such-and-such will take place,
—*fearing* that such-and-such will take place.

The characteristic feature of intentional states is that they are *about something* or *directed towards something*. This feature of *directedness* or *intentionality* distinguishes intentional states both from brute objects and from other mental phenomena such as qualia and feelings, none of which is about anything. The expressions 'intentional states' and 'cog-nitive states' denote the same class of mental states, but the two terms reflect different interests. The term 'intentionality' is employed primarily in philosophy, where it is used to denote specifically this directedness of certain mental states, a feature which is of importance in understanding several important philosophical problems, including opacity and transparency of reference and knowledge of extramental objects. The term 'cognition' is most commonly employed in psychology, where it is used to denote a domain for scientific investigation. As such, its scope and meaning are open to some degree of adjustment and change as the science of psychology progresses. A third term used to indicate this same domain is 'propositional attitude states'. This expression shows the influence of the widely accepted analysis of cognitive states as involving an *attitude* (such as believing or doubting) and a *content* that indicates the object or state of affairs to which the attitude is directed. Since the contents of mental states are often closely related to propositions, such attitudes are sometimes called *propositional* attitudes. These three expressions will be used interchangeably in the remainder of this book. In places where there is little danger of misunderstanding, the more general expression 'mental states' will also be used to refer specifically to intentional states.

1.2 MENTAL STATE ASCRIPTIONS IN INTENTIONAL PSYCHOLOGY AND FOLK PSYCHOLOGY

Attributions of intentional states such as beliefs and desires play an important role in our ordinary understanding of ourselves and other human beings. We describe much of our linguistic behavior in terms of the expression of our beliefs, desires, and other intentional states. We explain our own actions on the basis of the beliefs and intentions that guided them. We explain the actions of others on the basis of what we take to be their intentional states. Such explanations reflect a general framework for psychological explanation which is implicit in our ordinary understanding of human thought and action. A cardinal principle of this framework is that people's actions can often be explained by their intentional states. I shall use the term 'intentional psychology' to refer to any psychology that (*a*) makes use of explanations involving ascriptions of intentional states, and (*b*) is committed to a realistic interpretation of at least some such ascriptions.

This usage of the expression 'intentional psychology' should be distinguished from the common usage of the currently popular expression 'folk psychology'. The expression 'folk psychology' is used by many contemporary writers in cognitive science to refer to a culture's loosely knit body of commonsense beliefs about how people are likely to think and act in various situations. It is called "psychology" because it involves an implicit ontology of mental states and processes and a set of (largely implicit) assumptions about regularities of human thought and action which can be used to explain behavior. It is called "folk" psychology because it is not the result of rigorous scientific inquiry and does not involve any rigorous scientific research methodology. Folk psychology, thus understood, is a proper subset of what I am calling intentional psychology. It is a *subset* of intentional psychology because it employs intentional state ascriptions in its explanations. It is only a *proper* subset because one could have psychological explanations cast in the intentional idiom that were the result of rigorous inquiry and were not committed to the specific set of assumptions characteristic of any given culture's commonsense views about the mind. Many of Freud's theories, for example, fall within the bounds of intentional psychology, since they involve appeals to beliefs and desires; yet they fall outside the bounds of folk psychology because Freud's theories are at least attempts at rigorous scientific explanation and not mere distillations of commonsense wisdom. Similarly, many contemporary theories in cognitive psychology employ explanations in the intentional idiom that fall outside the bounds of folk psychology, in this case because the states picked out by their ascriptions occur at an infraconscious level where mental states are not attributed by commonsense understandings of the mind.

In understanding the importance of CTM in contemporary psychology and philosophy of mind, it would be hard to overemphasize this distinction between the more inclusive notion of intentional psychology, which embraces any psychology that is committed to a realistic construal of intentional state ascriptions, and the narrower notion of folk psychology, which is by definition confined to prescientific commonsense understandings of the mental. For CTM's advocates wish to defend the integrity of intentional psychology, while admitting that there may be significant problems with the specific set of precritical assumptions that comprise a culture's folk psychology. On the one hand, Fodor and Pylyshyn argue that the intentionally laden explanations present in folk psychology are quite successful,[1] that folk psychology is easily "the most successful predictive scheme available for human behavior" (Pylyshyn 1984: 2), and even that intentional explanation is indispensable in psychology.[2] On the other hand, advocates of CTM are often more critical of the *specific* generalizations implicit in commonsense understandings of mind. Folk psychology may provide a good starting point for doing psychology, much as animal terms in ordinary language may provide a starting point for zoological taxonomy or billiard ball analogies may provide a starting point for mechanics; but more rigorous research is likely to prove commonsensical assumptions wrong in psychology, much as it has in biology and physics.[3] Folk psychology is thus viewed by these writers as a protoscience out of which a scientific intentional psychology might emerge. One thing that would be needed for this transition to a scientific intentional psychology to take place is rigorous empirical research of the sort undertaken in the relatively new area called cognitive psychology.[4] Such empirical research would be responsible, among other things, for correcting such assumptions of common sense as may prove to be mistaken. What is viewed as the most significant shortcoming of commonsense psychology, however, is not that it contains erroneous generalizations, but that its generalizations are not united by a single theoretical framework.[5] CTM is an attempt to provide such a framework by supplying (*a*) an account of the nature of intentional states, and (*b*) an account of the nature of cognitive processes.

1.3 CTM'S REPRESENTATIONAL ACCOUNT OF INTENTIONAL STATES

The first thesis comprising CTM is a *representational account of the nature of intentional states* . Fodor provides a clear outline of the basic tenets of this account in the following five claims, offered in the introduction to *RePresentations,* published in 1981:

(*a*) Propositional attitude states are relational.
(*b*) Among the relata are mental representations (often called "Ideas" in the older literature).
(*c*) Mental representation[s] are symbols: they have both formal and semantic properties.
(*d*) Mental representations have their causal roles in virtue of their formal properties.
(*e*) Propositional attitudes inherit their semantic properties from those of the mental representations that function as their objects. (Fodor 1981: 26)

Claims (*a*) through (*c*) provide Fodor's views upon the nature of intentional states, while claims (*d*) and (*e*) provide the means for connecting this representational account of intentional states with a computational account of cognitive processes and an account of the intentionality of the mental, respectively.

Fodor supplies a more formal account of the nature of intentional states in *Psychosemantics,* published in 1987. There he characterizes the nature of intentional states (propositional attitudes) as follows:

Claim 1 (the nature of propositional attitudes):
For any organism O , and any attitude A toward the proposition P , there is a ('computational'-'functional') relation R and a mental representation MP such that

MP means that P, and

O has A iff O bears R to MP. (Fodor 1987: 17)

On Fodor's account, Jones's believing that two is a prime number consists in Jones being in a particular kind of functional relationship R to a mental representation MP. This mental representation MP is a symbol token, presumably instantiated in some fashion in Jones's nervous system. MP has semantic properties: in particular, MP means that two is a prime number. And Jones believes that two is a prime number when and only when he is relation *R* to *MP*.

There are some glaring unclarities about references to types and tokens of attitudes and representations in this formulation, but some of these are clarified when Fodor provides a "cruder but more intelligible" gloss upon his account of the nature of intentional states:

> To believe that such and such is to have a mental symbol that means that such and such tokened in your head in a certain way; it's to have such a token 'in your belief box,' as I'll sometimes say. Correspondingly, to hope that such and such is to have a token of that same mental symbol tokened in your head, but in a rather different way; it's to have it tokened 'in your hope box.' ... And so on for every attitude that you can bear toward a proposition; and so on for every proposition toward which you can bear an attitude. (Fodor 1987: 17)

On the basis of this gloss, it seems most reasonable to read Fodor's formulation as follows:

The Nature of Propositional Attitudes (Modified)

For any organism O, and any attitude-token a of type A toward the proposition P, there is a ('computational'-'functional') relation R and a mental representation token t of type MP such that

t means that P by virtue of being an MP-token, and

O has an attitude of type A iff O bears R to a token of type MP.[6]

While there are arguably some significant residual unclarities about Fodor's formulation in spite of these clarifications,[7] Fodor does make the main point adequately clear: namely, that it is the relationship between the organism and its mental representations that is to account for the fact that intentional states have the semantic properties and intentionality that they have. In the passage already quoted from *RePresentations,* for example, he writes that intentional states "inherit their semantic properties from those of the mental representations that function as their objects" (Fodor 1981: 26). And in that essay he also writes that "the objects of propositional attitudes are symbols (specifically, mental representations)" and that *"this fact accounts for their intensionality and semanticity"* (ibid., 25, emphasis added).[8]

The first thesis comprising CTM is thus a *representational account of the nature of intentional states*. On this account, intentional states are relations to mental representations. These representations are symbol tokens having both syntactic and semantic properties, and intentional

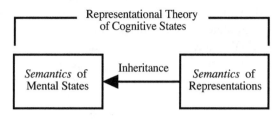

Figure 1

states "inherit" their semantic properties and their intentionality from the representations they involve (see fig. 1).

1.4 SEMANTIC COMPOSITIONALITY

An important feature of this account lies in the fact that the symbols involved in mental representation have both semantic and syntactic properties, and may be viewed as tokens in a "language of thought," sometimes called "mentalese." Viewing the system of mental representations as a language with both semantic and syntactic properties allows for the possibility of *compositionality of meaning*. That is, the symbols of mentalese are not all lexical primitives. Instead, there is a finite stock of lexical primitives which can be combined in various ways according to the syntactic rules of mentalese to form a potentially infinite variety of complex representations, just as in the case of natural languages it is possible to generate an infinite variety of meaningful utterances out of a finite stock of morphemes and compositional rules. Mentalese is thus viewed as having the same generative and creative aspects possessed by natural languages. So while the semantic properties of mental states are "inherited" from the representations they contain, those representations may themselves be either semantically primitive or composed out of semantic primitives by the application of syntactic rules.

1.5 COGNITIVE PROCESSES

If a representational account of the mind provides a way of interpreting the nature of individual thoughts, it does not itself provide any comparable account of the nature of mental *processes* such as reasoning to a conclusion or forming and testing a hypothesis, and hence does not provide the grounds for a psychology of cognition. For a psychology of cognition, something more is needed: a theory of mental processes that uses the properties of mental representations as the basis of a causal account of how one mental state follows another in a train of reasoning. Suppose, for example, that one wishes to explain why Jones has closed the window. An explanation might well be given along the following lines:

(1) Jones felt a chill.

(2) Jones noticed that the window was open.

(3) Jones hypothesized that there was a cold draft blowing in through the window.

(4) Jones hypothesized that this cold draft was the cause of his chill.

(5) Jones wanted to stop feeling chilled.

(6) Jones hypothesized that cutting off the draft would stop the chill.

so,   (7) Jones formed a desire to cut off the draft.

(8) Jones hypothesized that closing the window would cut off the draft.

so,   (9) Jones formed a desire to close the window.

so,   (10) Jones closed the window.

Here we have not a random train of thought, but a sequence of thoughts in which the latter thoughts are plausibly viewed as both (*a*) rational in light of those that have gone before them, and (*b*) consequences of those previous states—Jones formed a desire to close the window *because* he thought that doing so would cut off the draft. Moreover, a causal theory of inference would need to forge a close link between the semantic properties of individual states and their role in the production of subsequent states. It is changes in the *content* of Jones's beliefs and desires that we would expect to produce different trains of thought and different behaviors. If Jones had noticed the fan running instead of noticing an open window, we would expect him to entertain different hypotheses, form different desires, and act in a different way, all as a consequence of changing the content of his belief from "the window is open" to "the fan is running."

Now CTM's representational account of intentional states seems well suited to a discussion of the *semantic* relations between intentional states, since the semantic and intentional properties of intentional states are identified with those of the representations they involve.   But when it

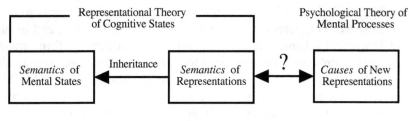

Figure 2

comes to the question of how intentional states can play a *causal* role in the etiology of a process that involves the generation of new intentional states, the notion of representation, in and of itself, has little to offer. Viewing intentional states as relations to representations allows us to locate the semantic relationships between intentional states in relationships between the representations they involve, but it does little to show how Jones's standing in relation R to a representation MP at time t can play a causal role in Jones coming to stand in relation Q to a representation MP^* at $t + \partial$.

This seems to present a problem. In order for a sequence of representations to make up a rational, cogent train of thought, the question of *which* representations should occur in the sequence should be determined by the meanings of the earlier representations. In order for the sequence of representations to *make sense,* the later representations need to stand in appropriate *semantic* relationships to the earlier ones. But in order for a sequence of representations to be a *causal* sequence, the question of what representations will occur later in the sequence must be determined by the causal powers of the earlier representations. Now intentional explanations pick out representations by their content—that is, by their semantic properties. But if such explanations are to be causal explanations, they must pick out representations in a fashion that individuates them according to their causal powers. But this can be done only if the semantic values of representations can be linked to, or coordinated with, the causal roles they can play in the production of other representations and the etiology of behavior. This has been seen by some as a significant stumbling block to the possibility of a causal-nomological psychology, as it is notoriously problematic to view semantic relationships as causal relationships or to equate reasons with causes.[9] The problem, then, for turning a representational theory of mental states into a psychological theory of mental processes is one of finding a way to link the semantic properties of mental representations to the causal powers of those representations (see fig. 2).

It is precisely at this point that the computer paradigm comes to be of interest. For computers are understood as devices that store and manipulate symbol tokens, and the manipulations that they perform are dependent upon what representations are already present, yet they are also completely mechanical and uncontroversially causal in nature. Machine computation provides a general paradigm for understanding symbol-manipulation processes in which the symbols already present play a causal role in determining what new symbols are to be generated. CTM seeks to provide an extension of this paradigm to *mental* representations, and thereby to supply an account of cognitive processes that can provide a way of discussing their etiology while also respecting the semantic relationships between the representations involved.

1.6 FORMALIZATION AND COMPUTATION

CTM's advocates believe that machine computation provides a paradigm for understanding how one can have a symbol-manipulating system that can *cause* derivations of symbolic representations in a fashion that "respects" their semantic properties. More specifically, machine computation is believed to provide answers to two questions: (1) How can semantic properties of symbols be linked to causal powers that allow the presence of one symbol token s_1 at time t to be a partial cause of the tokening of a second symbol s_2 at time $t + \partial$? And (2) how can the laws governing the causal regularities also assure that the operations that generate new symbol tokens will "respect" the semantic relationships between the symbols, in the sense that the overall process will turn out to be, in a broad sense, rational?

The answers that CTM's advocates would like to provide for these questions can be developed in two stages. First, work in the formalization of symbol systems in nineteenth- and twentieth-century mathematics has shown that, for substantial (albeit limited) interpreted symbolic domains (such as geometry and algebra), one can find ways of carrying out valid derivations in a fashion that does not depend upon the mathematician's intuition of the meanings of the symbols, so long as (*a*) the semantic distinctions between the symbols are reflected by syntactic distinctions, and (*b*) one can develop a series of rules, dependent wholly upon the syntactic features of symbol structures, that will license those deductions and only those deductions that one would wish to have licensed on the basis of the meanings of the terms. Second, digital computers are devices that store and manipulate symbolic representations.

Their "manipulation" of symbolic representations, moreover, consists in creating new symbol tokens, and the regularities that govern what new tokens are to be generated may be cast in the form of derivation-licensing rules based upon the syntactic features of the symbols already tokened in computer storage. In a computer, symbols play causal roles in the generation of new symbols, and the causal role that a symbol can play is determined by its syntactic type. Formalization shows that (for limited domains) the semantic properties of a set of symbols can be "mirrored" by syntactic properties; digital computers offer proof that the syntactic properties of symbols can be causal determinants in the generation of new symbols. All in all, the computer paradigm shows that one can coordinate the semantic properties of representations with the causal roles they may play by encoding all semantic distinctions in syntax.

These crucial notions of *formalization* and *computation* will now be discussed in greater detail. These notions are, no doubt, already familiar to many readers. However, how one tells the story about these notions significantly influences the conclusions one is likely to draw about how they may be employed, and so it seems worthwhile to tell the story right from the start.

1.6.1 FORMALIZATION

In the second half of the nineteenth century, one of the most important issues in mathematics was the *formalization* of mathematical systems. The formalization of a mathematical system consists in the elimination from the system's deduction rules of anything dependent upon the meanings of the terms. Formalization became an important issue in mathematics after Gauss, Bolyai, Lobachevski, and Riemann independently found consistent geometries that denied Euclid's parallel postulate. This led to a desire to relieve the procedures employed in mathematical deductions of all dependence upon the semantic intuitions of the mathematician (for example, her Euclidean spatial intuitions). The process of formalization found a definitive spokesman in David Hilbert, whose book on the foundations of geometry, published in 1899, employed an approach to axiomatization that involved a complete abstraction from the meanings of the symbols. The formalization of logic, meanwhile, had been undertaken by Boole and later by Frege, Whitehead, and Russell, and the formalization of arithmetic by Peano.

While there were several different approaches to formalization in nineteenth-century mathematics, Hilbert's "symbol-game" approach is of special interest for our purposes. In this approach, the symbols used in proofs are treated as tokens or pieces in a game, the "rules" of which govern the formation of expressions and the validity of deductions in that system. The rules employed in the symbol game, however, apply to formulae only insofar as the formulae fall under particular *syntactic* types. This ideal of formalization in a mathematical domain requires the ability to characterize, entirely in notational (symbolic and syntactic) terms, (*a*) the rules for well-formedness of symbols, (*b*) the rules for well-formedness of formulas, (*c*) the axioms, and (*d*) the rules that license derivations.

What is of interest about formalizability for our purposes is that, for limited domains, one can find methods for producing derivations that respect the meanings of the terms but do not rely upon the mathematician's knowledge of those meanings, because the method is based solely upon their syntactic features. Thus, for example, a logician might know a derivation-licensing rule to the effect that, whenever formulas of the form p and $p \supset q$ have been derived, he may validly derive a formula of the form q. To apply this rule, he need not know the interpretations of any of the substitution instances of p and q, or even know what relation is expressed by $\supset$, but need only be able to recognize symbol structures as having the syntactic forms p and $p \supset q$. As a consequence, one can carry out rational, sense- and truth-preserving inferences without attending to—or even knowing—the meanings of the terms, so long as one can devise a set of syntactic types and a set of formal rules that capture all of the semantic distinctions necessary to license deductions in a given domain.

1.6.2 A MATHEMATICAAL NOTION OF COMPUTATION

A second issue arising from turn-of-the-century mathematics was the question of what functions are "computable" in the sense of being subject to evaluation by the application of a rote procedure or algorithm. The procedures learned for evaluating integrals are good examples of computational algorithms. Learning integration is a matter of learning to identify expressions as members of particular syntactically characterized classes and learning how to produce the corresponding expressions that indicate the values of their integrals. One learns, for example, that integrals with the form $\int x^n dx$ have solutions of the form $(1/n+1)(x^{n+1})$, and so on.

Such computational methods are *formal,* in the sense that a person's ability to apply the method does not require any understanding of the meanings of the terms.[10] To evaluate , for example, one need not know what the expression indicates—the area under a curve—but only that it is of a particular syntactic type to which a particular rule for integration applies. Similarly, one might apply the techniques used in column addition (another algorithmic procedure) without knowing what numbers one was adding. For example, one might apply the method without looking to see what numbers were represented, or the numbers might be too long for anyone to recognize them. One might even learn the rules for manipulating digits without having been told that they are used in the representation of numbers. The method of column addition is so designed, in other words, that the results do not depend upon whether the person performing the computation knows the meanings of the terms. The procedure is so designed that applying it to representations of two numbers A and B will dependably result in the production of a representation of a number C such that $A + B = C$.

1.6.3 THE SCOPE OF FORMAL SYMBOL-MANIPULATION TECHNIQUES

It turns out that formal inference techniques have a surprisingly wide scope. In the nineteenth and early twentieth century it was shown that large portions of logic and mathematics are subject to formalization. And this is true not only in logic and number theory, which some theorists hold to be devoid of semantic content, but also in such domains as geometry, where the terms clearly have considerable semantic content. Hilbert (1899), for example, demonstrated that it is possible to formulate a collection of syntactic types, axioms, and derivation-licensing rules that is rich enough to license as valid all of the geometric derivations one would wish for on semantic grounds while excluding as invalid any derivations that would be excluded on semantic grounds.

Similarly, many problems lying outside of mathematics that involve highly context-specific semantic information can be given a formal characterization. A game such as chess, for example, may be represented by (1) a set of symbols representing the pieces, (2) expressions representing possible states of the board, (3) an expression picking out the initial state of the board, and (4) a set of rules governing the legality of moves by mapping expressions representing legal states of the board after a move m to the set of expressions representing legal successor states after move $m + 1$. Some games, such as tic-tac-toe, also admit of algorithmic strategies that assure a winning or nonlosing game. In addition to games, it is also possible to represent the essential features of many real-world processes in formal models of the sorts employed by physicists, engineers, and economists. In general, a process can be modeled if one can find an adequate way of representing the objects, relationships, and events that make up the process, and of devising a set of derivation rules that map a representation R of a state S of the process onto a successor representation R^* of a state S^* just in case the process is such that S^* would be the successor state to S. As a consequence, it is possible to devise representational systems in which large amounts of semantic information are encoded syntactically, with the effect that the application of purely syntactic derivation techniques can result in the production of sequences of representations that bear important semantic relationships: notably, sequences that could count as rational, cogent lines of reasoning.

1.6.4 COMPUTING MACHINES

The formalizability of limited symbolic domains shows that semantic distinctions can be preserved syntactically and that the application of syntactic derivation rules can result in a semantically cogent sequence of representations. In crude terms, formalization shows us how to link semantics to syntax. What is required, however, is a way of linking the semantic properties of representations with their ability to play a causal role in the generation of new representations to which they bear interesting semantic relationships (see fig. 3). In and of themselves, formal proof methods and formal algorithms do not provide such a link, since they depend upon the actions of the human computer who applies them. It is the paradigm of *machine* computation that provides a way of connecting the causal roles played by representations with their syntactic properties, and thus indirectly linking semantics with causal role.

The crucial transition from formal techniques dependent upon a human mathematician to mechanical computation came in Alan Turing's "On Computable Numbers" (1936). This paper was framed as an answer to the mathematical problem of finding a general characterization of the class of functions that admit of computational (i.e., algorithmic) soltions. Turing's approach to this problem was to describe a machine that was capable of scanning and printing symbols printed on a tape and governed in part by internal mechanisms and in part by the specific symbols found on the tape. Some of the details of this machine are described in chapter 5, but for present purposes it suffices to say that Turing

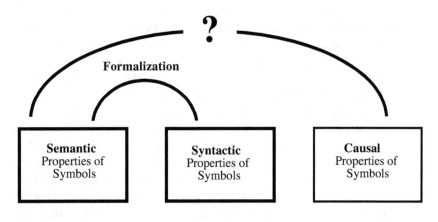

Figure 3

showed that *any computation that can be evaluated by application of a formal algorithm can be performed by a digital machine* of the sort he specifies. The original intent of Turing's article was to provide a general description of all computable functions: a function is computable just in case it can be evaluated by a Turing machine. But in providing this answer to a problem in mathematics, Turing also showed something far more interesting for psychologists and philosophers: namely, that it is possible to design machines that not only passively store symbols for human use, but also actively distinguish symbols on the basis of their shape and their syntactic ordering, and indeed operate in a fashion that is partially determined by the syntactic properties of the symbols on which they operate. In short, Turing showed that it is possible to link syntax to causal powers in a computing machine.

A *computing machine* is a device that possesses several distinctive features. First, it contains media in which symbolic representations can be stored. These symbols, like written symbols, can be arranged into expressions having syntactic structures and may be assigned interpretations through an interpretation scheme. Second, a computer is capable of differentiating between representations in a fashion corresponding to distinctions in their syntactic "shape." Third, it can cause the tokening of new representations. Finally, the causal regularities that govern *what* new symbols the computer will cause to be tokened are *dependent upon the syntactic form* of the symbols already stored by the machine.

To take a simple example, suppose that a computer is programmed to sample two storage locations *A* and *B* where representations of integers are stored and to cause a tokening of a representation at a third location C in such a fashion that the representation tokened at C will be a representation of the sum of the two numbers represented at A and B. The representations found at A, B, and C have syntactic structure: let us assume that each representation is a series of binary digits (1s and 0s). They also have semantic interpretations: namely, those assigned to them by the interpretation scheme employed by the designer of the program. Now when the computer executes the program, it will cause the tokening of a representation at C. Just *what* representation is tokened at C will depend upon what representations are found at A and B. More specifically, it will depend upon the *syntactic type* of the representations found at A and B—namely, upon what sequences of binary digits are present at those locations. What the computer does in executing this program is thus analogous to the application of a formal algorithm (such as that employed in column addition), which is sensitive to the syntactic forms of the representations at A and B. If the program has been properly designed, the overall process will accurately mimic addition as well, in the sense that what is tokened at C will always be a representation of the sum of the two numbers represented at A and B. That is, if the program is properly designed, the syntactically dependent operations performed by the machine will ensure the production of a representation at C that bears the desired semantic relations to the representations at A and B as well.[11] The semantic properties of the representations play no causal role in the process—they are etiologically inert. But since all semantic distinctions are preserved syntactically, and syntactic type determines what a representation can contribute causally, there is a *correspondence* between a representation's semantic properties and the causal role it can play.

This example illustrates three salient points. The first is the insight borrowed from formal logic and mathematics that at least some semantic relations can be reflected or "tracked" by syntactic relations. The second is the insight borrowed from computer science that machines can be made to operate upon symbols in such a way that the syntactic properties of the symbols can be reflected in their causal roles. Indeed, for any problem that can be solved by the application of a formal algorithm A, it is possible to design a machine M that will generate a series of representations corresponding to those that would be produced by the application of algorithm A. These two points jointly yield a third: namely, that it is possible for machines to operate upon symbols in a way that is, in Fodor's words, "sensitive solely to syntactic properties" of the symbols and "entirely confined to altering their shapes," while at the same time

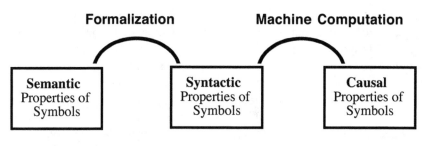

Figure 4

> The machine is so devised that it will transform one symbol into another if and only if the propositions expressed by the symbols that are so transformed stand in certain semantic relations—e.g., the relation that the premises bear to the conclusion of a valid argument. (Fodor 1987: 19)

In brief, "computers show us how to connect semantical with causal properties for *symbols*" (ibid.). And this completes the desired linkage between semantics and causality: for domains that can be formalized, semantic properties can be linked to causal properties by encoding semantic differences in syntax and designing a machine that is driven by the syntactic features of the symbols (see fig. 4).

1.7 THE COMPUTATIONAL ACCOUNT OF COGNITIVE PROCESSES

We have seen that the first thesis comprising CTM was a representational account of the nature of intentional states: namely, that such states are relations to mental representations. The second thesis comprising CTM is a computational account of the nature of cognitive processes: namely, that cognitive processes are *computations over mental representations,* or "causal sequences of tokenings of mental representations" (Fodor 1987: 17). Fodor writes,

> A train of thoughts, for example, is a causal sequence of tokenings of mental representations which express the propositions that are the objects of the thoughts. To a first approximation, to think 'It's going to rain; so I'll go indoors' is to have a tokening of a mental representation that means *I'll go indoors* caused, in a certain way, by a tokening of a mental representation that means *It's going to rain* . (ibid.)

This account may be broken down into several constituent claims. First, cognitive processes are *sequences* of intentional states. Now, according to CTM, to be in a particular intentional state is just to be in a particular functional relation to a mental representation. So if an organism is undergoing a cognitive process, it is passing through a sequence of functional relations to mental representations. Second, there are causal relationships between the intentional states that make up a cognitive process. Being in relation R to a representation of type MP at time t (say, believing at 12:00 noon that it is going to rain) can be a partial cause of coming to be in relation R^* to a representation of type MP^* at time $t + \partial$ (e.g., coming to a decision at 12:01 to go indoors). Third, the causal connection between the states picked out is not merely incidental, but depends in a regular way upon the *syntactic properties* of the mental representations. It is *because* the organism stands in relation *R to a token of (syntactic) type MP* at t that it comes to stand in relation R^* to a token of (syntactic) type MP^* at $t + \partial$, much as our adding program causes a particular representation to be tokened at C because representations with particular syntactic patterns are present at A and B. So just as the representations in computers can play a causal role in the generation of new representations, and do so by virtue of their syntactic form, so also "mental representations have their causal roles in virtue of their formal properties" (Fodor 1981: 26). Fourth, as in the case of a formal algorithm or a computer program, any semantic differences between mental representations are reflected by syntactic distinctions. So for any two mental representations MP and MP^* to which a single organism O is related, if MP and MP^* differ with respect to semantic properties, they must be of different syntactic types as well.

To view mental processes in this way is to treat the mind as being quite literally a digital computer. A computer is a device that performs symbol manipulations on the basis of the syntactic features of the symbols, and it can do so in a fashion that respects such semantic features as are encoded in the syntax. According to CTM, mental states involve symbolic representations from which they inherit their semantic properties. All semantic differences between representations are syntactically encoded, and the mind is a device whose causal regularities are determined by the syntactic properties of its representations.

This account of the nature of cognitive processes allows intentional state ascriptions to pick out intentional states by way of properties that are correlated with their causal powers. Intentional state ascriptions pick out intentional states by the semantic values of the representations they involve. These semantic values are not themselves causally efficient. But, according to CTM, the semantic properties of representations are correlated with their syntactic types. So when representations are picked out by their semantic value, their syntactic type is uniquely picked out as well. But the syntactic type of a representation *is* a determinant of the causal role it can play in causing tokenings of other representations and in the etiology of behavior. And so intentional state ascriptions can pick out causes, and indeed the semantic properties by which intentional states are picked out are correlated with the causal roles that they can play, because semantic properties are correlated with syntactic properties, and syntactic properties determine causal powers. This provides for the possibility of accounting for mental causation in a way that does not require semantic properties to be causally active, and yet correlates semantic value with causal role.

1.8 SUMMARY: THE COMPUTATIONAL THEORY OF MIND

In summary, we have now seen that CTM consists in two main theses. The first thesis is a representational account of the nature of intentional states. On this view, intentional states are relations between an organism and mental representations. These representations are physically instantiated symbol tokens having both semantic and syntactic properties. The second thesis is a computational account of the nature of cognitive processes. Cognitive processes, according to CTM, are computations over mental representations. That is, they are sequences of tokenings of mental representations in which the presence of one representation can serve as a partial cause of the tokening of a second representation. Just what causal roles a representation may play in the generation of other representations and the etiology of behavior is determined by its syntactic properties, and not by its semantic value. But while a representation's semantic value does not influence what causal roles it can play, the semantic value is nonetheless coordinated with causal role, because all semantic differences between representations are preserved syntactically, and syntax determines causal role.

CHAPTER TWO

Computation, Intentionality, and the Vindication of Intentional Psychology

The Computational Theory of Mind has received a great deal of attention in recent years, both in philosophy and in the empirical disciplines whose focus is cognition. On the one hand, the computer paradigm has inspired an enormous volume of theoretical work in psychology, as well as related fields such as linguistics and ethology. On the other hand, philosophers such as Fodor have claimed that CTM provides a solution to certain long-lived philosophical problems as well. The primary focus of this book is upon CTM's claims to solve philosophical problems. Two of these are of primary importance. The first is the claim that *CTM provides a philosophical account of the intentionality and semantics of intentional states* —in particular, that it does so in a fashion that provides thought with the same generative and compositional properties possessed by natural languages. The second is the claim that *CTM "vindicates" intentional psychology by providing a philosophical basis for an intentional psychology* capable of satisfying several contemporary concerns—in particular, concerns for (1) the compatibility of intentional psychology with materialistic monism, (2) the compatibility of intentional psychology with the generality of physics, and (3) the ability to construe intentional explanations as causal explanations based on lawlike regularities. Together, these claims imply that viewing the mind as a computer allows us to "naturalize" the mind by bringing both individual thoughts and mental processes within an entirely physicalistic world view.

It is important to note that the status of these distinctively philosophical claims is largely independent of the claim that the computer paradigm has been empirically fruitful in inspiring important theoretical work in psychology and other disciplines. On the one hand, the theory might ultimately prove to be philosophically interesting but empirically fallow. Such was arguably the case, for example, with representational theories of mind before CTM, and could turn out to be the case for computationalism as well if, in the long run, it goes the way of so many unsuccessful research programmes that initially showed such bright promise. On the other hand, it is possible to interpret psychological research inspired by the computer paradigm—"computational psychology" for short—in a fashion that is weaker than CTM. Fodor acknowledges this when he writes:

> There are *two*, quite different, applications of the "computer metaphor" in cognitive theory: two quite different ways of understanding what the computer metaphor *is*. One is the idea of Turing reducibility of intelligent processes; the other (and, in my view, far more important) is the idea of mental processes as formal operations on symbols. (Fodor 1981: 23-24)

The first and weaker view here is a *machine functionalism* that treats the mind as a functionally describable system without explaining intentional states by appeal to representations. On this view,

> Psychological theories in canonical form would then look rather like machine tables, but they would provide no answer to such questions as "Which of these machine states is (or corresponds to or simulates) the state of believing that P?" (ibid., 25)

The second and stronger application of the computer metaphor is Fodor's CTM, which adds the philosophically pregnant notion of mental representation to what is supplied by machine functionalism. As we shall see in the course of this chapter, Fodor's arguments for preferring CTM to functionalism turn largely upon its ability to "vindicate" intentional psychology and not merely upon factors internal to empirical research in psychology. And hence the strengths and weaknesses of the philosophical claims made on behalf of CTM are largely independent of the viability of computational psychology as an empirical research strategy.

2.1 CTM'S ACCOUNT OF INTENTIONALITY

The first philosophical claim made on behalf of CTM is that it provides an account of the intentionality of mental states. The basic form of this account was already introduced in chapter 1: namely, that mental states involve relationships to symbolic representations from which the states "inherit their semantic properties" (Fodor 1981: 26) and intentionality.

Or, in Fodor's words again, "Intentional properties of propositional attitudes are viewed as inherited from semantic properties of mental representations" (Fodor 1980b: 431). This claim that intentional states "inherit" their semantic properties, moreover, is intended to provide an *explanation* of the intentionality and semantics of intentional states. Beliefs and desires are about objects and states of affairs *because* they involve representations that are about those objects and states of affairs; intentional states are meaningful and referential because they involve representations that are meaningful and referential. In this chapter we will look at this account in greater detail, with particular attention towards (*a*) locating it within the more general philosophical discussion of intentionality and (*b*) highlighting what might be thought to be its strengths.

2.2 INTENTIONALITY

Since the publication of Franz Brentano's *Psychologie vom empirischen Standpunkt* in 1874, intentionality has come to be a topic of increasing importance in philosophy of mind and philosophy of language. While Brentano's own views on intentionality have not proven to be of enduring interest in their own right, his reintroduction of the Scholastic notion of intentionality into philosophy has had far-reaching ramifications. Brentano's pupil Edmund Husserl ([1900] 1970, [1913] 1931, [1950] 1960, [1954] 1970) made intentionality the central theme of his transcendental phenomenology, and the work of subsequent European philosophers such as Martin Heidegger, Jean-Paul Sartre, Jacques Derrida, and Michel Foucault has been articulated in large measure against Husserl's views about the intentionality of mind and language. In the English-speaking world, problems about intentionality have been introduced into analytic philosophy by Roderick Chisholm (1957, 1968, 1983, 1984a, 1984b), who translated and commented upon much of Brentano's work, and Wilfred Sellars (1956), who studied under Husserl's pupil Martin Farber.[1]

Several of the principal aspects of Brentano's problematic have been preserved in subsequent discussions of intentionality. Brentano's characterization of the directedness and content of some mental states has been adopted wholesale by later writers, as has his recognition that such states form a natural domain for psychological investigation and need to be distinguished both from qualia and from brute objects.[2] Recently, moreover, there has been a strong resurgence of interest in the relationship between what Brentano called "descriptive" (i.e., intentional) and "genetic" (i.e., causal, nomological) psychology. Brentano had originally thought that genetic psychology would eventually subsume and explain descriptive psychology, but subsequently concluded that intentionality was in fact an irreducible property of the mental and could not be accounted for in nonintentional and nonmental terms. This position is sometimes described as "Brentano's thesis." This discussion in Brentano is thus a direct forebear of current discussions of the possibility of naturalizing intentionality, with Brentano's mature position represented by writers such as Searle (1983, 1993).

On the other hand, later discussions have placed an increasing emphasis on several aspects of intentionality that are either given inadequate treatment in Brentano's account or missing from it altogether. Notable among these are a concern for relating intuitions about the intentional nature of mental states to other philosophical difficulties, such as psychophysical causation and the mind-body problem, and a conviction that intentionality is a property of *language* as well as of thought, accompanied by a corresponding interest in the relationship between the intentionality of language and the intentionality of mental states. This interest in the "intentionality of language" has taken two forms. On the one hand, writers such as Husserl (1900) and Searle (1983) have taken interest in how utterances and inscriptions come to be about things by virtue of being expressions of intentional states. On the other hand, Chisholm (1957) has coined a usage of the word 'intentional' that applies to linguistic tokens employed in ascriptions of intentional states.[3] This widespread conviction that language as well as thought is in some sense intentional has been paralleled by a similar conviction that some mental states can be evaluated in the same semantic terms as some expressions in natural and technical languages. Notably, it is widely assumed that notions such as *meaning, reference,* and *truth value* can be applied both (*a*) to occurrent states such as explicit judgments and (*b*) to tacit states such as beliefs that are not consciously entertained, in much the fashion that these semantic notions are applied to linguistic entities such as words, sentences, assertions, and propositions. Providing some sort of account of the intentionality and semantics of mental states is thus widely viewed to be an important component of any purported "theory of mind."

2.3 CTM, INTENTIONALITY, AND SEMANTICS

The motivation of CTM's account of intentionality found in Fodor (1981, 1987, 1990) plays upon several themes in the philosophical discussion of intentionality. In particular, it is an attempt to exploit the relationship between the semantics of thought and language in a fashion that provides a thoroughly naturalistic account of the intentionality of mental states—in other words, an account that is compatible with token physicalism and with treating beliefs and desires as things that can take part in causal relations. Fodor writes,

> It does seem relatively clear what we want from a philosophical account of the propositional attitudes. At a minimum, we want to explain how it is that propositional attitudes have semantic properties, and we want an explanation of the opacity of propositional attitudes; all this within a framework sufficiently Realistic to tolerate the ascription of causal roles to beliefs and desires. (Fodor 1981: 18)

Fodor begins his quest for such an account by making a case that intentional states are not unique in having semantic properties—symbols have them as well.

> Mental states like believing and desiring aren't... the only things that represent. The other obvious candidates are *symbols*. So, I write (or utter): 'Greycat is prowling in the kitchen,' thereby producing a 'discursive symbol'; a token of a linguistic expression. What I've written (or uttered) represents the world as being a certain way—as being such that Greycat is prowling in the kitchen—just as my thought does when the thought that Greycat is prowling in the kitchen occurs to me. (Fodor 1987: xi)

It is worth noting that Fodor assumes here that words such as 'represent' can be predicated univocally of intentional states and symbols. But his example also involves an even stronger claim: namely, that symbolic representations such as written inscriptions "represent the world as being a certain way... *just as* [my] thought does." Here the implication would clearly seem to be that there is just *one* sort of "representation" present in the two cases—an assumption that will be shown to have significant consequences later in this book.

The succeeding paragraph in *Psychosemantics* begins to reveal *what* Fodor takes to be common to what initially appear to be separate cases (i.e., mental states and symbolic representation):

> To a first approximation, symbols and mental states both have representational *content*. And nothing else does that belongs to the causal order: not rocks, or worms or trees or spiral nebulae. (Fodor 1987: xi)

It also reveals where his reasoning is headed:

> It would, therefore, be no great surprise if *the theory of mind and the theory of symbols were some day to converge.* (ibid., emphasis added)

There are, however, at least two directions that a convergence of the philosophy of mind and semiotics might take. On the one hand, philosophers like Husserl (1900) and Searle (1983) have argued that the intentional and semantic properties of symbols are to be explained in terms of the intentional and semantic properties of mental states. As we have already seen, however, Fodor's view is quite the reverse: namely, that it is the semantic and intentional properties of mental states which are to be explained, and they are to be explained in terms of the intentional and semantic properties of symbols—specifically, the symbols that serve as the objects of the propositional attitudes. While Fodor does acknowledge that *written* and *spoken* symbols get their semantic properties from the states that they express, he nonetheless holds that

> it is mental representations that have semantic properties in, one might say, the first instance; the semantic properties of propositional attitudes are inherited from those of mental representations and, presumably, the semantic properties of the formulae of natural languages are inherited from those of the propositional attitudes that they are used to express. (Fodor 1981: 31)

The resulting account of intentional states reduces the claim that a particular token intentional state has semantic or intentional properties to a conjunction of two claims to the effect that (*a*) a mental symbol token has semantic or intentional properties, and (*b*) an organism stands in a particular kind of functional relationship to that symbol token. As Fodor expresses it in the passage already cited from *Psychosemantics,*

> *Claim 1* (the nature of propositional attitudes):
> For any organism *O*, and any attitude *A* toward the proposition *P*, there is a ('computational'-'functional') relation *R* and a mental representation *MP* such that
>   *MP* means that *P*, and
>   *O* has *A* iff *O* bears *R* to *MP*. (Fodor 1987: 17)

It seems clear that questions about the *meaningfulness* and (putative) *reference* of intentional states are to be construed as questions about the symbolic representations involved. The same may be said for *truth value* in those cases where the concept applies, though the applicability of truthfunctional evaluation to a given intentional state would seem to depend upon the attitude involved, since most kinds of cognitive attitudes (e.g., desire, dread, etc.) are not subject to truth-functional evaluation.

2.4 THE VIRTUES OF THE ACCOUNT

There are several features of this account that render it attractive. First, the account locates the ultimate bearers of semantic properties in symbol tokens, and symbol tokens are among the sorts of things that everyone agrees can be physical objects. To the many who want intentionality and want materialism too, this is a substantial advance over previous theories that attributed intentionality either directly to minds (whose compatibility with materialism is in doubt) or directly to brain states (which are problematic as logical subjects of semantic predicates). The account also lends some clarity to the familiar analysis of intentional states in terms of intentional attitudes (such as belief and desire) and content. The attitude-content distinction is itself only a distinction of analysis. CTM fleshes this distinction out in a way that no previous theory had done. Attitudes are interpreted as functional relations between an organism and its representations, and content in terms of the semantic properties of the representations. CTM thus both retains and clarifies a central feature of the standard analysis of intentional states.

The account of intentionality and semantics offered by CTM also provides a way of understanding both narrow and broad notions of propositional content. According to CTM, what is necessary for an intentional state to have a particular content in the narrow sense—that is, what is necessary for it to be "about-X" construed opaquely, or in such a fashion as not to imply that there exists an X for the state to be about—is for it to involve a relationship between an organism and a symbol token of a particular formally delimited type. Whether the state is also contentful in the broad sense (i.e., "about X" under a transparent construal—one that *does* imply that there is an X to which the state is about) will depend upon how that symbol token is related to extramental reality: for example, whether it stands in the proper sort of causal relationships with X. While CTM does not provide an account of *what* relationships to extramental reality are relevant to the broad notion of content, the representational account of narrow content allows CTM to avoid several traditional pitfalls associated with the "hard cases" presented by illusions, hallucinations, false beliefs, and other deviant cases of perception and cognition. Notably, CTM escapes the Meinongian tendency to postulate nonexistent entities and the opposite inclination to identify the contents of intentional states with the extramental objects towards which they are directed.

Two features of CTM's account of intentionality, however, seem to be of utmost importance: its relation to CTM's account of cognitive processes and its ability to endow thought with a compositional semantics. It is perhaps an understatement to say that CTM's representational account of intentionality would be of little interest outside of narrowly philosophical circles if it were not coupled with a causal theory of cognitive processes. Locating the arcane property of intentionality in the equally mysterious meanings of hypothetical mental representations would cut little ice were it not for the fact that treating thoughts as relations to symbols provides a way of explaining mental processes as computations. Indeed, as writers like Haugeland (1978, 1981) have noted, it is the discovery of machine computation that has revitalized representational theories of the mind.

The other signal virtue of viewing thoughts as relations to symbolic representations is that this allows us to endow the mind with the same generative and creative powers possessed by natural languages. We do not simply think isolated thoughts—"dog!" or "red!" Rather, we form judgments and desires that are directed towards states of affairs and represented in propositional form. And our ability to think "The dog knocked over the vase" is in part a consequence of our ability to think "dog" in isolation. We are, furthermore, able to think new thoughts and to combine the ideas we have in novel ways. If I can think "The dog knocked over the vase" and I can think "cat," I can also think "The cat knocked over the vase." Therefore there is more to be desired from a theory of intentional states than an account of the meanings of individual ideas: there is also the fact that thought seems to be generative and systematic.

Viewing the mind as employing representations in a language of thought gives us this for free. For we already have a way of answering the corresponding questions in linguistics by employing the principle of compositionality. If a language is compositional, then the semantic values of complex expression are a function of (*a*) the semantic values of the lexical (or morphemic) atoms and (*b*) the syntactic structure of the expression. The generative and systematic qualities of languages are explained by the use of iterative syntactic structures and the substitution of known lexical items into the slots of known syntactic structures. So if the semantic properties of our thoughts are directly inherited from those of the symbols they involve, and the symbols involved are part of a language employing compositional principles, then these explanations from linguistics can be incorporated wholesale into our psychology. The mind has generative and systematic qualities because it thinks in a language that has a compositional semantics.

This is an important result because it is virtually impossible to make sense of reasoning by way of a representational theory except on the assumption that complex thoughts, such as "The cat knocked over the vase," are composed out of simpler parts, corresponding to "cat" and "vase." For when one has a thought of a cat knocking over a vase, this thought is immediately linked to all kinds of other knowledge about cats and vases and causality. One may infer, for example, that an animal knocked over the vase, that something knocked over an artifact, or that the vase is no longer upright. If mental representations were all semantic primitives, the ability to make such inferences on the basis of completely novel representations would probably be inexplicable. The simplest explanation for our ability to combine our knowledge about cats with a representation meaning "The cat knocked over the vase" is that the representation has a discrete component meaning "cat," and that the overall meaning of the representation is determined by how the component representations are combined. This, however, points to the need for a representational system in which syntax and semantics are closely connected. For the *only* known way of endowing a system of representations with this kind of compositionality is by way of supplying the representational system with *syntactic rules* that govern how to form semantically complex representations out of semantic primitives. CTM provides for this compositionality, and it is not clear that any account not based on an underlying system of languagelike representations would be able to say the same.

2.5 CTM AS THE BASIS FOR AN INTENTIONAL PSYCHOLOGY

The first important claim made on behalf of CTM is thus that it provides an account of the semantic and intentional properties of mental states. The second important claim made on behalf of CTM is that it provides a philosophical basis for intentional psychology. CTM's proponents believe that it provides a framework for psychological explanation that allows intentional state ascriptions to figure in such explanations, while also accommodating several contemporary concerns in philosophy of science. Three such concerns are of preeminent importance: (1) concerns that psychological explanations be causal explanations based on nomological regularities, (2) concerns that psychological explanations be compatible with the generality of physics (i.e., with the ability of an ideally completed physics to supply explanations for every token event), and (3) concerns that the ontology implicit in psychology be compatible with materialistic monism. Proponents of CTM thus view their project as one of "vindicating commonsense psychology" or "showing how you could have... a respectable science whose ontology explicitly acknowledges states that exhibit the sorts of properties that common sense attributes to [propositional] attitudes" (Fodor 1987: 10).

The perceived need for such a "vindication" was occasioned by the disrepute into which intentional psychology—and indeed mentalism in general—had fallen in the first half of the twentieth century. By the time that the notion of computation was available as a paradigm for psychology, many philosophers and psychologists believed that there could not be a scientific psychology cast in mentalistic or intentional terms. The roots of this suspicion of mentalism and intentional psychology may be traced to the views about the nature of science in general, and psychology in particular, associated with two movements: methodological behaviorism in psychology and Vienna Circle positivism in philosophy. In order to understand fully the significant emphasis placed upon "vindicating intentional psychology" in articulations of CTM (particularly early articulations), it is necessary briefly to survey these other movements which were so influential in the earlier parts of this century.

2.6 THE DISREPUTE OF MENTALISM—
A BRIEF HISTORY

The legitimacy of intentional psychology was seriously impugned in the first half of the twentieth century by ideas emerging from methodological behaviorists in psychology and from logical positivists in philosophy. Methodological behaviorism, as articulated by Watson (1913a, 1913b, 1914, 1924) and Skinner (1938, 1953), raised methodological concerns about explanations that referred to objects (mental states) that were not publicly observable and were not necessary (they argued) for the prediction and control of behavior.

Early logical positivism, as typified by Carnap's *Aufbau* (1928), adopted a "logical behaviorism" which Putnam describes as "the doctrine that, just as numbers are (allegedly) logical constructions out of *sets,* so *mental events* are logical constructions out of actual and possible *behavior events* " (Putnam [1961] 1980: 25). This interpretation of mental events is based upon a positivist account of the meanings of words, sometimes called the "verification theory of meaning." The criteria for verification of psychological attributes, the logical behaviorists argued, consist in observations of (*a*) the subject's overt behavior (gestures made, sounds emitted spontaneously or in response to questions) and (*b*) the subject's physical states (blood pressure, central nervous system processes, etc.). Since motions and emissions of sounds are straight-forwardly physical events, they argued, claims about psychological processes are reducible to statements in physical language.[4] The con-clusion, in Hempel's words, is that "all psychological statements which are meaningful, that is to say, which are in principle verifiable, are translatable into statements which do not involve psychological con-cepts, but only the concepts of physics.... Psychology is an integral part of physics" (Hempel [1949] 1980: 18).[5]

Vienna Circle positivism was characterized by a tension between *epistemological* concerns (with a concomitant tendency towards phe-nomenalism) and a commitment to *materialism*. Logical behaviorism emerged in the context of the *epistemological* concerns and radically empiricist (and even phenomenalistic) assumptions of early Vienna Circle positivism. As a consequence, it involved the assumption that "observational terms refer to subjective impressions, sensations, and perceptions of some sentient being" (Feyerabend 1958: 35). Carnap's *Aufbau* was the most significant work advocating this kind of logical reduction, though the influence of phenomenalism may be seen clearly in the early works of Russell and in the nineteenth-century German positivism of Mach.

Yet Carnap soon rejected the *Aufbau* account of the relationship be-tween physical and psychological terms and adopted a new understand-ing of science, emphasizing the materialist theme in positivism instead of the epistemological-phenomenalist theme. According to this view, observation sentences do not refer to the sense impressions involved in the actual observations, but to the (putative) objects observed, de-scribed in an intersubjective "thing-language."[6] Thus in 1936 Carnap writes, "What we have called observable predicates are predicates of the thing-language (they have to be clearly distinguished from what we have called perception terms... whether these are now interpreted sub-jectivistically, or behavioristically)" (Carnap [1936-1937] 1953: 69). And similarly Popper writes that "every basic statement must either be itself a statement about relative positions of physical bodies... or it must be equivalent to some basic statement of this 'mechanistic' kind" (Pop-per 1959: 103).

Oppenheim and Putnam's "Unity of Science as a Working Hypothe-sis" (1958) has become a locus classicus for this newer view, commonly called reductive physicalism—the view that every mental type has a cor-

responding physical type and all psychological laws are thus translatable into laws in the vocabulary of physics.[7] The ideal of science articulated by Oppenheim and Putnam shares with logical behaviorism and
Skinnerian operationalism a commitment to a "reduction" of mentalistic terms, including intentional state ascriptions, but the "reductions"
employed in the three projects differ both in nature and in motivation.[8]

Now while these three scientific metatheories differ with respect to
their motivations and their chief concerns, each contributed to a growing suspicion of intentional psychology. By the time the digital computer was available as a model for cognition, it was widely believed
that one could not have a scientific psychology that employed intentional state ascriptions. This skepticism about intentional psychology
reflected four principal concerns: (1) a concern about the nature of *evidence* for a scientific theory—particularly a concern that the evidence
for psychological theories be publicly or intersubjectively observable;
(2) a concern about the nature of scientific *explanation*—in particular, a
concern that scientific explanations be *causal* and *nomological;* (3) an
ontological concern about the problems inherent in dualism, and particularly a commitment to materialistic monism;[9] and (4) a commitment to
the generality of physics—that is, the availability of a physical explanation for every token event.

2.7 VINDICATING INTENTIONAL PSYCHOLOGY (1): MACHINE FUNCTIONALISM

The proponents of CTM believe that it has supplied a way of preserv-
ing the integrity of explanations cast in the intentional idiom while also
accommodating the concerns that had contributed to the ascendancy of
reductive approaches to mind in the first half of the century. Historically, the attempt to vindicate intentional psychology involved two distinct
elements: (1) the introduction of *machine functionalism* as a rigorous
alternative to behaviorism of various sorts and to reductive physicalism, and (2) CTM's combination of machine functionalism with the
additional notions of *computation* and *representation.*

In his 1936 description of computation, Alan Turing introduced the
notion of a computing machine. The machine, which has come to be called
a "Turing machine," has a tape running through it, divided into squares,
each capable of bearing a "symbol."[10] At any given time, the machine is
in some particular internal condition, called its "*m*-configuration." The
overall state of the Turing machine at a particular time is described by

"the number of the scanned square, the complete sequence of all symbols on the tape and the *m*-configuration" (Turing 1936: 232). A Turing machine is *functionally specifiable:* that is, the operations that it will perform and the state changes it will undergo can be captured by a "machine table" specifying, for each complete configuration of the machine, what operations it will then perform and the resulting *m*-configuration.

Machine functionalism is the thesis that intentional states and processes are likewise functionally specifiable—that is, that they may be characterized by something on the order of a machine table.[11] The thesis requires some generalizations from the computing machine described by Turing. In Putnam's 1967 articulation, for example, the tape of the machine is replaced by "sensory inputs" and "motor outputs," and a corresponding adjustment is made to the notion of a machine table to accommodate these inputs and outputs. Putnam also generalizes from Turing's deterministic case, in which state transitions are completely determined by the complete configuration of the machine, to a more permissive notion of a "Probabilistic Automaton," in which "the transitions between 'states' are allowed to be with various probabilities rather than being 'deterministic'" (Putnam [1967] 1980: 226). Since a single physical system can simultaneously be the instantiation of any number of deterministic automata, Putnam also introduces "the notion of a *Description* of a system." Of this he writes,

> A Description of *S* where *S* is a system, is any true statement to the effect that *S* possesses distinct states $S_1, S_2,..., S_n$ which are related to one another and to the motor outputs and sensory inputs by the transition probabilities given in such-and-such a Machine Table. The Machine Table mentioned in the Description will then be called the Functional Organization of *S* relative to that Description, and the S_i such that *S* is in state S_i at a given time will be called the Total State of *S* (at that time) relative to that Description. (ibid., 226)

This provides a way of specifying conditions for the type identity of psychological states in functional terms. As Block and Fodor articulate it, "For any organism that satisfies psychological predicates at all, there exists a unique best *description* such that each psychological state of the organism is identical with one of its machine states relative to that description" (Block and Fodor [1972] 1980: 240).

A psychology cast in functional terms possesses the perceived merits of behaviorist and reductive physicalist accounts while avoiding some of their excesses. First, a functional psychology founded on the machine analogy seems to provide the right sorts of explanations for a rigorous psychology. The machine table of a computer expresses relationships between types of complete configurations that are both *regular* and *causal*. If cognition is likewise functionally describable by something on the order of a machine table, psychology can make use of causal, nomological explanations.

Machine functionalism is also compatible with commitments to ontological materialism and to the generality of physics. A computing machine, after all, is unproblematically a physical object, all of its parts are physical objects, and all of its operations have explanations cast wholly in physical terms. If functional description is what is relevant to the individuation of psychological states and processes, the resulting functional psychology could be quite compatible with the assumptions that (*a*) all of the (token) objects in the domain of psychology are physical objects, and that (*b*) all of the token events explained in functional terms by psychology are susceptible to explanation in wholly physical terms as well.

While machine functionalism is compatible with materialism and token physicalism, it is incompatible with reductive or type physicalism, since functionally defined categories in a computer (e.g., AND-gates) are susceptible to indefinitely many physical implementations that are of distinct physical types. It is for this reason that much of the early computationalist literature focuses on comparing the merits of functionalism with those of reductive physicalism. For example, Fodor offers a general sketch of the case against reductive physicalism:

> The reason it is unlikely that every kind corresponds to a physical kind is just that (*a*) interesting generalizations...can often be made about events whose physical descriptions have nothing in common; (*b*) it is often the case that *whether* the physical descriptions of the events subsumed by such generalizations have anything in common is, in an obvious sense, entirely irrelevant to the truth of the generalizations, or to their interestingness, or to their degree of confirmation, or, indeed, to any of their epistemologically important properties; and (*c*) the special sciences are very much in the business of formulating generalizations of this kind. (Fodor 1974: 15)

Additional arguments for the benefits of functionalism over reductionism were marshaled on the basis of Lashley's thesis of equipotentiality, the convergence of morphological and behavioral features across phylogenetic boundaries, and the possibility of applying psychological predicates to aliens and artifacts (see Block and Fodor [1972] 1980). Advocates of functionalism thus see it as capturing the important insights of reductionists (compatibility with materialism and the generality of physics) while avoiding the problems of reductionism.

Advocates of machine functionalism view it as capturing the better side of behaviorism in similar fashion. Functional definition of psychological terms avoids appeals to introspection and private evidence, thereby satisfying one of the concerns of methodological behaviorists like Watson and Skinner. Any ontological suspicion of "the mental" is also avoided by machine functionalism, since computers are plainly objects that are subject to physical instantiation. Functionalism also permits the use of black-box models of psychological processes, much like behaviorism; and like the behaviorisms of Tolman and Hull (but unlike those of Watson and Skinner) it permits the models to include interactions *between* mental states and does not restrict itself to characterizations of states and processes in dispositional terms, thereby accounting for the intuition that psychological states can interact causally.

Machine functionalism is thus seen by its advocates as uniting the best features of behaviorism with those of physicalism. This, writes Fodor, allowed for the solution of

> a nasty dilemma facing the materialist program in the philosophy of mind: What central state physicalists seemed to have got right—contra behaviorists—was the ontological autonomy of mental particulars and, of a piece with this, the causal character of mind-body interactions. Whereas, what the behaviorists seemed to have got right—contra the identity theory—was the relational character of mental properties. *Functionalism, grounded in the machine analogy, seemed to be able to get both right at once.* (Fodor 1981: 9, emphasis added)

2.8 VINDICATING INTENTIONAL PSYCHOLOGY (2): SYMBOLS AND COMPUTATION

Despite its significant virtues, machine functionalism alone is not sufficient for vindicating intentional psychology. What machine functionalism establishes is that there can be systems which are characterized by causal regularities not reducible to physical laws. What it does not establish is that *physical objects* picked out by a functional description of a physical system can also be *mental* states or that functionally describable processes can also be rational mental processes. First, there is an ontological problem: functionalism alone does not show that the physical objects picked out by functional descriptions can be the very same things as the mental tokens picked out in the intentional idiom. As a consequence, explanations in intentional terms are still ontologically suspect, even if there can be *some* functionally delimited kinds which are unproblematic ontologically. The second problem is methodological: unless the kinds picked out by a psychology, even functional psychology, are the sorts of things susceptible to *semantic* relationships, the explanations given in that psychology do not have the characteristics that explanations in intentional psychology have.[12]

CTM seeks to rescue intentional psychology from this impasse by uniting functional and intentional psychologies through the notion of *symbol* employed in the computer paradigm. Computers, according to the standard account, are not merely functionally describable physical objects—they are functionally describable *symbol manipulators*. Symbols, however, are among the sorts of things that can have semantic properties, and computer operations can involve transformations of symbol structures that preserve semantic relationships. This provides a strategy for uniting the functional-causal nature of symbols with their semantic nature, and suggests that a similar strategy might be possible for mental states. Thus Fodor writes,

> Computation shows us how to connect semantical with causal properties for *symbols*. So, if having a propositional attitude involves tokening a symbol, then we can get some leverage on connecting semantical properties with causal ones for *thoughts*. (Fodor 1987: 18)

This, however, requires the postulation of mental *symbols:*

> In computer design, causal role is brought into phase with content by exploiting parallelisms between the syntax of a symbol and its semantics. But that idea won't do the theory of *mind* any good unless there are *mental* symbols: mental particulars possessed of both semantical syntactic properties. There must be mental symbols because, in a nutshell, only symbols have syntax, and our best available theory of mental processes—indeed, the *only* available theory of mental processes that isn't *known* to be false—needs the picture of the mind as a syntax-driven machine. (ibid., 19-20)

It is this addition of the notion of *symbol* that makes CTM stronger than machine functionalism. And it is in virtue of this feature that CTM can lay some claim to solving problems that functionalism was unable to solve. First, it can lay claim to solving the ontological problem. The ontological problem was that functionalism provided no warrant for believing that the functionally individuated (physical) objects forming the domain of a functional psychology could also be mental states—in particular, it seemed doubtful that they could have semantic properties. But if some of those functionally delimited objects are physically instantiated *symbols,* the computationalist argues, this difficulty is solved. Symbols can *both* be physical particulars *and* have semantic values. So if intentional states are relationships to physically instantiated symbol tokens, and the semantic and intentional properties of the symbol tokens account for the semantic and intentional properties of the mental states, then it would seem to be the case that mentalism is compatible with materialism.

The second problem for machine functionalism was that it was unclear how functionally delimited causal etiologies of physical events could also amount to rational explanations. But the computer paradigm also seems to provide an answer to this question. If we assume that (1) intentional states involve symbol tokens with semantic and syntactic properties, that (2) cognitive processes are functionally describable in a way that depends upon the syntactic but not the semantic properties of the symbols over which they are defined, and that (3) this functional description preserves semantic relationships, then (4) functional descriptions can pick out cognitive processes which are also typified by semantic relationships. Functional descriptions of computer systems are based in causal regularities, and so intentional explanations can pick out causal etiologies. And since the state changes picked out by the functional description are caused by the physical properties of the constituent parts of the system, intentional explanation is compatible with the generality of physics.

CTM thus purports to have accomplished a major tour de force. It claims to have vindicated intentional psychology by providing a model in which mentalism is compatible with materialism, and in which explanation in the intentional idiom picks out causal etiologies and is compatible with the generality of physics. The appeal of this achievement, moreover, has outlived the popularity of the movements in philosophy of psychology that originally motivated the desire for a "vindication" of intentional psychology. For while there are relatively few strict behaviorists or reductionists left on the scene in philosophy of science, it is still widely believed that a scientific psychology should employ causal-nomological explanations and be compatible with materialism and with the generality of physics. It is perhaps ironic that these desiderata emerged as consequences of particular short-lived theories in epistemology, philosophy of language, and the logic of science. The theories from which they emerged—the verification theory of meaning and the thesis that there are reductive translations between the languages of the various sciences—have largely been abandoned, but the suspicion of the mental they engendered has outlived them. And thus the "vindication" of intentional psychology will likely continue to be perceived as a virtue so long as this suspicion remains.

2.9 SUMMARY

This chapter has examined two major claims made on behalf of CTM: that it offers an account of the intentionality and semantics of intentional states, and that it provides a vindication of intentional psychology. These results are largely independent of one another, but both depend heavily upon computationalists' largely uncritical use of the notion of *symbol*. Each of these two results is highly significant in its own right, and if CTM can make good on either claim, it will have made a significant contribution to philosophy of mind and psychology. The next chapter will discuss some problems that have been raised about CTM's account of intentionality and semantics, and will argue that a proper evaluation of the account will require an examination of the notions of *symbol* and *symbolic representation*.

CHAPTER THREE

"Derived Intentionality"

If the computational theory has excited a maelstrom of interest in recent years, it has received a generous share of criticism as well. One important line of criticism, developed most notably by John Searle (1980, 1983, 1984, 1990) and Kenneth Sayre (1986, 1987), has centered around the suitability of the notions of *computation* and *symbolic representation* for explaining the semantic properties of mental states. I shall argue that there are several distinct lines of argument to be had here, but they share in common an intuition that there is something, either about the notion of symbolic representation in general or about representation in computers in particular, that makes it impossible to account for the semantic properties of mental states in representational terms.

What I shall do in this chapter is examine the criticisms offered by Searle and Sayre, and develop out of them three distinct lines of argument against CTM. The first, the "Formal Symbols Objection," locates the problem in CTM's attempt to wed the notion of representation with that of computation. It does this by claiming that computation is defined as "formal symbol manipulation," and hence is defined only over meaningless "formal symbols," with the consequence that, if the mind is a computer, it cannot operate upon meaningful symbols as required by CTM. The remaining arguments locate the difficulty for CTM more generally in the nature of symbolic meaning. More specifically, they locate the problem in the claim that symbolic meaning is "derived," whereas the meaningfulness of mental states is not derived, but "intrinsic." There are, however, two kinds of "derivativeness" that need to be explored here, as they provide the bases for two very different objections. The first, the "Causal Derivation Objection," agrees with CTM that there is a class of properties called "semantic properties" that can be predicated both of symbols and of mental states, but it claims that symbols must "derive" their semantic properties from preexisting meaningful mental states. The second, the "Conceptual Dependence Objection" makes a much more radical claim: namely, that the semantic vocabulary (i.e., the words used to express semantic properties) is systematically homonymous—or, more precisely, *paronymous*. On this view, words in the semantic vocabulary *express different properties* when applied (*a*) to symbols and (*b*) to mental states, and in such a fashion that an analysis of the meanings of these terms as applied to symbols will refer back to meaningful mental states. According to this objection, the "semantic properties" attributed to symbols are (*a*) distinct from and (*b*) conceptually dependent upon the "semantic properties" attributed to mental states.

The examination of these three lines of argument in this chapter will not itself yield a decisive verdict with respect to the viability of CTM. It will, however, make clear the questions that must be addressed in succeeding chapters in order to reach such a verdict. The main results of the chapter may be summarized as follows: CTM relies heavily upon the notion of *symbolic representation* as a notion that can be used to account for the meaningfulness of mental states. There is some question, however, about whether the notion of symbolic representation— and more precisely, symbolic *meaning* —may not itself be conceptually dependent upon the notion of meaningful mental states, and hence incapable of explaining them. In order to determine whether this is so, however, it will prove necessary to undertake a full-scale analysis of symbols and symbolic meaning (chapter 4) and apply the results of this to computers (chapter 5) and to CTM (chapter 7).

It will, moreover, prove necessary to examine an additional concern as well. This chapter and several of those that follow it will share in the assumption made by Searle and Sayre that when advocates of CTM speak of representations as "meaningful symbols," it is *symbolic* meaning that they have in mind—that is, the kind of semantic properties customarily attributed to symbols. It is this assumption that will undergird the examination of the nature of symbolic meaning and the attempt to apply this notion to an account of intentionality in chapter 6, and the thesis that will be advanced in this and the next three chapters—the paronymy of the semantic vocabulary—is somewhat radical. But in light of the suggestion that the semantic vocabulary might be systematically paronymous, it will prove necessary to investigate a second reading of CTM as well: namely, that words in the semantic vocabulary, such as 'meaningful,' do *not* express the same properties when applied to mental representations that they express when applied to garden-variety symbols—that words in the semantic vocabulary have a *special* use when applied to mental representations, a use whose analysis will differ from that of the analysis of terms such as 'meaningful' as applied to garden-variety symbols such as utterances and inscriptions. If this is the case, semiotics will prove irrelevant to the assessment of CTM, and the semantic properties of representations will have to be construed in some other way. In spite of the reasonableness of Searle's and Sayre's assumption that CTM attributes to mental representations the very same sorts of "semantic properties" that are attributed to inscriptions and utterances, this alternative reading must also be considered. Chapter 5 will develop an alternative interpretation of CTM's use of the words 'symbol' and 'syntax'. Chapters 7 and 8 will examine two distinct strategies for conceiving the *semantic* component of CTM in a way that does not depend on a semiotic analysis of representation.

3.1 SEARLE'S AND SAYRE'S CRITICISMS

In light of the key role that the notion of *symbol* plays in CTM, it is quite natural that some of the more important criticisms of the computational theory have been based upon objections to computationalists' use of that notion. John Searle and Kenneth Sayre have both articulated objections to CTM that are directed against (supposed) problems with the use to which writers like Fodor and Pylyshyn put the notion of *symbol,* especially as it occurs within the context of discussions of machine computation.

Searle and Sayre have argued that, whatever the virtues of CTM may be, one thing that it cannot provide is a model for understanding the intentionality and semantics of mental states. This, they argue, is a straightforward consequence of defining the notion of computation in terms of formal symbol manipulation. Sayre sums up the problem in this way:

> The heart of the problem is that computers do not operate on symbols with semantic content. Not even computers programmed to prove logical theorems do so. Hence pointing to symbolic operations performed by digital computers is no help in understanding how minds can operate on meaning-laden symbols, or can perform any sort of semantic information-processing whatever. (Sayre 1986: 123)

As Sayre sees it, the problem is that in order to provide a model for understanding cognitive (intentional) processes as manipulations of symbols, machine computation would have to provide a paradigm in which *meaningful* symbols were manipulated by a computer. But the very definition of computation as *formal* symbol manipulation, argues Sayre, prohibits this: "There is no purely formal system—automated or otherwise—that is endowed with semantic features independent of interpretation" (ibid.). And while the interpretation assigned by the programmer or user does, in *some* sense, lend semantic properties to symbols in computer memory, "whatever meaning, truth, or reference they have is derivative... tracing back to interpretations imposed by the programmers and users of the system" (ibid.).

The interpretations imposed by programmers and users are, in Sayre's view, quite irrelevant to the claims of CTM. For to say that a symbol in computer storage has some meaning (in virtue of an interpretation imposed by a programmer or user) is not to say something about what that symbol *is*, but rather to say something about how it is *used*. But computationalism attempts to explain human mental processes on the model of computation—that is, on the model of computers *just as computers,* not on the model of some use to which computers are or could be put. For Sayre, this seems to rule out the possibility of CTM providing a way of understanding the meaningfulness and intentionality of mental states: since computation is defined in formal terms, and claims about the meanings of computer symbols are claims about how computers are *used,* it seems to follow that "computers, just in and by themselves... do not exhibit intentionality at all" (Sayre 1986: 124). And hence, argues Sayre, thinking of mental activities as computations "is of no help in explaining the nature of the intentionality those activities exhibit" (ibid., 124-125).

A very similar case is made by John Searle in his 1984 book *Minds, Brains and Science.* Searle writes that "it is essential to our conception of a digital computer that its operations can be specified purely formally" (Searle 1984: 30). A consequence of this is that, in a computer, "the symbols have no meaning.... they have to be specified purely in terms of their formal or syntactic structure" (ibid., 31). Like Sayre, Searle deems this to be fatal to the ability to CTM to account for semantics and intentionality. He argues that "there is more to having a mind than having formal or syntactical processes. Our internal mental states, by definition, have certain sorts of contents.... That is, even if my thoughts occur to me in strings of symbols, there must be more to the thought than the

abstract strings, because strings by themselves can't have any meaning"
(ibid.).

3.2 THREE IMPLICIT CRITICISMS

The basic thread of criticism common to Searle and Sayre is clear
enough: symbolic representation in computers does not provide a fit
model for the intentionality of mental states. But if the general lines of
the criticism are plain enough, the exact details are a bit more difficult.
On the one hand, there seems to be some suggestion that the problem
lies specifically with symbols in *computers,* to the effect that these
symbols (unlike other symbols) are not meaningful at all, and hence are
poor candidates for explaining the meaningfulness of mental states. On
the other hand, other passages suggest a more general problem about
the very nature of symbolic meaning—namely, that the semantic properties of symbols, even symbols in computers, are somehow "derived"
from the meaning-bestowing acts and conventions of symbol users, and
that this somehow imperils the possibility of accounting for the meaningfulness of mental states in terms of the meaningfulness of symbols. I
shall argue, moreover, that there are in fact two different senses in
which symbolic meaning might be said to be "derivative," each of
which can serve as the basis of an attack upon CTM. In the following
sections, I shall discuss each of these variations upon Searle's and
Sayre's texts in turn. My concern here will, moreover, be with analysis
of the different lines of argument rather than with questions of exegesis.

3.3 THE FORMAL SYMBOLS OBJECTION

In some places, Searle and Sayre each seem to suggest that the problem
for CTM lies specifically in the fact that it tries to wed the notion of
computation to that of *symbolic meaning.* Sayre writes, for example,
that "computers do not operate on symbols with semantic content," and
he concludes from this that "pointing to symbolic operations performed
by digital computers is no help in understanding how minds can operate
on meaning-laden symbols" (Sayre 1986: 123). Similarly, Searle writes
that the symbols in a computer "have no meaning.... they have to be
specified purely in terms of their formal and syntactic structure" (Searle
1984: 31). This, according to Searle, is the crucial difference between
mental states and symbols in computers:

The reason that no computer program can ever be a mind is simply that a computer program is only syntactical, and minds are more than syntactical. Minds are semantical, in the sense that they have more than a formal structure, they have a content. (ibid.)

A natural way of reading these passages would be that Searle and Sayre believe that computation is defined only over a special class of "formal symbols" that are, by definition, devoid of semantic content. Indeed, Searle writes that a computer "attaches no meaning, interpretation, or content to the formal symbols it manipulates." On this reading, the problem with the computer paradigm is that it cannot be applied to symbols that have semantic and intentional properties, and hence cannot be applied to the kinds of mental representations postulated by CTM. I shall call this objection the "Formal Symbols Objection."

It is easy to see how such a line of criticism might arise. If computers are defined as "formal symbol manipulators," it is tempting to conclude that this means that they are devices that manipulate some class of entities called "formal symbols"—that is, symbols devoid of semantic content. Moreover, this interpretation of CTM is not without textual support from important advocates of CTM. Pylyshyn, for example, speaks of computation as operation upon "meaningless symbol tokens," and goes so far as to bemoan the fact that (even) computationalists sometimes experience an "occasional lapse" in which "semantic properties are... attributed to representations" (Pylyshyn 1980: 114-115). It is, however, easy enough to find passages in expositions of CTM that are in contradiction with this interpretation as well. Fodor consistently insists that computers *do* operate upon symbols that have meanings, though he claims that computers have *access* only to the syntactic properties.[1] And even Pylyshyn takes a line similar to Fodor's in his book *Computation and Cognition*.[2] (There are some writers [e.g., Stich 1983] who have advocated a purely "syntactic" theory based on the computer metaphor, but their views are significantly at odds with CTM.)

Now if computation *were* restricted by definition to meaningless symbols, there would indeed be a problem with extending the computer paradigm to provide an account of the intentionality of mental states in the fashion indicated by CTM. For CTM requires that the mind be a system that performs computations over meaning*ful* representations. As a consequence, if computation is *defined* as applying only to meaning*less* symbols, CTM demands the impossible, and the claims it makes are self-contradictory and hence false.

This criticism, however, seems based upon a dubious understanding of the notion of computation, and more particularly upon a dubious parsing of the expression 'formal symbol manipulator'. If the Formal Symbols Objection turns upon the claim that computers are, by definition, capable of manipulating only "meaningless formal symbols," the objection is deeply flawed and reflects a misunderstanding of the use of the word 'formal'. If this is Searle's and Sayre's point, they seem to have confused questions about the *formal specifiability* of symbol *systems* with questions about the *meaningfulness* of *symbol tokens.* Symbol tokens, strictly speaking, are neither formal nor informal. *Derivation techniques* are said to be formal if they do not depend upon the meanings of the symbols, and *systems* that employ symbol structures, such as logic or geometry, are said to be formal if they involve only formal derivation techniques. But formal logic and formal geometry generally *do* involve some degree of semantic interpretation. Indeed, it is only because the systems have interpretations that they can be regarded as *logic* and *geometry*. When one speaks of "formal symbol manipulations," the word 'formal' modifies the word 'manipulations', not the word 'symbol'; and when one speaks of computers as "formal symbol systems" one does not thereby imply that the symbols lack interpretations, but only that the symbol *manipulations* performed by the machine do not *depend upon* the interpretations of the symbols. There is thus no contradiction in saying that the mind is a computer that operates on meaningful symbols.

There is, of course, a much milder sort of objection that might be voiced about *which* symbols in computers *do in fact* have meanings, or indeed if *any* do. To the extent that symbols in computers might turn out to be meaningless, computers become less appealing as a model for the mind. But really this poses no significant threat to CTM. CTM's claim, after all, is not that production-model computers provide a good metaphor for the mind, but that the exact mathematical notion of computation provides the right sort of resources for supplementing a representational account of intentionality with a computational account of cognitive processes. And this claim requires only the possibility of consistently combining computation with representation *in the case of mental states,* regardless of whether it takes place in production-model computers. However, the *persuasive force* of the arguments marshaled in favor of CTM depend in large measure on the claim that the paradigm of machine computation shows that it is possible to combine symbolic meaning with syntactically driven computation in the desired fashion, and upon the assumption that this same union can be made to work in the case of mental representations, and it remains to be seen whether an investigation of symbols and computation in computing machines will bear these assumptions out. An examination of symbolic representation in general, and representation in computers in particular, thus seems desirable.

3.4 DERIVED INTENTIONALITY

If the Formal Symbols Objection does not seem to present serious problems for CTM, Searle's and Sayre's discussions raise what would seem to be a more serious objection as well. For while both writers sometimes speak as though the problem with CTM lies specifically with symbols in *computers,* each also says things that suggest that the problem is a problem concerning symbolic meaning generally. The nub of the problem is that symbolic meaning is "derived" from the meanings of mental states and from conventions governing the use of symbols, and thus CTM has the relationship between symbolic meaning and mental meaning precisely reversed.

Searle develops this view in his discussion of the relationship between intentional states and illocutionary acts. Searle holds that the sense in which intentional and semantic properties may be attributed to symbols in computers is precisely analogous to the sense in which they may be attributed to illocutionary acts such as assertions and promises. Illocutionary acts, according to Searle, have their intentional properties because they are *expressions* of intentional states: "In the performance of each illocutionary act with a propositional content, we express a certain Intentional state with that propositional content.... The performance of the speech act is *eo ipso* an expression of the corresponding Intentional state" (Searle 1983: 9). The intentionality of illocutionary acts (and other linguistic tokens) is *derived from* the intentionality of mental states. Indeed, illocutionary acts are said to be "intentional" in two ways, each of which depends upon the intentionality of a mental state. First, since a speech act *derives its content* from the intentional state of which it is the expression, it is intentional in the sense of *having a content* in virtue of its relationship to an intentional state with that same content. (An assertion is about Lincoln, for example, because it is an expression of a belief about Lincoln.) What unites the utterance with the intentional state it expresses, however, is a second intentional state—namely, the intention of the speaker that his utterance be an expression of a particular intentional state (see Searle 1983: 27).

The intentionality of symbols in computers, according to Searle, can be explained in just the same fashion. Symbols in a computer, like marks on paper or vocalized sounds, are not intrinsically meaningful. Meaning is *imputed* to symbols by some being who has intentional states. In the case of language, it is the speaker or writer. In the case of symbols in computers, it is the designer, programmer, or user. Intentional states have intentionality *intrinsically;* symbols have it only *derivatively.*

Sayre makes a case against the extendability of the computer paradigm in a similar fashion. Like Searle, he admits that symbols in computers may in *some* sense be said to have meanings and intentionality, but he maintains that "whatever meaning, truth, or reference they have is derivative... tracing back to interpretations imposed by the programmers and users of the system" (Sayre 1986: 123). Sayre argues that treating computers as dealing with meaningful symbols involves talking *not* just about the *computer,* but about the uses to which it is put and the interpretations imposed upon its symbols and its operations by the user. "Computers, just in and by themselves... do not exhibit intentionality at all" (ibid., 124). Since intentional states *do* have intentionality "in and by themselves"—that is, independently of impositions of interpretations from outside sources—the computer paradigm is ill suited to providing an understanding of the intentionality and semantics of mental states.

3.5 THE AMBIGUITY OF
"DERIVED INTENTIONALITY"

The notion of "derived intentionality" to which both Searle and Sayre appeal is of crucial importance in assessing CTM. Yet it is also ambiguous and admits of two significantly different interpretations. On one interpretation, the word 'derived' indicates something about *how* an object that *has* intentional or semantic properties *got* them. On this interpretation, words such as 'intentionality' and 'meaning' express the same properties when applied to symbols and mental states, and an object has *derived* intentionality just in case it received or inherited its intentional properties from another object having intentional properties by way of some causal connection. This will be called "causally derived intentionality." On the second interpretation, the "derivativeness" of the intentional properties of symbols is a *logical* feature of the *way intentional properties can be ascribed to symbols.* On this view, terms such as 'meaningful' and 'intentional' cannot be predicated univocally of symbols and mental states, and hence any theory that depends upon the univocal application of such terms is conceptually confused. This will be called the "conceptually dependent intentionality" of symbols. These two notions have significantly different impacts upon CTM, and so will receive independent development. The Causal Derivation Objection assumes with CTM that there is one kind of property called "intentionality," and claims that there is a one-way inheritance relationship between mental states and symbols. The Conceptual Dependence Objection claims that what words in the semantic vocabulary signify when applied to symbols is (*a*) distinct from and (*b*) logically dependent upon what they signify when applied to mental states.

3.6 CAUSALLY DERIVED INTENTIONALITY

The first way of interpreting the expressions 'intrinsic' and 'derived intentionality' is to construe them as pointing to differences in the *sources* of the intentional properties of mental states on the one hand, and symbols and illocutionary acts on the other. On this view, there is one *property* called *intentionality* which cognitive states, symbols, and illocutionary acts can each possess, but they come by it in different ways. Thus Searle writes,

> An utterance can have Intentionality, just as a belief has Intentionality, but whereas the Intentionality of the belief is *intrinsic,* the Intentionality of the utterance is *derived.* The question then is: How does it derive its Intentionality? (Searle 1983: 27)

This way of phrasing the problem reflects Searle's views on the nature of the problem of linguistic or symbolic meaning:

> Now the problem of meaning in its most general form is the problem of how do we get from the physics to the semantics; that is to say, how do we get (for example) from the sounds that come out of my mouth to the illocutionary act? (ibid.)

Searle's answer is that utterances come to have semantic properties because the person making the sounds intends "their production as the performance of a speech act" (ibid., 163). This is an instance of what Searle calls "Intentional causation": the speaker's "meaning intention" that the sounds express an intentional state is a *cause* of the fact that the utterance comes to have intentionality.[3]

If the expression 'derived intentionality'—or more generally, 'derived

semantic properties'—is meant to signify this sort of causal dependence, we may clarify the usage of the expression in the following way:

Causally Derived Semantics

X has semantic property P derivatively iff

   (1) X has semantic property P

   (2) There is some Y such that

      (*a*) $Y \neq X$,

      (*b*) Y has semantic property P, and

      (*c*) Y's having P is a (perhaps partial) cause of X's having P.

X may be said to have semantic property P *intrinsically* just in case X
has P and X does not have P derivatively.

Now I take it that both Searle and Sayre would wish to claim that
symbolic meaning is causally derivative, whereas the intentionality of
mental states is intrinsic. That is, they would claim that the semantic
properties of symbols are causally derived from the semantic properties
of the mental states of symbol users, while there is no Y such that a
mental state M's meaning P is causally dependent upon Y's meaning P.
If this is correct, then CTM errs in two respects: first, it assumes that
the intentionality of mental states is derived (i.e., from mental representations) rather than intrinsic; second, it assumes that the intentionality
of symbols can be accounted for without recourse to causal derivation
from mental states.

3.7 ASSESSING THE CAUSAL DERIVATION OBJECTION

In order for the Causal Derivation Objection to prevail against CTM, it
would be necessary to establish two claims:

   (S1) *The Derivative Character of Symbolic Meaning:* Necessarily, all
        symbols with semantic properties have those properties by way
        of causal derivation.

   (S2) *The Intrinsic Character of Mental Semantics:* All semantic
        properties of mental states are intrinsic (i.e., not causally de-
        rived).

It seems quite clear that, if these two claims are correct, CTM is fundamentally flawed. Unfortunately, the arguments provided by writers like

Searle and Sayre do not establish the derivative character of symbolic meaning, but merely the weaker claim that certain familiar *kinds* of symbols—inscriptions, utterances, symbols in computers—have *their* semantic properties by way of causal derivation. As for the claim that all semantic properties of mental states are intrinsic, no argument has been given for that claim at all.

Moreover, both of these claims have come under fire from proponents of CTM and other writers in cognitive science. I shall explore two lines of response, which I shall call the "Fodor move" and the "Dennett move."

3.7.1 THE "FODOR MOVE"

It is not too surprising to find that Fodor is a dissenter with respect to claim (S1). It is, however, illuminating to note how far he is willing to go along with Searle's analysis of symbolic meaning. Fodor largely agrees with the analysis Searle gives of the semantic properties of *discursive* symbols such as those involved in illocutionary acts, and he even seems to sympathize with the notion that one must give a similar sort of account of semantics for symbols in computers. But he also thinks that mental representations *differ* from discursive symbols precisely in this regard:

> It is mental representations that have semantic properties in, one might say, the first instance; the semantic properties of propositional attitudes are inherited from those of mental representations and, presumably, the semantic properties of the formulae of natural languages are inherited from those of the propositional attitudes that they are used to express. (Fodor 1981: 31)

Thus Fodor agrees with Searle that there are certain properties called "semantic properties" that can be possessed by mental states and by discursive symbols. He agrees, additionally, that discursive symbols get their semantic properties from the intentional states they are used to express. He simply adds that there are these *other, non* discursive symbols in the mind that (*a*) do not get their semantic properties the way discursive symbols do, but in some other fashion, and (*b*) are the ultimate source of the semantic properties of intentional states. I shall call this reply to the Causal Derivation Objection the "Fodor move."

There seems to be little in Searle's texts to militate against the Fodor move. Searle agrees with Fodor that there is a class of properties such as "intentionality" that can be predicated indifferently of symbols and mental states. And, while he has argued convincingly that certain familiar classes of symbols—perhaps all of the familiar classes—acquire their semantic properties by way of causal derivation, he has offered no reason to draw the stronger conclusion that there cannot be other classes of symbols that can acquire semantic properties by other means. There should, of course, be a burden of proof upon CTM's advocates to justify the claim that there *is* some other way for symbols to acquire meaning (and to do so without equivocating on the notion of "meaning"), but arguably this is precisely what they are doing when they seek to provide "theories of content" for representations.[4]

3.7.2 THE "DENNETT MOVE"

Claim (S2), which asserts the intrinsic character of the intentionality of mental states, has likewise met with disagreements. The very assertion of CTM involves its explicit denial, since CTM claims that mental states "inherit" their semantic properties from representations. Moreover, the claim of intrinsicality seems to swim against a strong current within cognitive science of attempts to explain high-order cognitive phenomena by breaking them down into (hypothetical) lower-order cognitive phenomena, sometimes explaining the semantic properties of the high-order phenomena by appeal to the semantic properties of their components. Dennett (1987) has perhaps taken this move as far as anyone, claiming that if sociobiology provides legitimate explanations, then the semantic properties of our mental states are ultimately traceable to the intentions of our genes. (Call this the "Dennett move.")

Now there are two ways of taking the Dennett move. On the strong reading, Dennett is really claiming that genes are in fact the ultimate source of intentionality—that they have intentionality intrinsically, and that everything else that has intentionality, including mental states and discursive symbols, has it derivatively. (On this view, many mental representations would not have intentionality intrinsically either, but the intentionality of intentional states could still be causally derived from that of mental representations.) This version of the Dennett move is only as plausible as (*a*) the claims of sociobiology, and particularly (*b*) the assumption that semantic properties can sensibly be ascribed to entities such as genes. I think that most readers find these claims, especially (*b*), to be more than a little suspect. But there is also a weaker way of treating the Dennett move: namely, to take it as a kind of reductio of Searle's Causal Derivation Objection to theories like CTM. On this reading,

Dennett's real point is that once you let in the notion of causal derivation of intentionality, there is no reason to stop with intentional states,
since cognitive scientists regularly explain beliefs and desires by appeal
to infraconscious states to which they also impute semantic properties.
Perhaps the chain does not go back so far as genes, but why assume
that it stops with intentional states?

Now it may well be possible to muster an adequate reply to the Dennett move, but doing so would seem to require something beyond what
is present in the Causal Derivation Objection. That objection already
acknowledges that several diverse kinds of things (symbols and mental
states) have semantic properties, and that the presence of semantic
properties in one can cause semantic properties to be present in the other. What is to prevent the possibility of other kinds of entities having
such properties as well, or to prevent them from causing the semantic
properties of mental states? The answer one *wants* to give, perhaps, is
that there is something simply *outrageous* about attributing semantic
properties to genes or nerve firings—that these just are not among the
sorts of things of which properties such as "meaning" and "intentionality" may sensibly be predicated. They are not proper logical subjects for
belief attributions. To make a case for this, however, would require
something more than the notion of causally derived intentionality provides—namely, an analysis of the nature of meaning and intentionality.
It is here that the next line of criticism will find much of its appeal.

3.7.3 PROSPECTS OF THE CAUSAL DERIVATION OBJECTION

All in all, it seems that the notion of causally derived intentionality may
serve to place a significant burden of proof upon would-be advocates of
CTM, but it does not reveal any fundamental flaw with that theory. The
burden of proof arises from the fact that mental representations would
have to differ from all known kinds of symbols in a fundamental respect: namely, in how they come to have semantic properties. The only
way we know of for symbols to acquire meaning is through interpretive
acts. This does not preclude the possibility of symbols acquiring meaning in some other way, but it is likewise unclear that there is any other
way for them to acquire meaning. Since CTM *requires* that there be
another such way, its advocates had best show that there can be such a
way. This, however, is arguably just what CTM's proponents are doing
when they discuss "theories of content" for mental representations.

3.8 THE CONCEPTUAL DEPENDENCE OBJECTION

While I rather expect that something on the order of *causal* dependence was what Searle and Sayre had in mind when they spoke of "derived intentionality," there is another way of interpreting the expression 'derived intentionality' which may damage CTM in a more fundamental manner. On the *causal* construal of Searle's expression 'derived intentionality', the term 'intentionality' could be predicated univocally of both mental states and symbols. The difference between cognitive states and symbols lay in how they *came to have* this one property. This is probably what Searle had in mind in his discussion of derived intentionality. But one *might* read the expression 'derived intentionality' in quite another way. One might read it as meaning "intentionality in a derivative *sense.*" On this reading, attributions of intentional and semantic properties to cognitive states and attributions of intentional and semantic properties to symbols *are not attributions of the same properties.*

3.8.1 THE HOMONYMY OF 'HEALTHY':
AN ARISTOTELIAN PARADIGM

In setting up the problem, it may be helpful briefly to recall Aristotle's discussion of homonymy. Aristotle points out that some words, such as 'healthy', are used in different ways when they are applied to different kinds of objects. We say that there are healthy people, healthy food, healthy exercise, healthy complexions, and so on. But when we say that some kind of food is healthy, we are not predicating the same thing of the food that we are predicating of a person when we say that he is healthy. If I say that fish is a healthy food, I mean that *eating* fish is *conducive* to health in humans. But if I say that Jones is healthy, I mean that he is *in good health.*[5] (Individual fish can be healthy in the same sense that Jones is healthy, but they have ceased from being in good health by the time they are healthy food for Jones.) Yet words like 'healthy', according to Aristotle, are not *merely* homonymous. Rather, there is *one* meaning which is *primary* or *focal,* and the other meanings are all to be understood in terms of how they relate to the primary meaning. In the case of 'healthy', the primary meaning is the one that applies to people (or, more generally, to living bodies): to be healthy in the primary sense is to be in good health. Things other than living beings are said to be "healthy" in other senses because of the way that they relate to being in good health: for example, because of the way they *contribute* to being in good health (e.g., a healthy diet or regimen), or because of the way they *indicate* good health (e.g., a healthy complexion). Aristotle calls this kind of homonymy *paronymy* or "homonymy *pros hen*." The sense of 'healthy' that applies to food is dependent upon, and indeed *points* to the sense of 'health' that applies to bodies. "Healthy food" *means* "food that is conducive to bodily health." And similarly the senses of 'healthy' that apply to exercise, appearance, and so on point to the notion of bodily health. As a result, questions about the "healthiness" of a particular food amount to questions about *how it contributes to bodily health*. Someone who thought that bodily health was derived from the "health" contained in the food one ate would simply be mistaken about how the word 'health' is used. And it would betray conceptual confusion if one were to say, "I don't want to know how broccoli contributes to bodily health, I just want to know why it is *healthy*."

Of course, someone *could* use the word 'healthy' in some *new* manner. For example, someone convinced that vitamins were the source of bodily health might start applying the word 'healthy' in a way that just meant "full of vitamins." This use of the word 'healthy' (to be indicated 'healthy$_v$') would no longer be conceptually dependent upon the notion of bodily health. But 'healthy$_v$ ' would not mean what 'healthy' is normally understood to mean either. In particular, one could not draw upon any implications of the *normal* use of the word 'healthy' in reasoning about things that are healthy$_v$. For example, presumably things are healthy$_v$ in proportion to the number and quantity of vitamins present in them. A meal with ten thousand times the recommended daily allowance of all vitamins would be very healthy$_v$. But one cannot infer from this that such a meal would be healthy (conducive to health). First, since 'healthy$_v$ ' no longer bears a semantic connection to the notion of bodily health, the analytically based inference does not go through. Second, the conclusion happens to be empirically false. Massive doses of some vitamins are not conducive to health, but toxic. Old words can be given completely new meanings, but then what you have is homonymy plain and simple. And not all of the things that may be said of things that can be said to be healthy can also be said of things that are healthy$_v$.

3.8.2 "DERIVED INTENTIONALITY" AS THE HOMONYMY OF 'INTENTIONAL'

Now one could interpret the expression 'derived intentionality' as pointing to a conceptual dependence between different uses of words such as 'intentional', 'intentionality', 'meaningful', 'referential'—in short, of all words used in attributing intentional and semantic properties. And the nature of the dependence would be along the following lines: both symbols and mental states are *said to be* intentional, meaningful, referential, and so on. But words such as 'intentional' and 'meaningful' are not used in the same way when they are said of symbols as when they are said of mental states. Intentional and semantic terms are homonymous. But they are not *merely* homonymous. Rather, it is a case of *paronymy,* or homonymy *pros hen,* where there is a *primary* or *focal* sense of each term: specifically, the sense that is applied to cognitive states. The sense that is applied to symbols is "derivative" or *conceptually dependent* because it refers back to the sense that is applied to cognitive states. For example, when we say that a speech act is intentional, what we *mean* is that it is an *expression of an intentional state.* On this view, there would be no sense in which a speech act could be said to be intentional that did not point to an intentional state in similar fashion.

Now I believe that a view of this sort is *implicit* in some of the things written by Sayre and Searle, but I do not see that it is ever explicitly articulated in this form, or marshaled as an explicit objection to CTM.[6] Searle's analysis, moreover, is confined almost exclusively to illocutionary acts, and is not developed more generally for *symbols.* Since CTM does not posit that mental representations are illocutionary acts, Searle's analysis would at very least have to be broadened if it is to provide a criticism of CTM. It is quite possible, however, that Searle has in mind something like this notion of conceptual dependency of symbolic intentionality and meaning when he blames the inadequacies of CTM upon the fact that the "meaningfulness" and "intentionality" of symbols in computers is "dependent" upon the intentions of users and programmers.

Sayre's analysis of the shortcomings of CTM might also be read as relying upon the premise that the kind of intentionality that symbols may be said to have is conceptually dependent upon the kind of intentionality that cognitive states may be said to have. Sayre places more stress than does Searle upon the role that computer users and programmers play in *imbuing* symbols in computers with meaning and intentionality. He writes, for example, that

> none of the representations internal to the machine has meaning, or truth, or external reference, just in and by itself. Whatever meaning, truth, or reference they have is derivative... tracing back to interpretations imposed by programmers and users of the system....
>
> ... My point is that computers, *just in and by themselves,* no matter how programmed, do not exhibit intentionality at all. (Sayre 1986: 123, 124)

If assertions which appear to be just about the meaningfulness of symbols in computers turn out to be (covert) assertions about the actions and intentions of computer users and programmers, then the computer paradigm does involve symbols with "intentional" and "semantic" properties, but only in the sense that it involves a *human-computer system* in which the humans impute semantic and intentional properties to the symbols in the computer. If this be the case, there may well be problems about extending the model to account for intentionality in humans.

Unlike Searle, Sayre also touches more broadly upon the semantic features of symbols in general. In discussing the semantic properties of symbols in a natural language, he stresses the point that natural language symbols have semantic properties only because of interpretive conventions:

> Inasmuch as the English word "cat" refers to cats, the word consists of more than can be uttered or written on paper. It consists of the symbolic form CAT (which can be instantiated in many ways in speech and writing) *plus interpretive conventions by which instances of that form are to be taken as referring to cats*. Similarly, the symbolic form Go means the opposite of STOP (or COME, etc.) by appropriate interpretive conventions of English, while by those of Japanese it means a board game played with black and white stones. *But without interpretive conventions it means nothing at all.* (Sayre 1986: 123, emphasis added)

If this passage is read with the notion of conceptual dependence in mind, it is extremely suggestive. If talk about the meaningfulness of symbols is necessarily (covertly) talk about linguistic conventions, then the meaningfulness of symbols is conceptually dependent upon conventions. And if *this* is the case, CTM may be in very serious trouble indeed.

3.8.3 THE POTENTIAL OF A "CONCEPTUAL DEPENDENCE OBJECTION"

While Searle's and Sayre's criticisms of CTM may well include the kernel of a "Conceptual Dependence Objection," no full-scale development of such an objection has yet been offered. Developing such an objection will, among other things, involve a careful examination of the notion of *symbol* and the ways that symbols of various sorts may be said to have semantic and intentional properties. Such an analysis will be undertaken in chapter 4.

But even prior to such an analysis, it is possible to see, in general terms, what force such an objection would have. CTM's representational account of cognitive states consists primarily in the claim that these involve symbolic representations which have semantic properties. If the Conceptual Dependence Objection can be made to stick, however, all attributions of semantic and intentional properties to symbols refer to something more than the symbol: they refer to the beings who are responsible for the symbol's having an interpretation. This would present two kinds of problems for CTM. First, like the Causal Derivation Objection, it calls the *credibility* of the computer paradigm into question: it just seems incredible to postulate that there *is* some being (or beings) responsible for interpreting mental symbols. But there is also a more fundamental problem: if *all* attributions of symbolic meaning are (covertly) attributions of the *imposition* of meaning, then attributions of intentional and symbolic properties to *any* symbol would have to involve attributions of intentional states of some agent or agents responsible for the imposition of meaning upon that symbol. And this would seem to involve CTM in regress and circularity: CTM explains the intentionality and semantics of cognitive states in terms of the intentionality and semantics of symbols. But if the intentionality and semantics of symbols are, in turn, cashed out in terms of cognitive states, there is circularity in the interexplanation of cognitive states and symbols, and a regress of cognitive states responsible for the intentionality and semantics of other cognitive states. Such an objection would be far more damaging than the Causal Derivation Objection.

3.9 THE NEED FOR SEMIOTICS

This chapter has been devoted to a discussion of attacks marshaled against CTM that are directed specifically against its use of the notion of *symbol*. The upshot of the chapter is that the notion of *symbol* needs further elucidation. The first objection, the Formal Symbols Objection, turned upon claims about the symbols stored in and manipulated by computers—specifically, the claims that computers only *could* store or only *do* store meaningless "formal symbols." It was suggested that this objection rested upon a confusion about the meaning of expressions such as 'formal symbol manipulation'. The "formality" of formal systems and computers, it was argued, consists in the fact that the techniques through which derivations of symbols are effected are blind to the semantic properties of the symbols. A mathematical system or a computer can be formal in this sense and still operate upon meaningful symbols. Indeed both formalized mathematical systems (such as Hilbert's geometry) and digital computers often do involve meaningful symbols—that is, symbols that are assigned interpretations by the mathematician, the programmer, or the computer user. The Formal Symbols Objection is nonetheless very alluring, and the literature on computers and the mind is replete with suggestions that computers operate upon some special class of "meaningless formal symbols." The ambiguity of expressions such as 'formal symbol manipulation' and the difficulty of characterizing the semantic status of symbols in computers gives us reason to inquire more carefully into the nature of attributions of semantic properties to symbols in general and to symbols in computers in particular.

A general examination of the notion of *symbol* is also made necessary by our development of the notion of "derived intentionality." The Causal Derivation Objection consisted in the claim that the account one would give of the intentional and semantic properties of symbols cannot also be used as an account of the intentional and semantic properties of cognitive states, because symbols have their intentional and semantic properties only by virtue of causal derivation. But all that was really shown was that illocutionary acts and symbols in computers do not have intentionality intrinsically. Computationalists now generally agree that (*a*) CTM does not provide a full-fledged semantic theory, and (*b*) mental representations do not come by their intentionality the way symbols in computers do. The question of whether some *other* kinds of symbols *might* have intentionality and semantics intrinsically remains open.

Finally, the Conceptual Dependence Objection argues that the very *notion* of *symbol* makes essential reference to cognizers who are responsible for the imposition of meaning upon symbols, and upon the cognitive states which are involved in this imposition of meaning. This objection might well undercut CTM completely, but it has yet to be developed in detail and requires a careful examination of the nature of symbols as a prerequisite.

A further issue also arises here: if the semantic vocabulary *does* turn out to be systematically homonymous, it may turn out additionally that the kind of "meaning" that is to be attributed to mental representations is not the same kind of "meaning" that is attributed to symbols. So, in addition to assessing the question of whether symbolic meaning can explain mental meaning, it may prove necessary to examine whether there might be other kinds of "meaning" possessed by mental representations (i.e., other properties expressed by a distinct usage of the word 'meaning').

This sets some agenda for the remainder of this book. Chapter 4 will undertake the task of clarifying the notion of *symbol*—specifically, it will examine what it is to be a symbol, what it is to have syntactic properties, and what it is to have semantic properties. This analysis will be applied to CTM in chapter 7 in order to assess the force of the objections marshaled by Sayre and Searle. Meanwhile, chapter 5 will explore an alternative way of interpreting the use of the words 'symbol' and 'syntax' by CTM's advocates, and chapters 8 and 9 will examine two ways of articulating a notion of "semantics" that is in important ways discontinuous with the usage of that word as applied to symbols.

PART II

Symbols, Computers, and Thoughts

CHAPTER FOUR

Symbols—An Analysis

The preceding chapter has brought us to a crucial juncture in assessing the merits of the Computational Theory of Mind. The criticisms raised by Searle and Sayre point to some potentially serious problems for computationalism. But the exact nature and force of the problems cannot be judged without first undertaking an analysis of the notion of *symbol*, which figures prominently both in the claims made by computationalists and in the criticisms leveled by their opponents. In particular, we must ask what it is to be a symbol and how symbols may be said to have syntactic and semantic properties. This chapter will present answers to these questions, and will offer a rich set of terminology for talking about symbols, syntax, and semantics.

The terminology makes two important kinds of distinctions. First, the ordinary usage of the word 'symbol' as a sortal term is ambiguous. Sometimes the word is used precisely to denote utterances or inscriptions that have semantic interpretations—things that symbol*ize* something. But in other contexts the word is used to denote things which do not have semantic properties: there are purely formal symbol games, for example, in which the tokens have syntactic but not semantic properties, and there are even symbols such as letters on eyecharts which have *neither* syntactic *nor* semantic properties. To distinguish these different senses of the word 'symbol', three sortal terms will be developed. The term 'marker' will be used to capture the road usage of 'symbol' which includes letters on eyecharts. To be a marker is just to be a token of a conventional type, and does not have any necessary semantic or syntactic consequences. The term 'signifier' will be used to denote markers that have semantic interpretations. An object is a signifier just insofar as it is a token of a marker type which has an interpretation. The term 'counter' will be used to pick out markers on the basis of their syntactic features within some language game. To be a counter is to be a token of a marker type which has a particular set of conventionally determined syntactic properties in a particular language game.

The terminology developed in this chapter will also reflect a second, and equally important, distinction. For there are several different senses in which an object may be said to be a marker, a counter, or a signifier. For example, if we say that a marker token "has an interpretation," we might mean one of four things: (1) that there is a linguistic convention that associates the marker's type with that interpretation, (2) that the author of the marker meant it to have that interpretation, (3) that someone who apprehended the marker *took* it to have that interpretation, or merely (4) that there is an interpretation scheme available in principle that associates that marker's type with that interpretation. The terminology developed in this chapter disambiguates expressions like 'is a signifier' or 'has semantic properties' by offering different locutional schemas for each of the four legs of the ambiguity. In the four cases above, for example, the marker token would be said, respectively, to be (1) *interpretable (under convention C) as* signifying *X*, (2) *intended (by its author S) as* signifying *X*, (3) *interpreted (by some Y) as* signifying *X*, and (4) *interpretable-in-principle as* signifying *X*. These locutions point to four *modalities* under which an object may be said to have properties dependent upon conventions or intentions, and these modalities also apply to the sortals 'marker' and 'counter', as well as 'signifier', in ways that will be made clearer in the course of the chapter. The result is a terminology that reflects four different ways in which an object may be said to be a marker (a symbol in the barest sense), four ways a marker may be said to take on syntactic properties, and four ways it may be said to take on semantic properties. The remainder of this chapter will be devoted to a more detailed development of these distinctions.

4.1 SYMBOLS: SEMANTICS, SYNTAX,
   AND TOKENING A TYPE

It should come as no surprise that the word 'symbol' is used in widely differing ways by writers with different research interests. When a linguist studying the development of the set of graphemic characters used to represent English words speaks of the graphemes as "symbols," he will very likely mean something different from what a Jungian psychologist means when he expresses an interest in finding out what "symbols" are important to a patient. But even if we restrict our attention to the *linguistic* notion of *symbol* that is relevant to the analysis of natural, technical, and computer languages, there are still ambiguities that need to be unraveled.

First, the word 'symbol' is sometimes used precisely to indicate objects that symbol*ize* something else. An object is a symbol in this sense *just in case it has a semantic interpretation.* This usage of the word 'symbol' is found quite frequently in discussions of computation and the philosophy of mind. Fodor, for example, uses the word 'symbol' in this way in the introduction to *RePresentations,* where he repeatedly glosses the word 'symbol' with the phrase "semantically interpreted object[s]" (Fodor 1981: 22, 23, 30) and claims that the objects of propositional attitudes "are symbols... and that this fact accounts for their intensionality and semanticity" (ibid., 24). Haugeland likewise uses the word 'symbol' in this way when he writes, "Sometimes we say that the tokens in a certain formal system *mean* something—that is, they are 'signs,' or 'symbols,' or 'expressions' which 'stand for,' or 'represent,' or 'say' something" (Haugeland 1981: 21-22).

But not all writers who discuss the tokens employed in formal systems follow Haugeland's practice of applying the word 'symbol' only to objects having semantic interpretations. Pylyshyn, for example, distinguishes between "a system of formal symbols (data structures, expressions)" and a scheme of interpretation "for interpreting these symbols" (Pylyshyn 1984: 116). Here Pylyshyn uses the word 'symbol' in a way which clearly and explicitly does not have semantic overtones, since the "symbols" of which he speaks are purely "formal" and are only imbued with meaning through the additional imposition of a scheme of interpretation. Logicians interested in formal systems likewise use the word 'symbol' to denote the characters and expressions employed in those systems, even though by definition semantics falls outside of the purview of formal systems.

Such a practice is also justified by ordinary usage: it is quite acceptable, for example, to use the word 'symbol' to refer to graphemic characters such as letters, numerals, punctuation marks, and even to such characters as those employed in musical notation. To merit the application of *this* use of the word 'symbol', an object need not have any semantic interpretation. For example, individual letters employed in inscriptions in a natural language seldom have semantic values, and yet there is nothing strange about referring to them individually as symbols.

Here it might seem tempting to follow Haugeland's terminological practice and to contrast "symbols" (things with interpretations) with "*formal* tokens"—or, alternatively, to join Pylyshyn in using the expression 'formal symbols' when referring to such entities as character strings without reference to their semantic properties. But to do so would be to risk running afoul of a further distinction. For the word 'formal' has weaker and stronger uses. In its weaker use, it means "not semantic"; in its stronger use, it means "syntactic." This distinction is important because *entities such as letters and phonemes fall into types quite independently of their syntactic properties*. The same set of letter types, for example, is employed in the written forms of most of the European languages, and the *same* letters take on *different* syntactic properties in different languages. Now if letter types were *determined* by the syntactic positions that their tokens could occupy in a symbol game, then symbol games with different syntactic rules would, by definition, have to be construed as employing different symbol types. For example, given that the spelling rules of English and French allow different combinations of letters to occur, one would have to say that English and French employ different letters. But surely such a conclusion would be misguided: there is good reason to say that written French and written English employ the *same* symbol types (i.e., the same letter types), but that symbols of the same types take on different syntactic properties when used in inscriptions in different languages. It is surely more natural, for example, to say that the letter *y* can stand alone as a word in French but not in English than to say that French and English have distinct symbol types which happen to look alike, just because the English *y* can occur only within a larger word while the French *y* can occur alone. Or, to take a different example, it seems natural to say that base-2 notation and base-10 notation both employ the numerals zero and one, even though those numerals take on different combinatorial properties in the two systems. (This is trivially true, since the digits 0 and 1 can be combined in base-10 notation with digits that are not employed in base-2 notation.)

We thus stand in need of *three separate sortal terms* to play the different roles played by the ordinary term 'symbol'. First, we need a term that designates objects like letters and numerals quite apart from any considerations about what syntactic or semantic properties they might take on in a particular context. Second, we need a term that designates objects just insofar as they are assigned a semantic interpretation. Finally, we need a term that designates objects just insofar as they are of a particular syntactic type.

4.2 MARKERS, SIGNIFIERS, COUNTERS

I propose to use three existing words in new and technical ways in order to supply the necessary sortal terms. I propose to use the word 'marker' to replace the term 'symbol' in its broadest sense, the usage that can be applied to letters and numerals and carries no syntactic or semantic connotations. There are marker *types* (e.g., the letter P) and marker tokens (a particular inscription of the letter P). Marker types such as letter types and numeral types are a particular class of conventionally established types. *And so an object is a marker token just insofar as it is a token of such a conventional type.* Sometimes markers are used in such a fashion that they carry semantic values. The complex marker type 'dog', for example, has a conventional interpretation in English, but does not have one in French. *Insofar as an object is a marker that carries a semantic value, it will be called a signifier.* Finally, markers can be employed in symbol games in such a fashion that they have syntactic properties. The lower-case letters, for example, take on no syntactic properties when they are used on an eyechart, but take on one set of syntactic properties when used as proposition letters in the propositional calculus, and take on a *different* set of syntactic properties when used as variable letters in the predicate calculus. The syntactic rules of a symbol game serve to partition the markers employed in that game according to the syntactic positions they can occupy. These syntactic types will be called *counter* types, and *a marker will be said to be a counter just insofar as it takes on syntactic properties within a symbol game.*

These three sortal terms—'marker', 'signifier', and 'counter'—will play a significant role in the discussion of the nature of symbols and symbolic representation that is to follow. Although this book does not undertake to develop a thoroughgoing semiotics, it will prove helpful to undertake a brief discussion of each of these three terms.

4.3 MARKERS

The first and most basic of the three sortal terms is 'marker'. Thus far the development of the term 'marker' has consisted of the citation of a few paradigm examples (letters, numerals, characters employed in musical notation) and a negative claim to the effect that being a marker has nothing to do with syntax or semantics. To come to a better understanding of markers, it will be useful to employ a thought experiment.

4.3.1 THE "TEXT FROM TANGANYIKA" EXPERIMENT

Suppose that the noted Victorian-age explorer and linguist Sir Richard Francis Burton, while traversing central Africa in search of the source of the Nile, comes upon a lost city. There he finds a clay tablet on which there are inscriptions of unknown origin and meaning. One line of the script reads as follows:

L⌐⟨ ⟩◊▢| ⟨ ⟩·L⟨L ·|\◊   ▢|/▢·| ◊⟨Υ   •Υ

What assumptions can Burton reasonably make about the inscription? First, he can probably proceed upon the assumption that what he has come upon is an inscription in a written language, which he dubs "Tanganyikan." He can assume that, like other written languages, Tanganyikan will employ symbols, that it will have a syntactic structure, and that at least some of the symbols will be used meaningfully. At this point, however, he most emphatically does not know what any of the symbols mean. *Nor does he even know what symbolic units are meaningful.* What he encounters may be a phonetically based script like that used in written English, in which case few if any of the individual characters will be meaningful. On the other hand, it might be an ideographic notation like that employed in written Chinese, in which case individual ideogram types are correlated with specific interpretations. Or it might be like Egyptian or Coptic script, in which characters can function as ideograms in some contexts and function as indications of phonemes in others. (If English were to be represented in a similar way, for example, we might have a character   to represent the word 'heart, and then represent the word 'hearty' by the string ♥-y.) And of course it *could* be the case that what he sees is not writing at all, but mere ornamentation or doodlings.

Now there is a great deal that Burton can do without an interpretation scheme for this writing. Notably, he can begin by making a list of the atomic characters employed, and on the basis of this he can do such things as compare them with characters used in other African languages to see if Tanganyikan may be related to any of these. For example, if the writing found at Timbuktu contains a character ⌊, then Burton might postulate that the symbol ⌊ found in Tanganyikan script is a variant of ⌊, and that Tanganyikan is related to Timbuktuni. And he can do all of this without knowing anything about the syntax or semantics of Tanganyikan. Indeed, even if it turns out that what he has found are a child's handwriting exercises or an ancient eyechart—in which case what he sees does not have either syntactic structure or semantic interpretation—his conclusions about the character type need not be imperiled. And the reason for this is that the characters themselves can be understood as falling into types quite independently of the linguistic uses to which they are put.

Once Burton has made this observation, he begins to realize that it is not only atomic graphemic character types that can be studied apart from their syntactic and semantic properties. On the one hand, *strings of characters that function together* can be treated as a single unit, and hence Burton can make some guesses about what sequences of characters make up words.[1] On the other hand, graphemic characters are not the only tokens whose membership in a type can be understood apart from syntax and semantics. The very same kind of analysis can be applied to nonvisual units, such as phonemes, Morse code units, or ASCII units in computer storage locations. If, for example, Burton had a tape recording of someone speaking Tanganyikan, he might undertake a very similar analysis of the phonemes employed in the language, even without knowing where the breaks between words fall or what anything in the language means. Or, if he were in a position to intercept an electronically transmitted message such as a transmission in Morse code, he might be able to figure out the basic units (e.g., dots and dashes) and how they were instantiated in a telegraph wire or through modulations of radio waves. In light of these realizations, of course, he would come to realize that he could no longer employ the term 'character' to cover all of the relevant cases, and would be in search of a suitably neutral term: for example, the term 'marker.'

4.3.2 WHAT IS ESSENTIAL TO THE NOTION OF A MARKER?

If 'marker' is to serve as a generic term for phonemes, graphemes, units of Morse code, and other such entities, it is worth asking just what is involved in being an entity of one of these kinds. And the best way of answering is by making a series of observations.

(1) *Markers are tokens of types.* The type-token distinction is applicable to all markers—to letters, numerals, Morse code units, ASCII code units, phonemes, and so on.

(2) *Marker types are conventional.* To say that a graphite squiggle on

a sheet of paper is a letter P is to say that it is a token of a particular type that is employed by particular linguistic communities. To say that it is a rho is to say that it is a token of a different particular type employed by a different community. And to claim that a particular squiggle is a P is not the same thing as to claim that it is a rho, even if it is the case that an object has the right shape to count as a P if and only if it has the right shape to count as a rho. This is because the claim that the squiggle is a P (or a rho) makes reference to more than the shape of the object: it makes reference to specific conventions of a specific linguistic community as well. Likewise, the claim that the squiggle is a P (or a rho) is *not* equivalent to a claim about its shape—for example, that it is composed of a vertical line on the left and a half-oval attached to the right side of the upper half of the line.

When I say that marker types are *conventional,* what I mean is merely that marker types are established by the beliefs and practices of language users. In particular, I wish to emphasize that marker types are not *natural kinds*. To be sure, sounds and squiggles may *also* fall into natural kinds on the basis of physical patterns present in them, such as their waveforms or their shapes: a sound wave is a sine wave at 440 kHz just because of its physical characteristics, and an inscribed rectangle is a rectangle just because of the distribution of graphite on paper. But when we say that an object is a marker—for example, an inscription of the letter P or an utterance of the word 'woodchuck'—we are not picking it out just by its sound or its shape, but by the way it fits into established linguistic practices in some community of language users. To determine what marker types an object falls into, we need to know more than what patterns are present in the object: we need to know what marker types there are as well, and what kinds of objects can count as tokens of those types. And to answer *those* questions, we need to know what linguistic communities there are and what shared understandings and practices members of those communities have about using sounds and inscriptions communicatively. An object can only be a P-token if there is a letter type P, and there can only be a letter type P if there is some community of language users who have a set of shared beliefs and practices to the effect that there is a marker type whose tokens are shaped in certain ways and may be employed in certain activities. So when I say that marker types are conventional, I mean that the existence of the type is determined by the beliefs and practices of language users.

(3) *The conventions that establish marker types involve criteria gov -*

erning what can count as tokens of those types. So while the assertion that a squiggle is a rho involves more than claims about its shape, it does entail things about the shape of the squiggle as well. The letter type rho is established by the conventions employed by writers of Greek, but part of what is involved in those conventions is a set of criteria governing what a squiggle has to look like in order to count as a rho.

(4) *The criteria governing what can count as a token of a marker type pick out a set of (physically instantiable) patterns such that objects having those patterns are suitable to count as tokens of that type.* In the case of letters, numerals, and other graphemes, the patterns are two-dimensional visible spatial patterns. In the case of phonemes, they are acoustic patterns distinguishable by the human auditory system. In the case of Morse code and computer data storage they are abstract patterns made up, respectively, of dots and dashes or binary units which can be instantiated in various ways in different media. One can also have complex marker types that are formed from arrangements of simple marker types: written words, for example, are complex markers composed of sequences of atomic markers (letters).

(5) *The criteria for a marker type may be flexible and open-ended, and need not be subject to formulation in terms of a rule.* This is clearest in the case of graphemic symbols. As Douglas Hofstadter (1985) has argued, letter types seem to permit an indefinite number of stylistic variations. A reader who has not foreseen these can nonetheless quickly recognize them as such when presented with them. It is by no means clear that one could provide a rule (e.g., in the form of a computer program) that could, for example, distinguish all of those patterns that a person could recognize as stylistic variants of the letter *P* from those patterns which a person would not recognize as such.

(6) *Marker types are often found in groups or clusters that are employed in the same symbol games.* Thus we speak of different sets of graphemic characters such as "the letters," "the numbers," "the punctuation symbols," and so on.

(7) *Criteria for marker types may overlap, both within groups and across groups.* Thus the same squiggles that count as letter o's can count as zeroes and omicrons as well. And indeed, as anyone who has had trouble reading another person's handwriting knows, handwritten letters are often interpretable in a number of different ways.

(8) *Language users possess a repertoire of marker types, which can be used in various ways.* Mathematicians, for example, are in the busi-

ness of developing new symbol games. In doing so, they commonly employ existing marker types such as letters and numerals whose origins may be traced to various linguistic communities. Mathematicians use existing marker types, but put them to new uses in new symbol games. Similarly, one can use one's knowledge of phonemes and the rules for combining them into words in one's language in order to coin a new word if one is needed.

(9) *Marker types can be added to or deleted from an individual's repertoire.* That is, a person can learn marker types and also forget them.

(10) *Marker types can be added or deleted from the repertoire of a linguistic group.* New words (complex markers) are coined, new atomic markers are invented (as in the case of the integration sign used in the calculus or the missionary St. Cyril's invention of the Cyrillic alphabet) and imported (as in the case of Europe's adoption of the Arabic numerals). Markers also disappear from usage. Many of the complex markers (Middle English words) one finds in Chaucer's writings, for example, are no longer in use; and the Old English letter thorn has survived only in the guise of a y on the signs of anglophilic innkeepers.[2]

(11) *The boundaries of a "linguistic group" and the extent to which conventions are shared within a group are highly flexible.* In the case of natural languages, for example, there are often significant differences in dialect and idiolect which involve differences in the conventions for pronunciation, inscription, and so on. It is not always fully clear when one should say that one is faced with separate linguistic groups and when one is faced with a variety of practices within a single group. Moreover, there may be groups within groups: all topologists may observe certain notational practices, but topologists who work in a particular topological specialty (e.g., surgery theory) may all observe an additional set of practices not shared by other topologists, and an individual mathematician who has developed his own techniques for a particular problem may be the only person employing his new conventions. Similarly, an individual may find the need for a new word in a natural language and may therefore choose a phonetic sequence (a complex marker type) that is not currently used in his language and then employ it as a marker type. The new marker type is *conventional* in the sense that it is *established by a human convention* and not simply by a natural pattern, even though the convention that establishes it is not (yet) a convention *of English,* but merely a convention within some individual's idiolect. (Of course, it can become a convention of English; new words are introduced into languages, and they all start out as someone's idiosyncrasies.)

4.4 SIGNIFIERS

While the conventions that establish marker types function independently of the particular uses to which the markers are put in actual practice, it is nonetheless part of the nature of markers that they can be used meaningfully. So while there is nothing, for example, about the marker type consisting of the sequence of letters *d-o-g* that binds it to a particular meaning, the marker type—just by virtue of being a marker type—is the sort of thing that *can* be associated with a meaning in such a fashion that its tokens can count as carrying or expressing that meaning. And within a linguistic community—such as the community of English speakers—there are conventions that set up an association between marker types and meanings. When we speak of something as a *meaningful word* in a natural language such as English, for example, we refer to it as a token of a marker type (be it typified phonetically, graphemically, or both) that is associated with a meaning by English semantic conventions, and we pick it out *both* by its marker type *and* by the associated meaning.[3] (Hence we can distinguish between different words with the same spelling but different meanings, and vice versa.) Similarly, when we speak of a written number, we refer to a marker string and to its associated meaning. The marker string 1-1-0-1 can be used in the representation of various numbers: thirteen in base-2 notation, thirty-seven in base-3 notation, and so on. In the technical terminology introduced in this chapter, insofar as an object is a marker that is associated with a meaning, it may be called a *signifier*.

It will prove useful to think of symbols as things that can be examined at *several different levels of analysis*. Thus the inscription

dog

can be seen at several levels. First, it can be examined at what might be called the "marker level," at which it is a sequence of letters from the Roman alphabet, and also a complex marker employed in English. The atomic marker types are established by the conventions of a linguistic community, and the complex type is licensed for use in English by similar conventions. But the above inscription may also be examined at a second or "signifier level." At the signifier level, the inscription is a token of a signifier type employed in English. That signifier type is established by a linguistic convention that associates a complex marker type with a meaning.

The conventionality of signifier types is again a matter of there being

certain shared understandings and practices within a linguistic community. Individual inscriptions of the word 'dog' mean *dog* because they are
tokens of a particular signifier type employed in English. That signifier
type exists by virtue of a convention: in this case, a shared understanding
among speakers of English that tokens of the complex marker type *d-o-g*
can be used to express the meaning *dog,* and a shared practice of using
tokens of that complex marker type to express that meaning.

There are thus at least two levels of conventionality involved in being a signifier token. First, anything that is a signifier must also be a
marker, and marker types are established by conventions. Second, signifier types are established by conventions that associate marker types
with interpretations. The first kind of conventionality appears at the
marker level, the second connects the marker level to the signifier level.
It is by virtue of marker conventions that objects bearing patterns can
count as markers, and it is by virtue of signifier conventions that markers can count as signifiers.

4.5 COUNTERS

Markers can, of course, take on syntactic as well as semantic properties.
But like semantic properties, syntactic properties are extrinsic to the
marker type. That is, there is nothing about the marker type P that implies anything about the syntactic properties of P-tokens. P's can be used
in symbol games without syntactic rules—for example, on eyecharts.
They can also be used in games that have syntactic rules, such as written
English, written French, algebraic topology, and predicate logic. Just
what syntactic properties a P-token can take on depends on what symbol
game it is used in, what syntactic categories are involved in that symbol
game, and which syntactic slots can be occupied by P-tokens.

Now all of this implies that there is more to syntax than marker order—that the syntactic properties of a marker token are intimately connected with the role it plays in larger linguistic activities, and are not just
a matter of the marker's combinatorial properties. One could, of course,
use the word 'syntax' so broadly as to include *all* arrangements of markers—or, indeed, to include all arrangements of *objects,* since all objects
can, in principle, serve as markers. But the word 'syntax' has some paradigm uses in which it is applied to specifically linguistic structures, and
there is arguably a great deal about linguistic structure that falls under the
rubric of syntax that goes beyond combinatorial features. There is, for
example, a sense in which we should say that a sentence has a syn-

tactic structure while the order found in other entities (e.g., the sequence of cars in a traffic jam, the sequence of philosophy courses taken by an undergraduate major) is not plausibly regarded as syntactic.

Let us briefly inquire as to how the syntactic structure of a string of markers is dependent upon the symbol game in which it is employed. Consider, for example, the marker string

F a d

What is the syntactic structure of this sequence of markers? The answer depends entirely upon the symbol game that is operative. If the letters appear on a line of an eyechart, one would be inclined to say that the string of markers has no syntactic structure: there is an order to the markers, to be sure, but it is not a *syntactic* order. But if the markers make up the English word 'fad' with a capitalized *f*, the story is quite different. It has both *internal* syntactic structure, since spelling rules can plausibly be called "syntactic" (even if spelling is not the kind of syntax that comes most quickly to mind). It also has *external* or *relational* syntactic properties, since the word 'fad' is of a grammatical type that can occupy certain slots in English sentence structure, but not others. For example, sentence (1) is grammatically permissible in English, while sentence (2) is not:

(1) The hula hoop was a fad.

(2) * The hula hoop fad was.

The string *F-a-d* could also be used as an expression in the predicate calculus, with *F* being a predicate letter and *a* and *d* its arguments. Here once again the string would have both internal and relational syntactic properties, but very different ones from the previous case. The difference, of course, lies in the fact that the same marker string can be used in several different language games, but those games have different syntactic rules, and the role that the markers play in the different games is correspondingly different. Moreover, the kinds of syntactic *categories* in terms of which one can analyze a marker string are closely related to kinds of symbol games. Natural languages have nouns, verbs, adjectives, and so on. Some natural languages also have syntactic features that others do not: articles, plural suffixes, case indicators, privative prefixes, and so on. (Greek has all of these features; Chinese has none of them.) Technical languages may have very different categories: predicate logic, for example, has no nouns or verbs but does have quantifiers, predicate letters, variable letters, and connectives, while the propositional calculus has only sentence letters and connectives.

When we are interested precisely in the syntactic role that a marker or marker string plays in a particular symbol game, it is useful to be able to refer to it precisely as an object of a type distinguished by its syntactic role in that symbol game (as a predicate letter, for example, or as a count noun). Each symbol game has some set of syntactic categories. (It may be the empty set, as in the case of the eyechart.) These are established by the conventions governing the symbol game—that is, the set of beliefs and practices, shared by those who have mastered the game, that govern how symbols may be combined within the game. These conventions also govern what markers and marker strings can be employed in the symbol game, and which syntactic slots they may occupy.

Sometimes, as in the case of the predicate calculus or the Fortran programming language, the stock of markers is set up from the very beginning to fall into categories such that one can tell from the marker type itself what syntactic roles it can play. In the predicate calculus, capital letters can be predicate letters but not variables, while lower-case letters can be variables but not predicate letters. In Fortran, variables with names beginning with the letter i can only store integer values, while variables with names beginning with the letter n can only store floating-point values. But other symbol games are more complicated. In English, the marker string h-o-u-s-e can be used either as a verb or as a noun, and one cannot tell just from the string of symbols which it will be in a given instance. The language has conventions establishing both 'house' the noun and 'house' the verb; and there is no reason that the marker string could not be used as an adjective as well. Likewise, in the computer language Pascal, virtually any string of ASCII characters can be used as a name for a variable that can store any kind of value. One simply has to specify elsewhere what kind of variable it is, and that will have consequences for its syntactic properties. (A variable that stores a boolean value, for example, cannot appear immediately before a slash indicating division.)

The word 'counter', as it is being developed here, will be used to indicate a marker as it takes on particular syntactic properties in a specific language game. Thus, for example, 'house' the noun and 'house' the verb are of separate counter types; for while they employ the same marker string, they have different syntactic properties in English. When we are attending specifically to syntax, we may say that we are working at the counter level. Like the marker and signifier levels, the counter level has its uses. Notably, the study of formal systems, for example, takes place almost exclusively at the counter level, since it brackets semantics and treats differences in what markers are employed as "notational variants." Likewise, much of computer science is devoted to work at the counter level.

4.6 THE RELATIONSHIP OF THE MARKER, SIGNIFIER, AND COUNTER LEVELS

Since marker types are independent of the syntactic and semantic properties that their tokens can take on in different symbol games, while counter and signifier types presuppose the existence of marker types, there is a hierarchic relationship between the marker level and the signifier and counter levels. Analyzing a complex of sounds or squiggles as counters presupposes dividing them into markers, and so both the counter and signifier levels are dependent upon the marker level.

There is not, however, any absolute dependence between the counter and signifier levels. One can, for example, assign interpretations to marker types without situating them within a syntactically structured symbol game, and one can concoct "purely formal systems" for which there is no interpretation scheme. This does not mean, however, that syntax and semantics are absolutely *independent,* either. The semantic values of some marker complexes, such as sentences, are dependent upon the syntactic structure of the complexes as well as the interpretations of the signifying terms. Such structures are subject to *compositional analysis.* But there is no *absolute* dependence of either the counter or the signifier level upon the other in the way that both are dependent upon the marker level.

The marker level is similarly related to lower levels of analysis. An entity's ability to count as a marker, after all, depends not only upon conventions but upon the fact that it bears a physically instantiated pattern satisfying the criterion for its type. One might see such patterns as abstract physical features that are literally present in objects, and one might thus speak of a "pattern level" which is connected to the marker level above it by marker conventions and to other physical descriptions below it by various kinds of abstraction. These abstractions bracket those features of an object that are not relevant to its having a pattern, rendering it suitable to count as a token of a marker type. We might represent the resulting structure of levels of analysis graphically as in Figure 5, with the nodes representing the objects appearing at a level and the arrows

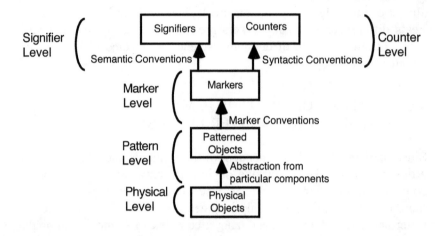

Figure 5

between nodes representing what relates the objects appearing at one level to those appearing at the next.

Now it is important to note that the sortal terms 'marker', 'signifier', and 'counter' designate conventional rather than natural kinds, and that they can pick out the same objects under different aspects. Indeed, any object that is a signifier or a counter *must* also be a marker, and objects that are markers may very well be signifiers and counters as well. The need for the sortal terms arises not because there are three mutually exclusive classes of particulars, but because there are different sorts of questions about symbols that call for classifications based on different features. (There are, for example, questions about orthography, syntax, and semantics.) The distinction between markers, signifiers, and counters is also useful for discussing certain aspects of language, such as ambiguity, homonymy, homophony, and certain kinds of performance errors. One kind of ambiguity occurs, for example, when one has marker strings that admit of multiple semantic interpretations. Homonymy occurs when a single graphemic marker string is associated by different signifier conventions with two or more meanings. Homophony occurs when a single auditory marker string is associated with multiple meanings. Performance errors such as slips of the tongue, malapropisms, and spoonerisms are ways of producing a marker token that is not compatible with the semantic interpretation that one intended one's utterance to have.

The different sortal terms also license different kinds of inferences about the objects they pick out. From the fact that an object is a counter in some language game, it follows that it is a marker and that it has syntactic properties. Nothing follows, however, about whether it has semantic properties. Similarly, if an object is a signifier, it follows that it is a marker and that it has semantic properties; but nothing follows about whether it is used in a syntactically structured symbol game. And from the fact that an object is a marker, nothing follows about whether it has either syntactic or semantic properties.

4.7 FOUR MODALITIES OF CONVENTIONAL BEING

This concludes the first part of the disambiguation of the notion of *symbol* —the separation of 'symbol' into three separate sortal terms. But there is also a need for a second disambiguation, a disambiguation of the senses in which a thing can be said to "be" a symbol. And the ambiguity that is of concern here is reflected in the technical terms 'marker', 'signifier', and 'counter', as well as the original term 'symbol'. I intend to present a case that, because each of these categories is convention-dependent, there are four ways in which an object can be said to be a token of one of the types, corresponding to four ways it can be related to human conventions and intentions. Once again, the distinctions are best motivated by a series of thought experiments.

4.7.1 CASE 1: THE OPTOMETRIST

A man named Jones goes to an optometrist for an eye examination. The examination involves a test which requires the patient to look through a device containing a number of movable lenses. The device is pointed at an eyechart, and is so positioned that just one character on the chart can be seen through the eyepiece. The examination begins with the device being pointed at the single character on the uppermost line of the chart, in this case a letter *P*. Jones looks into the eyepiece and sees the following image:

P

The optometrist asks Jones, "What letter do you see?" Jones responds, "The letter *P*." For purposes of this example, assume that Jones has correctly identified the character. One of the things that Jones has accomplished is the successful identification of a physical particular as a token of a particular conventionally sanctioned marker type. To do this, Jones need not impute any syntactic or semantic properties to the marker token he sees. Indeed, if the doctor were to ask Jones "What does that symbol mean?" or "What is its truth value?" or "What are its syntactic properties?" Jones would likely perceive the questions as very queer indeed. Letters on eyecharts simply do not have syntactic or semantic properties. Moreover, it would be possible for Jones to learn to identify the symbol correctly even if he had never used the Roman letters in the representation of meaningful discourse, much as he might learn to distinguish Chinese ideograms without learning their meanings or the syntactic rules for Chinese—and even without learning that the ideograms were used by the Chinese as a form of writing. Even with such a poverty of competence with written language, Jones could still be said to have recognized and identified what he saw as a token of the type P.

4.7.2 CASE 2: THE BILINGUAL OPTOMETRIST

Yet if we adjust the circumstances in the right ways, it quickly becomes more difficult to characterize what Jones has and has not accomplished. Suppose that Jones goes to a second optometrist, Dr. Onassis. Dr. Onassis lives and works in a Greek neighborhood and has a number of clients who speak and read only Greek, and so he has two sets of eyecharts—one with Greek letters, one with English letters. When Jones looks through the eyepiece of Dr. Onassis's instrument, he sees the following pattern:

P

Dr. Onassis asks Jones, "What letter do you see?" And Jones responds, "The letter P." At this, however, Dr. Onassis casts Jones a very puzzled look. He then looks at the eyechart and laughs. "Oh, I see," he says. "I made a mistake, and put up the Greek eyechart instead of the English one, and then I was puzzled, because the English chart begins with the letter Q and does not even contain a letter P. What you see, by the way, isn't a P but a rho."

This example differs from the first in that our natural intuitions about what Jones has and has not accomplished no longer serve us as well as they did in the first case. Indeed, they may tend to lead people towards two opposite extremes. To continue the story: Jones, upon being told that what he is looking at is not a P at all, becomes quite indignant. "Of course it's a P," he says. "I know what a P looks like, and I can see this one as plain as day, and it's a *P* if ever I've seen one!" This, however, is taken by the doctor as a challenge to his professional competence. "Look here," he says, "I made this chart myself, so I know darned well what the letters are. I made it for my Greek patients, and meant this symbol to be a rho, so a rho is what it is!"

Jones and the doctor are each partially correct in their claims, and each is partially mistaken as well. The most important thing to see, however, is that they are both making the implicit assumption that there is just one univocal meaning to the locutional schema 'is a *P*' (or 'is a rho'), while in fact there are several ways a particular may be said to be a token of a conventional type. The necessary distinctions are easily missed, however, because the same English locution can be used to express each of the several ways. Yet the distinctions may be formulated out of fairly ordinary English locutions, and are easily mastered if one attends to the nature of the situation rather than the form of the ordinary locutions.

4.7.3 INTERPRETABILITY

First, consider Jones's line of reasoning: Jones is a competent user of the letters employed in the representation of English. (Here they will be called "the Roman letters.") The pattern he sees meets the spatial criteria for counting as a token of the marker type *P*. Under the conventions governing the Roman letters, the pattern Jones sees can count as a *P*, and cannot count as a token of any of the other marker types which form the set of Roman letters. There is thus a sense of "being a *P*" which does apply to the mark on the eyechart.

Notice, however, that the exposition of how the character Jones sees can be said to be a *P* has required an appeal to several things in addition to the mark and the marker type—notably, it has required an appeal to (*a*) a community which employs a certain set of marker types which includes *P*, and (*b*) conventions within that community which govern what can count as a token of that marker type. The sense of "being a *P*" that is operative here, then, turns out to be more complex than is suggested by the locution used to express it. To put it differently, the predicate indicated by this usage of the locutional schema 'is a *P*' is more complex than one might assume. To spell out entirely the sense in which Jones might be right in saying that what he sees is a *P*, one would have to say something like the following: "This mark t has a pattern p_i which is a member of the set P of patterns suitable for tokening the marker type T employed by linguistic community L."

We may capture and codify this sense of "being a symbol" by coining the technical expression 'is interpretable as a token of type T' (e.g., 'is interpretable as a rho'). The rules for the application of this predicate may be articulated as follows:

(M1) An object X may be said to be *interpretable as a token of marker type T* iff

    (1) there is some linguistic community L which employs marker type T,

    (2) the conventions in L which govern what can count as a token of

        type T allow any object having any pattern $p_i \in P: \{p_1,..., p_n\}$ to
        be suitable to count as a token of type T,

    (3) X has a pattern p_j, and

    (4) $p_j \in P$.

This sense of "being a P" points to a relationship between (1) a physical particular, (2) a pattern present in that particular, (3) a convention linking that pattern to a marker type, and (4) a linguistic community using that marker type and employing that convention. An object X related in such a fashion to a marker type T will be said to be *interpretable as a (marker) token of type T (under convention C) (for linguistic group L)*. The parentheses are used here to separate a short form of the new technical term—'interpretable as a token of type T'—from its complete form. In many cases it will prove unnecessary to allude specifically to a convention or a linguistic group, and so the shortened locution 'interpretable as a token of type T' can purchase some measure of simplicity with little cost in terms of exactitude. The items in parentheses, however, are *not optional* —any claim that a physical pattern is interpretable as a token of a marker type involves at least implicit reference to a convention and to a linguistic community, even if these are not specified.

It is, of course, quite possible for a single object X to be interpretable as a token of a number of different marker types $\{T_1,..., T_n\}$. In each of the optometrist examples, the object Jones sees is interpretable under the conventions for Roman letters as a P and interpretable under the conventions for Greek letters as a rho. It may be subject to interpretation as a token of other marker types as well. There is no inconsistency in saying that a mark is interpretable as a token of a variety of different types. Such illusion of an inconsistency as there may be is quickly dispelled if one looks at the long way of describing interpretability. If one says "X

is interpretable as a token of type T" and "X is interpretable as a token of type U," the long versions of the two statements will always reveal additional differences which will explain how it is that X is multiply interpretable. These will be differences in what linguistic community's conventions are involved (as in the case of the bilingual optometrist), or differences in the particular conventions of a single community which are operative in the different cases (as in the case of the numeral zero and the letter o in our community), or differences in what pattern in each particular mark is relevant to its interpretability as a marker of that type.[4]

4.7.4 INTENTIONAL TOKENING AND AUTHORING INTENTIONS

If Jones has something of a point, the doctor does as well. The doctor's line of argument is that he drew the eyechart himself, and as a consequence he is in a special position to say what the characters are. Indeed, he might go so far as to say that he is in a position to *stipulate* what they are. The mark on the chart was, after all, made with the intention that it be a token of a particular marker type—in this case that it be a token of the Greek letter rho. There is thus a sense in which it seems right to say that the doctor inscribed a rho. And in this sense it would not be correct to say that he inscribed a P, because he did *not* intend it to be a P.

Hence, in distinction with the *interpretability* of a particular object X as a token of type T, one may also develop another technical locution:

(M2) An object X may be said to have been *intended (by S) as a token of marker type T* iff

(1) there is some linguistic community L that employs marker type T,

(2) there is a language user S who is a member of L (or is otherwise able to employ the conventions in L governing marker type T),

(3) S inscribed, uttered, or otherwise "authored" X, and

(4) S intended what he authored to count as a token of type T by virtue of conventions in L governing marker type T.

Several clarifications and caveats are immediately in order. First, the term 'intended' is meant very broadly here. Notably, it need not imply that the author of the mark must have a conscious, linguistically formulated characterization of what he is doing in producing the marker. When someone writes a sentence without any explicit awareness of making marks with a pen, he would, according to this usage, "intend" his marks to be letters of particular types.[5] This use of 'intend' is also meant to allow a great deal of latitude in how direct a causal chain there is between the intention of the author of the marker and its ultimate production. Notably, it is intended to be broad enough to cover at least some instances of the printing of stored representations of text by a computer. The explanation of how the marks on a printed page—a page of a book, for example—are said to count as letters (and how conglomerations of them are to count as words, sentences, statements, and arguments) will need to appeal in part to the intentions of the author. (It may also need to appeal to the intentions of the various engineers and programmers who designed the hardware, software, and coding schemes which mediate the process which begins with the author's striking keys on a keyboard and ends with the production of a printed page.)

A second clarification which needs to be made is this: the author of a marker token may intend it to be a token of more than one type. Within the story about the bilingual optometrist, one should say that the mark which Jones saw was interpretable as a *P and* interpretable as a rho, but that it was *intended* as a rho and *not* intended as a *P*. In *devising* the two scenarios used in this thought experiment, however, the visible pattern that was chosen—namely,

P

—was deliberately chosen precisely for its susceptibility to multiple interpretations. One could devise more complex enterprises which turn upon such ambiguities, such as acrostics which make sense in two languages, or which make sense in one language vertically and another horizontally. (In spoken language, puns might well fall into this category. Take for example the case of Lewis Carroll's "We called him the Tortoise because he *taught us,* " which works in British but not American English because the expressions 'tortoise' and 'taught us' sound the same in British English, but different in American English.)

From these two clarifications a third emerges—namely, that there is room for some very different ways of intending an utterance or inscription to count as a token of more than one type. Here are a few examples: (1) In devising the *P*/rho example, the intention was to find an inscription that could clearly count as a token of either of two marker types which might be presumed to be familiar to those likely to read these pages. (2) In legal, political, and diplomatic enterprises, it is often deemed prudent to choose what one says or writes so that it has multiple interpretations—in particular, so that it has one natural interpretation that is likely to appeal to the hearer or reader, and another more exacting interpretation which can be offered as what was "really meant" at a later date. (For example, promising "no new taxes" does not, strictly speaking, involve promising that existing taxes will not be raised by 10 percent or even 1000 percent.) This kind of intentional ambiguity is most important on the semantic level, but could occur at the level of marker interpretability as well. (3) A slightly different form of ambiguity is present when what is said or written is intended to be interpretable in more than one way, all of which are intended to be understood by the hearer or reader, who then chooses which leg of the ambiguity to treat as operative. An expression of interest in doing business together in the future, for example, can be treated as an opening move in negotiations to do business or as a mere expression of good will. Properly used and properly understood, such ambiguous expressions can allow two parties to explore one another's interests without risk of "losing face." (This practice is reportedly expected by Japanese in business dealings to a degree seldom appreciated by American businessmen.)

4.7.5 ACTUAL INTERPPRETATION

In addition to the *interpretability* of a marker token and its *intended interpretation,* one may identify two additional relationships between a particular marker token and a marker type. The first of these is *(actual) interpretation* of the figure as a marker of some particular type. In both of the optometrist examples, Jones interprets the figure he sees as a letter of a familiar type—he identifies each as a P-token. One might want to say there is a sense in which he was *right* in so identifying each (because each is interpretable under English conventions as a P) or that there is a sense in which he was *wrong* in his identification of the second figure (because its author *intended it* to be a rho and did not intend it to be a P). But neither of these facts alters one fact about what Jones did: namely, he placed an interpretation upon a figure he saw; he *interpreted it as* or *took it to be* a P-token.

Once again, the new terminology has hidden references to marker types, conventions governing what can count as tokens of the types, and linguistic communities which use the types. To interpret a figure as a token of type T is to be familiar with marker type T employed by some linguistic community L, which in turn involves understanding (not necnessarily perfectly) how to apply the criteria for interpretability as a token of that type (though the "understanding" does not necessarily involve the ability to form or consciously articulate a rule for what can and cannot count as a *P*, but is better understood as a kind of *competence*).

This notion of actual interpretation may once again be expressed by a more technical definition:

(M3) An object *X* may be said to have been *interpreted (by H) as a token of marker type T* iff

   (1) there is some linguistic community *L* which employs marker type *T*,

   (2) there is a language user *H* who is a member of *L* (or is otherwise able to employ the conventions in *L* governing marker type *T*),

   (3) *H* saw, heard, or otherwise apprehended *X*, and

   (4) *H* construed *X* as a token of type *T*.

Now it is important to see the distinction between authoring intentions and mere interpretations. For while authoring intentions do, in a sense, involve interpretation, the author of a marker's intention is not "just another interpretation." There is a significant difference between Dr. Onassis's original interpretation of the figure on his eyechart—the one that was involved in its authoring—and Jones's interpretation of it, and this leads to our strong intuition that there is a sense in which the figure "is a rho" and "is not a *P*." The difference between intended interpretation, or *authoring interpretation,* and other interpretations of the same figure also applies to Dr. Onassis's *own* later interpretations of what he has inscribed. The author of a marker token is certainly likely to be in a unique epistemic position with regard to what the token was meant to be, even long after he has brought it into being, and hence he is usually accorded unique authority in clarifying any ambiguities which might be spotted. But the reason for this is precisely that he is believed to know better than anyone else what he *intended* to write or utter, and it is what he *intended* that determines "what it is" in one sense—namely, in the sense captured by the technical locution 'intended to be a token of type *T*.' (Note, for example, that the author's [current] interpretation of his words and actions is *not* accorded the same respect if its fidelity to his *original* intent is in question—if he is a defendant in a libel suit, for example, or if he has suffered a loss of memory.)[6]

4.7.6 INTERPRETABILITY-IN-PRINCIPLE

There is one way of "being a symbol" that is yet to be discussed. It is most easily developed for signifiers—and will be shortly—but can be developed for markers as well, albeit with less intuitive appeal. Again let us perform a thought experiment. Assume that there is a sandstone cliff in the Grand Canyon that bears certain dark patterns against a lighter background. Let us assume, moreover, that there are no actual orthographic conventions, past or present, by virtue of which these patterns would be interpretable as marker tokens. The patterns are not now interpretable as marker tokens. But consider the future. It could be the case that some future culture will develop an orthography whose conventions will be such that the patterns on the sandstone cliff would then be interpretable as markers in that orthography. It could even be that members of that culture would naturally perceive the cliff as bearing a meaningful message in their language. Let us call this scenario "Future A." Now of course it could also be the case that such a culture will not arise—that it will never be the case that there is a culture anywhere that will employ conventions by virtue of which the patterns on the cliff face would be rendered interpretable as marker tokens. Call this scenario "Future B."

Now it would seem to make some sense to say that the patterns on the cliff face are already *suitable* to count as markers, given the existence of the right sorts of conventions. It seems right to say that, if only the right sorts of conventions were adopted—for example, the conventions that are eventually adopted in Future A but not in Future B—those patterns would then be interpretable as markers. To put it slightly differently, we might say that, while those patterns are not *in fact* interpretable (under any *actual* conventions) as markers, they are nonetheless *interpretable-in-principle* as markers under conventions that could be (or could have been) adopted, and their being so interpretable-in-principle is independent of which future—A or B—actually comes about.

This notion of *interpretability-in-principle* can be developed more exactly as follows:

(M4) An object X may be said to be *interpretable-in-principle as a token of a marker type T* iff

    (1) a linguistic community could, in principle, employ conventions governing a marker type T such that any object having any pattern $p_i \in P: \{p_1,..., p_n\}$ would be suitable to count as a token of type T,

(2) X has a pattern p_j, and

(3) $p_j \in P$.

That is, for any object X one might consider, if X has some pattern that could, in principle, be used as the criterion for a marker type, then X is interpretable-in-principle as a marker. (One could, for example, establish a convention whereby spherical objects could count as markers of a particular type, and hence globes, oranges, and planets are interpretable-in-principle as markers.)

Now it should be immediately evident that this notion of *interpretability-in-principle* is extremely permissive. For while the range of patterns that human beings can easily employ for marker types is rather limited, and the range of patterns they do in fact employ is more limited still, this is more a consequence of the nature of our bodies than of the nature of markers. The patterns we use for markers are chosen for the ease with which we can perceive and implement them. Thus until very recently marker types were confined largely to those distinguished by patterns that could be easily seen or heard. With the aid of instruments, however, humans can deal with markers that are distinguished by patterns of voltage levels in a wire or across a field of circuits, or by patterns of magnetic activity, or by various other kinds of patterns. And there is no reason why a being with very different powers and senses could not use very different sorts of things as markers. (To take an extreme example: an all-powerful God might use configurations of stars as criteria for marker types employed in storing messages for very large angels, and use patterns of electron activity in a single atom as criteria for marker types used to send messages to very small angels.) As a consequence, it would seem that everything whatsoever is interpretable-in-principle as a marker token.

4.7.7 THE FOUR MODALITIES

The expression 'is a marker' has been replaced by four locutional schemas that have been given technical definitions:

— 'is *interpretable* (under convention C of linguistic group L) as a marker of type M'

— 'was *intended* (by its author S) as a marker of type M'

— 'was *interpreted* (by some H) as a marker of type M'

— 'is *interpretable-in-principle* as a marker'

To these four locutional schemas correspond what might be called *four modalities of conventional being,* or four ways in which an object can be related to a conventionally established type (though in the case of interpretability-in-principle, the conventions and the type need not be actual). These four modalities can be applied not only to markers, but to other conventionally established types as well, as we shall see presently. These locutional schemas, moreover, are intended to capture and distinguish four different senses in which one might speak of an object "being" a marker (e.g., a letter or a Morse code dot) or "being" of one of the other conventionally established types. These different senses are, to some extent, already operative in ordinary and technical uses of the word 'symbol', but existing terminology is not subtle enough to distinguish the different senses.

4.8 FOUR WAYS OF BEING A SIGNIFIER

Just as it is important to distinguish four senses of "being a marker," it is likewise important to distinguish four different senses in which a marker may be said to "have" or "bear" semantic properties, and hence four ways in which a marker may be said to be a signifier. In order to clarify these four senses, we shall employ another thought experiment. The great detective Sherlock Holmes has been called in to solve a murder case. The victim, a wealthy but unpleasant lawyer, has been poisoned. Before dying, however, he managed to write a single word on a piece of paper. The inscription is

PAIN

Inspector Lestrade of Scotland Yard has concluded that the deceased was merely expressing the excruciating agony that preceded his death. Holmes, however, makes further investigations and discovers that the victim's French housekeeper is also his sole heir. It occurs to Holmes that 'pain' is the French word for bread, and upon inquiring he discovers that the housekeeper did indeed do the baking for the household. Perhaps, reasons Holmes, the deceased was poisoned by way of the bread, and has tried to indicate both the means by which the poison was conveyed and the identity of his murderess by writing the French word for bread.

Which was inscribed on the dead lawyer's stationery—the English word 'pain' (meaning a particular kind of sensation) or the French word 'pain' (meaning bread)? To put it differently, what does the inscription *mean*—pain or bread? It should immediately be evident that this question is very much like the question about the figure on the bilingual optometrist's eyechart. First, there is a sense in which what is on the paper is *interpretable* (under English conventions) *as meaning pain*. In this very same sense the mark on the paper is *interpretable* (under French conventions) *as meaning bread*. That is, the sequence of Roman letters on the stationery is used by English speakers to carry one meaning and used by French speakers to carry a different meaning.

Yet there is also a sense in which the inscription can be said to mean one thing and not the other, provided that one assumes that the victim *intended* what he wrote to mean one thing rather than the other. If Holmes's hypothesis is correct, for example, the lawyer meant to write the French word for bread and did not mean to write the English word for pain. Assuming that this was the case, there is a sense in which the inscription can be said to mean bread but *not* to mean pain.

This distinction between two ways a marker token can be related to a meaning should seem familiar, as it parallels the first two ways an object could be said to "be" a marker token—namely, *interpretability* and *intended* (or *authoring*) *interpretation*.

(S1) An object X may be said to be *interpretable as signifying (meaning, referring to) Y* iff

    (1) X is interpretable as a marker of some type T employed by linguistic group L, and

    (2) there is a convention among members of L that markers of type T may be used to signify (mean, refer to) Y.

(S2) An object X may be said to be *intended (by S) to signify (mean, refer to) Y* iff

    (1) X was produced by some language user S,

    (2) S intended X to be a marker of some type T,

    (3) S believed that there are conventions whereby T-tokens may be used to signify Y, and

    (4) S intended X to signify Y by virtue of being a T-token.

Two observations should perhaps be noted about these definitions. First, neither of them is intended to correspond precisely to what is meant by the vernacular usage of the words 'meaning' or 'reference'. Indeed, the whole enterprise of specifying new terms such as these is necessary only because ordinary usage is ambiguous and imprecise. In assuming that the inscription meant *pain,* Lestrade was probably (implicitly) assuming *both* that the inscription was interpretable under English conventions as carrying the meaning pain *and* that the deceased had intended the inscription to mean pain. But his assumption would be implicit in that he has probably never made the distinction under discussion. It is only when someone like Holmes notices that the ordinary assumptions do not always hold that distinctions can be made, and at such a point it is of little interest to the specialist (be he detective or philosopher) to argue about whether interpretability or authoring intention or the combination of the two best captures the "real" (i.e., the vernacular, precritical) use of the term 'meaning' (or 'reference'). It is the new, more refined terms that are needed. The determination of vernacular usage may be left to the descriptive linguist.

Yet there is most definitely no intention here to imply that ordinary usage is irrelevant in the pursuit of philosophy. Attention to ordinary usage can often be of great help in solving philosophical problems, especially when those problems are themselves caused by an impoverished understanding of language on the part of the philosopher. The point here is that language points to the phenomena to be studied, and sometimes it points too vaguely and indistinctly to serve the purposes of the theorist. When this happens, terminology must be refined to capture distinctions the specialist needs but the ordinary person does not. The enterprise is far more risky when the process proceeds in the opposite direction—that is, when ordinary terms are *extended* instead of *refined.* The application of the terms 'symbol' and 'representation' to the contents of intentional states is a case in point. (This entire book is an examination of what has gone wrong in the extension of such ordinary terms as 'symbol' and 'representation'.)

The second observation about these definitions is that the definition of authoring intention allows for the possibility that the speaker is wildly idiosyncratic in his use of language. If, for example, Jones believes that the word 'cat' is used to refer to newspapers, and utters "The cat is on the mat" to express the belief that the newspaper is on the mat, we may nonetheless say that Jones *intended to signify* the newspaper. In particular, he uttered a token of the marker type 'cat', which he believed could be used to signify newspapers, and intended to signify the newspaper by uttering the word 'cat'. Of course, there is no convention of *English* that allows the word 'cat' to be used to signify newspapers. (Utterances of 'cat' are not interpretable, under English conventions, as signifying newspapers.) But Jones nonetheless intended to refer to the newspaper by uttering the word 'cat'. And of course there could be subgroups of English speakers who employ semantic conventions that are not conventions of English, but only of a dialect of English (as, for example, some Baltimoreans refer to street vendors as "Arabs" [pronounced *ay*-rabz], or Bostonians refer to submarine sandwiches as "grinders"). And indeed one might even wish to speak of idiolects in terms of the special semantic conventions of a linguistic subgroup consisting of one member, in which case Jones correctly believes that there is a convention licensing the use of 'cat' to refer to newspapers, but incorrectly believes that it is a convention of English rather than of his own idiolect. One might wish to use the term 'convention' in such a case because there are beliefs and practices that can govern how a marker may be used. These beliefs and practices are, in principle, public and shareable, even though *in fact* only one person possesses them. (Because they are essentially public, and the fact that they are possessed by only one person is merely incidental, Wittgenstein's concerns about a private language do not arise here.)

Third, it should be noted that the semantic features to which these definitions are relevant are meaning and reference. The truth value of a signifier is undetermined by the relationships between the token, linguistic conventions, and the intentions of its speaker or inscriber. (There are some exceptions, such as analytic truths, but here the interest is in a general characterization of ways objects can be said to have semantic properties.)

In addition to interpretability (under conventions employed by some linguistic group) and intended interpretation, one may distinguish two additional ways in which a thing may be said to carry a semantic value. These correspond to the two remaining ways that a figure could be said to count as a marker token: namely, *actual interpretation (by someone apprehending the signifier)* and *interpretability-in-principle*. Regardless of what the deceased lawyer intended his inscription to mean, it is nonetheless the case that it was *interpreted* by Lestrade as meaning pain and *interpreted* by Holmes as meaning bread. These actual acts of interpretation are, indeed, independent of whether the lawyer intended his inscription to mean *anything at all* —they would be unaltered if, for example, he had been scribbling random letters. The notion of *actual interpretation* may be defined for signifiers as follows:

(S3) An object X may be said to have been *interpreted (by H) as signifying (meaning, referring to) Y* iff

(1) some language user H apprehended Y,

(2) *H* interpreted *X* as a token of some marker type *T*,

(3) *H* believed there to be a linguistic convention *C* licensing the use of *T*-tokens to signify *Y*, and

(4) *H* construed *X* as signifying *Y* by virtue of being a *T*-token.

Finally, it is notorious that any symbol structure (i.e., any marker, simple or complex) can be used to bear any semantic interpretation whatsoever. Haugeland, for example, writes of a set of numerical inscriptions he supplies as examples in *Mind Design* that "formally, these numerals and signs are just neutral marks (tokens), and many other (unfamiliar) interpretations are possible (as if the outputs were in a code)" (Haugeland 1981: 25). And Pylyshyn writes of symbols in computers,

> Even when it is difficult to think of a coherent interpretation different from the one the programmer had in mind, such alternatives are, in principle, always possible. (There is an exotic result in model theory, the Lowenheim-Skolem theorem, which guarantees that such programs can always be coherently interpreted as referring to integers and to arithmetic relations over them.) (Pylyshyn 1984: 44)

In the terminology developed in this chapter, what this means is that there is nothing about markers that places intrinsic limits upon what interpretations they may be assigned, and so it is possible for there to be conventions which assign any interpretation one likes to any marker type one likes. Now there are two different ways in which we might wish to formulate this insight. One way of formulating it would be to say that, for any marker type *T* and any interpretation *Y*, it is possible for there to be a semantic convention to the effect that *Y*-tokens are interpretable as signifying *T*. In terms of a technical definition:

(S4) An object *X* may be said to be *interpretable-in-principle as signifying Y* iff

(1) *X* is interpretable-in-principle as a token of some marker type T, *and*

(2) there could be a linguistic community *L* that employed a linguistic convention *C* such that *T*-tokens would be interpretable as signifying Y under convention *C*.

That is, to say of some *X* and some *Y* that "*X* is interpretable-in-principle as signifying *Y*" is to say (1) that one could, in principle, have a marker convention whereby *X* would be interpretable as a marker of some type T, and (2) that one could, in principle, have a semantic convention C whereby T-tokens would be interpretable as signifying Y.

One might, however, wish to characterize semantic interpretability-in-principle in a different manner. All that is necessary for an object X to be interpretable-in-principle as signifying Y is the availability of an interpretation scheme that maps X's marker type onto Y. And all that this requires is that X be interpretable-in-principle as a marker, and that there be a mapping available from a set of marker types to a set of interpretations that takes X's marker type onto Y. In terms of a technical definition:

(S4*) An object X may be said to be *interpretable-in-principle as signifying Y* iff

  (1) X is interpretable-in-principle as a token of some marker type T,

  (2) there is a mapping M available from a set of marker types including T to a set of interpretations including Y, and

  (3) $M(T) = Y$.

Definitions (S4) and (S4*) are extensionally equivalent for real and counterfactual cases. Under either definition, for any object X and any interpretation Y that one might specify,[7] X is interpretable-in-principle as signifying Y. First, we have already seen that every object is interpretable-in-principle as a marker token of some type T. Now, according to definition (S4), all that is additionally necessary for X to be interpretable-in-principle as signifying Y is that one could, in principle, have a convention licensing T-tokens as signifying Y. But one could, in principle, have such a convention for any type T and any Y. Similarly, according to definition (S4*), what is necessary for X to be interpretable-in-principle as signifying Y (over and above X's being interpretable-in-principle as a marker of some type T) is the availability of a mapping M from marker types to interpretations such that Y is the image of T under M. Such a mapping is merely an abstract relation between two sets, however, and there is such a mapping, for any type T and any Y, that maps T onto Y. So both (S4) and (S4*) license the conclusion that every object is interpretable-in-principle as signifying anything whatsoever. This conclusion may seem bland in and of itself, but it is important to distinguish this sense of "having a meaning" or "having a referent" from other, more robust senses. It is all the more important to do so since computationalists seem at times to be interested in this sort of "having a meaning," but do not always make it adequately clear what role (if any) it plays in their accounts of semantics and intentionality for cognitive states.

4.9 FOUR MODALITIES FOR COUNTERS

The four conventional modalities are also applicable to counters. Returning to the optometrist examples, suppose that the optometrist's instrument is adjusted so that Jones can see more than one symbol at a time. Suppose, moreover, that what he sees is the following image:

$$p \ \& \ q$$

Jones has just come from his logic class, and so, when asked what he sees, says "*p* and *q*." The optometrist, however, is ignorant of the conventions of logic. To him, this is just a line of three characters: the letter *p*, an ampersand, and the letter *q*. As the doctor sees it, the symbols on the eyechart are not related to one another syntactically, *because the "eyechart game" does not bare any syntactic rules*.

   Once again, both Jones and the doctor are partially right, and in much the same ways they were each partially right in the original examples. Jones has a point in that the figures he sees are interpretable as markers of familiar types (and are in this case intended to be of the types that Jones guesses), and he is furthermore right in seeing that they are arranged in a fashion that is *interpretable,* under the conventions he has been taught for the propositional calculus, as having a certain syntactic form in the propositional calculus. Yet the optometrist has a point as well: the chart at which Jones is looking was designed as an eyechart. (We may, if we like, assume once again that the doctor drew the chart himself, and knows quite well what he meant to draw.) It was not intended to contain formulas in the notation employed in propositional logic, and the fact that some symbols in the eyechart are interpretable as forming such a formula is quite accidental. Similarly, if a diagonal sequence of letters should be interpretable as a sentence in Martian, that fact would be quite accidental. When the author of the eyechart drew it, Martian language played no role in his activity, and neither did the propositional calculus. To use terminology developed earlier, syntactic relationships did not form part of the *authoring intention* with which the

Figure 6

chart was created, and so the symbols in question would not rightly be said to have been *intended* to have a particular syntactic form.

It is, of course, possible that someone should devise an eyechart or some other display of symbols with more than one symbol game in mind. Someone who believed in subliminal suggestion, for example, might devise a display of symbols so that parts of it were interpretable under standard English conventions in a fashion that was not supposed to be consciously recognized by the reader. Thus a greedy optometrist might try to sell extra pairs of glasses by designing his eyechart like that shown in figure 6. In this case, the figures on the chart can be interpreted in two ways: (1) as characters on an eyechart, and (2) as letters forming English words that make up the sentence "Buy an extra pair now." As they are employed in the "eyechart game," the markers on the display do not enter into syntactic relationships, because syntactic relationships are always *relative to a system with syntactic rules,* and the "eyechart game" has no syntactic rules. As markers used in the formation of an English sentence token, however, they are counters having syntactic properties, because the English language does have syntactic rules. In this example, moreover, the markers on the chart are not only (*a*) *interpretable as* syntactically unstructured tokens in the eyechart game and (*b*) *interpretable as* syntactically structured tokens in a written English sentence, they are also (*c*) *intended as* syntactically unstructured tokens in the eyechart game and (*d*) *intended as* syntactically structured tokens in a written English sentence. *Both* "games" are intended by the author of the chart in this case—unlike the earlier case, in which the optometrist did *not* intend the line of symbols p-&-q to count as a formula structured by the rules of propositional logic, even though the line of symbols was nonetheless *interpretable* as such.

In that example, moreover, the line of symbols was also interpreted (by Jones) as a formula in propositional calculus notation, and was interpreted in such a fashion that Jones imputed to it a certain syntactic structure that is provided for by propositional logic. This interpretation is not affected in the least by the fact that the eyechart was not designed with it in mind, or even by the fact that the author of the chart was unfamiliar with propositional logic. Finally, as in the case of markers and signifiers, there is infinite latitude in the ways a display of markers could, in principle, be interpreted as counters of various sorts, because any given marker type can be employed in an indefinite number of systems characterizable by syntactic rules. For any arrangement of markers, one could, as Haugeland says, "imagine any number of (strange and boring) games in which they would be perfectly legal moves" (Haugeland 1981: 25).

It is now possible to provide definitions for the four ways of being a counter. These definitions will not be employed directly in the argumentation that follows, but are provided for the sake of exactitude and balance in the development of semiotic terminology. They may safely be skimmed over by the reader who is not interested in the definitions for their own sake, but only in their contribution to the main line of argument.

(C1) An object X may be said to be *interpretable as a counter of type C* iff

    (1) X is interpretable as a marker of type T,

    (2) the marker type T is employed in some language game G practiced by a linguistic community L,

    (3) G is subject to syntactic analysis,

    (4) there is a class C of markers employed in G sharing some set F of syntactic properties, and

    (5) the conventions of G are such that tokens of type T fall under class C.

(C2) An object X may be said to be *intended (by S) as a counter of type C* iff

    (1) there is a language user S who is able to apply the conventions of a language game G,

    (2) the marker type T is employed in G,

    (3) G is subject to syntactic analysis,

(4) there is a class C of markers employed in G sharing some set F of syntactic properties,

(5) the conventions of G are such that tokens of type T fall under class C,

(6) S intended X to be a marker of type T,

(7) S intended X to count as a move in an instance of language game G, and

(8) S intended X to fall under class C.

(C3) An object X may be said to be interpreted (by H) as a counter of type C iff

(1) some language user H apprehended X,

(2) H interpreted X as a token of type T,

(3) H is able to apply the conventions of language game G,

(4) the marker type T is employed in G,

(5) G is subject to syntactic analysis,

(6) there is a class C of markers employed in G sharing some set F of syntactic properties,

(7) the conventions of G are such that tokens of type T fall under class C,

(8) H interpreted X as counting as a move in an instance of G, and

(9) H interpreted X as falling under class C in game G.

(C4) An object X may be said to be interpretable-in-principle as a counter of type C iff

(1) X is interpretable-in-principle as a token of marker type T,

(2) there could be a language game G employing markers of type T,

(3) that game G would be subject to syntactic analysis,

(4) these conventions would be such that there would be a class C of markers employed in G sharing some set F of syntactic properties, and

(5) the conventions of G would be such that tokens of type T would fall under class C.

4.10 THE NATURE AND SCOPE OF THIS
SEMIOTIC ANALYSIS

The preceding sections of this chapter have been devoted to the development of an analysis of symbols and their semantic and syntactic properties. In the ensuing chapters this analysis will be applied towards an assessment of CTM's claims about the nature of cognition. Before proceeding to that assessment, however, it is important to clarify the nature and status of the semiotic analysis that has been presented here.

The new terminology is intended to resolve perilous ambiguities in the uses of (*a*) the word 'symbol' and (*b*) expressions used to predicate semantic and syntactic properties of symbols (for example, 'refers to', 'means', 'is a count noun'). For purposes of careful semiotic analysis, the technical terms are meant to *replace* the ordinary locutions rather than to *supplement* them. Thus the sortal terms 'marker', 'signifier', and 'counter' do not name different *species* of symbol, nor do they signify different objects than those designated by the word 'symbol'. Rather, these terms serve collectively as a disambiguation of the word 'symbol' as it is applied to discursive signs, and each sortal term is designed to correspond to one usage of the word 'symbol'.

Similarly, the modalities of *interpretability* (under a convention), *authoring intention, actual interpretation,* and *interpretability-in-principle* have been referred to as "ways of being" a marker, signifier, or counter. But this does not mean that there is such a thing as just being a marker, signifier, or counter, and—over and above that—*additional* properties of being *interpretable* as one, being *intended* as one, and so on. For there is no such thing as simply *being* a symbol. *Symbol* is not a natural but a conventional kind, and to say that something "is a symbol" (a marker) is to relate it in some way to the conventions that establish marker types.

But there are several ways in which an object can be related to such conventions: it can be *interpretable* as a token of a type by virtue of meeting the criteria for that type, it can be *intended* by its author as being of that type, it can be *interpreted* as being of that type, or it can simply be such that one could have a convention that would establish a type such that this object would be interpretable as a token of that type. The case is much the same for semantics and syntax: there is no such thing as a marker simply *being* meaningful or simply *referring* to an object. To say that a marker has a meaning, or that it refers to something, is to say something about interpretation and interpretive conventions. We can say that

the marker is of such a type that it is interpretable, under English semantic conventions, as referring to Lincoln. We can say that its author
intended it to refer to Lincoln, or that someone who apprehended it
construed it as referring to Lincoln. And we can say that one could, in
principle, have a convention whereby it would be interpretable as referring to Lincoln. But there is no *additional* question of whether a symbol *just plain refers* to Lincoln. Expressions such as 'refers to Lincoln',
'is a marker', 'is a *P* ', or 'is an utterance of the word *dog* ' are ambiguous. The process of disambiguation consists of substituting the four
expressions, 'is interpretable as', 'was intended as', 'was interpreted
as', and 'is interpretable-in-principle as' for 'is'.

So, for example, if someone asks of an inscription, "What kind of
symbol is that?" we should proceed by supplying four kinds of information: (1) We should provide a specification of how it is interpretable
as a marker token by virtue of meeting the criteria for various marker
types. For example, we might point out that it is interpretable under
English conventions as a *P* or under Greek conventions as a rho. (2) If
the mark was in fact inscribed by someone, we should say what kind of
marker it was intended to be: for example, that it was intended as a *P*,
or that it was intended as a rho, or that it was intended precisely to meet
the criteria for both *P* and rho. (3) If someone has interpreted the inscription as a marker token, we should say who did the interpreting and
what they took it to be. We might say, for example, that Jones took the
symbol to be a *P*, while Mrs. Mavrophilipos took it to be a rho. (4) We
should point to the fact that such a mark might be used in all kinds of
ways—namely, that one could, for example, develop new marker conventions whereby that mark might count as a token of some new type.

Similarly, if someone asks what an inscription means, a full response
would involve the following: (1) A list of the meanings that the inscription could be used to bear under the semantic conventions of various
linguistic groups. (For example, English speakers use the marker string
p-a-i-n to mean pain while French speakers use it to mean bread.) (2) A
specification of what the author of the inscription intended it to mean.
(The deceased lawyer in the thought experiment, for example, might
have used it to mean bread, while I, the author of the example, intended
precisely that it be interpretable as meaning either bread or pain.) (3) A
specification of how anyone who apprehended the symbol interpreted
it. (For example, Lestrade took it to mean pain and Holmes took it to
mean bread.) (4) A reference to the fact that one could, in principle, use
markers of that type to refer to anything whatsoever.

And similarly for counters, if one were to inquire as to the syntactic properties of an inscription such as 'p & q', a complete answer would require four kinds of information: (1) A list of ways that string could be interpreted as bearing a syntactic structure in different symbol games. (It could be a series of syntactically unrelated markers on an eyechart, for example, or a conjunction in the sentential calculus.) (2) A specification of how the inscription was intended by its author. (For example, the optometrist intended those markers as items on an eyechart, and did not intend them to bear any syntactic relation to one another.) (3) A specification of how such persons as apprehended the symbols took them to be syntactically arranged. (Say, Jones took them to constitute a propositional calculus formula of the form 'p and q', while Mrs. Mavrophilipos took them to just be individual letters.) Finally, (4) an allusion to the fact that one could devise any number of symbol games with quite a variety of syntactic structures such that this inscription would be interpretable as being of the syntactic types licensed by the rules of those games.

Now there *are* other uses of the term 'symbol'—for example, those employed in Jungian psychology and cultural anthropology. Similarly, there are other senses in which a marker might be said to "mean something." Holmes's companion Dr. Watson might, for example, inquire of Holmes, "What does the deceased attorney's inscription mean?" and Holmes might reply, "What it means, Watson, is that the housekeeper is a murderess." In this case, Watson's query, "What does it mean?" amounts to asking "What conclusions about this case can we draw from it?" and Holmes's answer supplies the relevant conclusion.

Yet it is important to emphasize that there is no *general* sense of "being a symbol" or "meaning such-and-such" over and above those captured by our technical terms. For suppose that someone were to ask what the first mark on the eyechart was, and we told him about how it was interpretable under various conventions, how it was intended by the doctor who drew it, how it was interpreted by various people who saw it, and pointed out, finally, that one could develop all sorts of conventions that could apply to marks with that shape. Suppose, however, that our questioner was not satisfied with this, but insisted upon asking for more. Suppose he said, "I don't want to hear what conventional types it meets the criteria for, or how it was intended, or how anyone construed it, or how it could, in principle, be interpreted. I just want to know what kind of symbol it *is*." Suppose that it was clear from the way that he spoke that he thought that there was just some kind of brute fact about an object that consisted in its being a marker of a particular type, quite apart from how it met the criteria for conventionally sanctioned types, how it was intended, and so on. How would we construe such a question?

There are, I think, two basic possibilities. The first is that the questioner is just confused, and does not realize that the relevant uses of the expression 'is a symbol' have effectively been replaced by our technical terminology. If this is the case, he would seem to be suffering from a misunderstanding of what is meant when we say that something is a rho, or a *P*, or a token of some other marker type. He is much like the person who misunderstands the use of the word 'healthy' when it is applied to food and demands of us that we tell him what "makes vitamin *C* healthy" without telling him how it contributes to the health of a body.

The second possibility is that the questioner has some *special* use of the expression 'is a symbol' in mind. He might, for example, be asking for an answer cast in the vocabulary of some particular psychological or anthropological tradition. (We might, for example, respond to a query about something on the wall of an Irish church in the following fashion: "This is the Celtic cross, a fine example of syncretic symbolism. In it one finds the Christian cross, the symbol of salvation through the death of Christ, cojoined with the Druidic circle, symbolizing the sun, the source of life and light.") Or he might have some more novel use of words in mind. He might, for example, just *use* the word 'symbol' in a way that did not make appeals to conventions. Allen Newell, for example, apparently *identifies* symbols with the physical patterns that distinguish them. Newell writes, "A physical symbols system is a set of entities, called symbols, *which are physical patterns* that can occur as components of another type of entity called an expression (or symbol structure)" (Newell and Simon [1975] 1981: 40, emphasis added). In another place, Newell (1986: 33) speaks of symbols systems as involving a physical medium and writes that "the *symbols* are patterns in that medium."

I shall discuss the proper interpretation of Newell's usage at length in chapter 5, but the basic point I wish to make may be summarized as follows: In characterizing symbols in this way, Newell is using the word 'symbol' differently from the way it is normally used in English, not unlike the way someone might just use the word 'healthy' to mean "full of vitamins." (By the same token, one *could* use the word 'symbol' to designate all and only objects that have odors pleasing to dogs. Why one should wish to abuse a perfectly good word in such a fashion, however, is quite another matter.) This kind of idiosyncratic use of words may be confusing, but it need not be pernicious so long as the writer (*a*) does

not draw inferences that are based upon a confusion between his idiosyncratic usage of the word and its normal meaning (e.g., inferring that
food that is healthy$_v$ [i.e., full of vitamins] must be healthy [i.e., conducive to health]), and (b) makes his own usage of the word adequately
clear that his readers are not drawn into such faulty inferences. Thus
there is nothing troublesome about using the word 'charm' to denote a
property of quarks because (a) physicists have an independent specification of the meaning of 'charm' as applied to quarks, and (b) no one is
likely to mistakenly infer that quarks would be pleasant guests at a soi-
rée.

Similarly, it is possible to use words such as 'means' and 'refers to'
in novel ways. One could, for example, become so enamored of causal
theories of reference that one began to *use* sentences like "The word
'dog' refers to dogs" to *mean* something like "Tokens of 'dog' stand in
causal relation R to dogs." This would, of course, be an enterprise involving linguistic novelty: the locutional schema 'refers to' is not generally used by English speakers to report causal relationships per se.
But the idiosyncratic usage of the locutional schema might be an efficient way of expressing something that is important and for which there
is no more elegant means of expression. So long as the writer makes his
usage of words clear and does not make illicit inferences based on non-
operative meanings of words, his idiosyncrasy need not be construed as
being pernicious. But if, for example, someone uses 'refers to' to mean
"is larger than," he cannot draw an inference like that below from (A)
to (B) just by virtue of the meanings of the sentences

(A) The title 'Great Emancipator' refers to Abraham Lincoln.

(B) Abraham Lincoln is also known as the Great Emancipator.

If one used such a novel definition to try to show that one could derive
"X is known as Y" from "X is greater than Y," one would be arguing
fallaciously.

Nor can the inference from (A) to (B) be drawn by virtue of the
meanings of the sentences if one just *defines* 'refers to' in causal terms.
That is, if one uses (A) to *mean* "Tokens of 'Great Emancipator' stand in
causal relation R to Abraham Lincoln," one cannot infer from (A) that
Abraham Lincoln is also known as the Great Emancipator. One might,
however, be able to infer (B) from the conjunction of the *two* claims
(A*): "Tokens of 'Great Emancipator' stand in causal relation R to Lincoln" and (C): "For every signifier token M and every object N, if M
stands in causal relation R to N, then M refers (in the ordinary sense) to N."

But (A*) and (C) jointly entail (B) only because (A*) and (C) jointly entail (A), and (A) entails (B). (A*) alone does not entail (A), however, even if there is a causal relation R that always in fact holds between signifiers and their referents.

4.11 THE FORM OF ASCRIPTIONS OF INTENTIONAL AND SEMANTIC PROPERTIES

One of the motivations for undertaking this analysis of symbols was an objection to CTM that was suggested in chapter 3. This objection, called the Conceptual Dependence Objection, involved two important claims about ascriptions of semantic and intentional properties. The first claim was that terms used in ascriptions of semantic and intentional properties are ambiguous: ascriptions of semantic and intentional properties to *symbols* and ascriptions of semantic and intentional properties to *mental states* have different logical forms and indeed involve attributions of different properties. The second claim was that ascriptions of semantic and intentional properties to symbols are conceptually dependent upon attributions of cognitive states. In Aristotelian terms, the homonymy of semantic and intentional terms is an example of homonymy *pros hen,* and the focal meaning of the terms is that which applies to cognitive states. These claims were offered only provisionally in chapter 3, however, and a major reason for undertaking this analysis of the nature of symbols was to provide resources for investigating the claims.

I shall argue in the next chapter that the analysis that has been offered here bears out both claims. For present purposes, I shall confine myself to commenting on the logical form of ascriptions of semiotic properties to symbols. We have discovered that the surface form of ascriptions of semantic values and intentionality to symbols is misleading. When we say, for example, "(Inscription) I refers to X," it looks as though the verb phrase 'refers to' expresses a two-place predicate with arguments I and X. This way of reading the sentence, however, is wrong in two respects. First, the locutional schema 'refers to' is ambiguous, and may be used to express four very different propositions. More perspicuous expressions of these propositions are supplied by our technical terminology:

(1) I is interpretable (under convention C of linguistic group L) as referring to X.

(2) I was intended (by its author A) to refer to X.

(3) *I* was interpreted (by some reader *R*) as referring to *X*.

(4) *I* is interpretable-in-principle as referring to *X*.

Second, on *none* of these interpretations does 'refers to' turn out to be a two-place predicate linking a symbol and its referent. The first interpretation, an attribution of *semantic interpretability,* involves implicit reference to a linguistic community and the semantic conventions of that community. The second interpretation, an attribution of *semantic authoring intention,* involves implicit reference to the cognitive states (namely, the authoring intentions) of the author of the symbol. The third interpretation, an attribution of *actual semantic interpretation,* involves implicit reference to the cognitive states of an individual who apprehends *I*. Finally, if we look at the definition of interpretability-in-principle, we see that the fourth interpretation involves implicit reference as well, either to the availability of a mapping that takes *I* 's marker type onto an interpretation, or to possible conventions. What has been said of ascriptions of reference may be said of ascriptions of meaning and intentionality as well. In each case, there are four ways of interpreting such ascriptions, and these involve covert reference to intentions and conventions in just the same ways as ascriptions of reference to symbols involve it.

4.12 SUMMARY

This chapter has developed a set of terminology for dealing with attributions of syntactic and semantic properties to symbols. The terminology involves the disambiguation of the term 'symbol' into three sortal terms—'marker', 'signifier', and 'counter'—and a distinction between four ways in which an object may be said to be a symbol (a marker) and to have syntactic or semantic properties. The analysis has already produced a significant conclusion: once we have rendered ascriptions of semantic properties to markers more perspicuous by employing the terminology that has been developed here, it becomes apparent that the logical forms of these expressions involve complex relations with conventions and intentions.

This analysis provides the basis for an investigation of the claims of CTM. The next chapter will investigate the implications of this analysis for the nature of semantic attributions to minds and to symbols in computers. The one that follows it will examine an objection to the analysis presented in this chapter and articulate an alternative reading of the semiotic vocabulary as employed by advocates of CTM. Afterwards, we shall examine the implications of this analysis for CTM's representational account of the nature of cognitive states and its attempt to vindicate intentional psychology by claiming that cognitive processes are computations over mental representations.

CHAPTER FIVE

The Semantics of Thoughts and of Symbols in Computers

The preceding chapter presented an analysis of the nature of symbols, syntax, and symbolic meaning. The upshot of this analysis was that symbolhood, syntax, and symbolic meaning are all conventional to the core. There is no such thing as simply *being* a *P*, a count noun, or a referring term. Words used to attribute semiotic categories do not express simple one- or two-place predicates, but hide complex relationships involving conventions and intentions. I shall refer to the analysis presented in chapter 4 as the "Semiotic Analysis."

The ultimate reason for undertaking this Semiotic Analysis was to assess a particular kind of attack upon CTM: namely, the claim, urged on us by Sayre and Searle, that the notions of *symbol* and *symbolic meaning* were somehow unsuited to the tasks of explaining the intentionality of mental states and of "vindicating" intentional psychology. We shall begin to develop some definitive answers to this question in chapter 7. Before doing that, however, it is necessary to address two issues, which will be the task of this chapter and the one that follows. To take things in reverse order, the next chapter will address an important kind of objection to the Semiotic Analysis: namely, that it conflates a "purely semantic" element of languages that is nonconventional with conventional features that accrue to natural languages only because they are used for communication. On this opposing view, often identified (rightly or wrongly) with Tarski and Davidson, semantic analysis is applied to things called "abstract languages" that are nonconventional in nature, while conventions come into play only in our *use* or *adoption* of such languages for communication.

The present chapter will draw out some consequences of the Semiotic Analysis in two very separate areas, both of which will prove important to the larger argument. First, the Conceptual Dependence Objection sketched in chapter 3 claims that the semantic vocabulary is paronymous, in that (1) the semantic vocabulary *expresses different properties* when applied (*a*) to symbols and (*b*) to mental states; and (2) the usage that is applied to symbols is *conceptually dependent upon* the usage that is applied to mental states, and not vice versa. In the first part of this chapter it will be argued that the Semotic Analysis gives us what we need to justify this claim of conceptual dependence.

Second, it will be useful and interesting to examine the application of the Semiotic Analysis to symbols in computers. On the one hand, applying this analysis makes it quite clear that, *pace* the Formal Symbols Objection described in chapter 3, computers can and do store and operate upon entities that are symbols with syntactic and semantic properties in all of the ordinary senses. That is, we can speak of syntax and semantics for symbols in computers in exactly the same ways we speak of them for utterances and inscriptions. On the other hand, it will become clear upon closer inspection that the *functional* analysis of computers is a completely separate matter from their *semiotic* analysis. Computers can be analyzed in functional terms *and* in semiotic terms, and computer designers take great pains to make these two descriptions line up with one another in practice. But it is not the functional properties of the computer that make things inside it *count as* markers, counters, and signifiers (or vice versa). Contrary to some writers, the study of computation adds *nothing* to our understanding of symbols per se.

5.1 SEMIOTICS AND MENTAL SEMANTICS

The Semiotic Analysis was an analysis of the properties of symbols. A part of this analysis considered what it is we are imputing to symbols when we impute to them meaning or reference or intentionality. This involved looking both (1) at the logical form of such utterances, and (2) at the conditions for their satisfaction. Semantic terms like 'means' and 'is about' turned out to be both ambiguous and surprisingly complex. If, for example, I say,

"Symbol *X* means *P*"

I could be asserting one or more of the following:

(1) *X* is interpretable under convention *C* of language *L* as meaning *P*.

(2) *X* was intended by its author *S* to mean *P*.

(3) *X* was interpreted by some observer *H* as meaning *P*.

(4) *X* is, in principle, interpretable as meaning *P*.

Each of these locutional schemas has a distinct logical form and a distinct set of argument slots that can be filled by different kinds of objects. Some of these arguments do not always appear in the surface grammar of attributions of semantic properties in ordinary language, though they are likely to be filled in when a speaker is called upon to clarify her utterance. In the first three cases—that is, conventional interpretability, authoring intention, and actual interpretation—some of these suppressed arguments refer either to conventions of a linguistic community or to intentions of those who produce or apprehend the symbol tokens. It also turned out that there was a plausible reading of interpretability-in-principle which construed it as a modal variation upon interpretability in a language.

Here, I think, we have all we need to justify two claims sketched in the articulation of the Conceptual Dependence Objection in chapter 3:

(1) Terms in the semantic vocabulary express different properties when applied to mental states from those they express when applied to symbols.

(2) The applications of semantic terms to symbols are conceptually dependent upon their applications to mental states.

5.1.1 THE HOMONYMY OF THE SEMANTIC VOCABULARY

The case for homonymy is fairly straightforward. First, if a natural language verb *V* is used in two contexts, *A* and *B*, and the logical form of *V*-assertions in *A* differs from that in *B*, then *V* is used to express different predicates in the two contexts. Moreover, *predicates with different logical forms express different properties*. In particular, two predicates expressing relational properties can only express the same property if they relate the same number of relata. Predicates with different numbers of argument slots in their logical form relate different numbers of relata, and hence express different properties.[1] Likewise, two predicates with the same number of arguments do not express the same property if the things that can fill their argument slots come from different domains.

Second, *the predicates used to express semantic properties of symbols have argument slots that must be filled by references to conventions and intentional agents.* They thus express complex relational properties that essentially involve conventions and agents. This much is a straightforward consequence of the Semiotic Analysis. One may contest the analysis on other grounds; but if one accepts it, one has already bought into this consequence.

Third, *the logical forms of attributions of semantic properties to minds and mental states do not contain argument slots to be filled by references to conventions or intentions, and hence they attribute properties distinct from the semiotic properties.* I have always thought that this was pretty self-evident, but sometimes people have accused me of just asserting this point without arguing it. The only way I see to demonstrate this point is to test each of the schemas developed in chapter 4 as a possible interpretation of attributions of semantic properties to mental states and see whether any of them seem intuitively plausible. So suppose John is having a thought T at time t, and what he is thinking about is *Mary*. We say (in ordinary English)

"John is thinking about Mary,"

or (in awkward philosophical jargon)

"John's thought T is about Mary."[2]

Now can this be analyzed in terms of conventional interpretability in a language? That is, could the logical form of this utterance possibly be the following?

(1) John's thought T is interpretable under convention C of language L as being about Mary.

The answer, I think, is clearly *no.* The more you think about "meanings" of *utterances,* the clearer it becomes that there is a notion of conventional meaning in a language that applies there, and that it is part of what we meant all along when we spoke of meaning for symbols. But it is hard to see how that notion could apply to *thoughts.* Except for special cases like the interpretation of dreams, there are no conventions for interpreting thoughts. Nor do thoughts *require* such conventions for interpretation: thoughts come with their meanings already attached. You cannot separate the thought from its meaning the way you can separate the marker from its meaning. This is why some writers describe the semantics of mental states as *intrinsic* to them.

Likewise for the other semiotic modalities:

(2) John intended that this thought T be about Mary.

(3) H apprehended John's thought T and interpreted it as being about Mary.

We do sometimes have thoughts as a result of intentions to have thoughts, as suggested in (2). John might, for example, deliberately think about his wife Mary while he is away on a business trip on their anniversary. Or, dealing with transference on a therapist's couch, he may *intend* to think about Mary but end up thinking about someone else instead. These things happen, but they are surely not what we are talking about when we say John's thought T is about Mary. Usually the intentionality of our thoughts is unintentional.

As for (3), there is some question about whether we apprehend one another's thoughts at all. We surely *guess* at one another's thoughts, and may rightly or wrongly surmise that John's thought at a given time is about Mary. But this is very different from *seeing* a marker as a marker and then interpreting it. We never apprehend thoughts as markers. And more to the point, even if we do apprehend people's thoughts and interpret them, this is not what we mean when we attribute meaning to their thoughts. I might say of a *symbol,* "It means X to Jim." But it surely makes no sense to say, "John's thought means 'Mary' to *Jim* " (or for that matter, "John's thought means 'Mary' to John"). Finally, when we say that John's thoughts are about Mary, we certainly do not mean merely to assert the existence of a mapping relationship (i.e., interpretability-in-principle) from John's thought to Mary. If we try to apply the logical form of the semiotic vocabulary to our attributions of meaning to mental states, the results are nonsensical.

So semantic terms like 'means' and 'is about' have a different logical form when applied to mental states. It *does* seem reasonable to construe the logical form of *these* attributions as involving a three-place predicate relating subject, thought-token, and meaning, as the surface grammar suggests. There are no hidden references to conventions and intentions. As a consequence, the semantic vocabulary also expresses distinct *properties* when applied to mental states. When applied to symbols, it expresses relational properties in which some of the relata are conventions or producers and interpreters of symbols. But these relata are missing in the case of mental states. In short, differences in logical form point to differences in properties expressed.

It thus behooves us to differentiate between two classes of properties that are expressed using the same semantic vocabulary: there is one set of *semiotic-semantic properties* as described by the Semiotic Analysis in chapter 4, and a separate set of *mental-semantic properties* attributed to mental states.

5.1.2 CONCEPTUAL DEPENDENCE

It is also quite straightforward to show that attributions of semiotic-semantic properties are conceptually dependent upon attributions of mental-semantic properties. In the case of authoring intentions and actual interpretation, the analysis of semantic attributions alludes to meaningful mental states on the part of the author or interpreter: their intentions and acts of interpretation. Claims about authoring intention and actual interpretation are built upon a more fundamental stratum of attributions (or presuppositions) of meaningful mental states to human individuals. In the case of conventional interpretability the case is only slightly less direct. For a large part of what linguistic conventions consist of is the shared beliefs and practices of members of a linguistic group. Thus any appeal to conventions assumes a prior stratum of meaningful mental states as well. It thus turns out that the semantic vocabulary is ambiguous and indeed paronymous as claimed by the Conceptual Dependence Objection. Words like 'means' and 'is about' are used differently for mental states and for symbols, and the usage that is applied to symbols is conceptually dependent upon the usage that is applied to mental states.

It remains to be seen, however, what impact this will have on CTM's claims to explain mental-semantics in terms of the semantics of mental representations. Indeed, in light of this distinction between different uses of the semantic vocabulary, it will turn out that we have to clarify what kinds of "semantic" properties are even being *attributed* to mental representations. Are they semiotic-semantic properties (the natural assumption)? Or are they some other kind of properties that add a new ambiguity to the semantic vocabulary? Chapter 7 will examine the prospects of semiotic-semantic properties for explaining mental-semantics, and chapters 8 and 9 will explore two different strategies for attributing a distinct kind of "semantics" to mental representations.

5.2 SYMBOLS IN COMPUTERS

At this point, I wish to shift attention to a second application of the Semiotic Analysis. In the remainder of this chapter, I shall consider the applications of the Semiotic Analysis to symbols in *computers*. There are really two parts to this exercise. First, I shall argue (against the Formal Symbols Objection articulated in chapter 3) that it is quite unproblematic to say that computers do, in fact, both store and operate upon objects that may be said to be symbols, and do have syntactic and semantic properties in precisely the senses delineated by the Semiotic Analysis. To be sure, the story about how signifiers are tokened in microchips is a bit more complicated than the story about how they are tokened in speech or on paper, but it is in essence the same *kind* of story and employs the same resources (namely, the resources outlined in the Semiotic Analysis). Second, I shall address claims on the opposite front to the effect that there is something special about symbols in computers, and that computer science has in fact revealed either a *new kind* of symbol or revealed something *new and fundamental* about symbols in general. I shall argue that this sort of claim, as advanced by Newell and Simon (1975), is a result of an illegitimate conflation of the functional analysis of computers with their semiotic properties. Or, to put it another way, Newell and Simon are really using the word 'symbol' in two different ways: one that picks out semiotic properties and another that picks out functionally defined types. Neither of these usages explains the other, but both are important and useful in understanding computers.

5.2.1 COMPUTERS STORE OBJECTS THAT ARE SYMBOLS

In light of the centrality of the claim that computers are symbol manipulators, it is curious that virtually nothing has been written about how computers may be said to store and manipulate symbols. It is not a trivial problem from the standpoint of semiotics. Unlike utterances and inscriptions (and the letters and numerals on the tape of Turing's computing machine), most symbols employed in real production-model computers are never directly encountered by anyone, and most users and even programmers are blissfully unaware of the conventions that underlie the possibility of representation in computers. Spelling out the whole story in an exact way turns out to be cumbersome, but the basic conceptual resources needed are simply those already familiar from the Semiotic Analysis. I have divided my discussion of symbols in computers into two parts. I shall give a general sketch of the analysis here and provide the more cumbersome technical details in an appendix for those interested in the topic, since the details do not contribute to the main line of argumentation in the book.

The really crucial thing in getting the story right is to make a firm distinction between two questions. The first is a question about semiotics: *In virtue of what do things in computers count as markers, signifiers, and counters?* The second is a question about the design of the machine: *What is it about computers that* allows them to *manipulate symbols in ways that "respect" or "track" their syntax and semantics?* Once we have made this distinction, the basic form of the argument that computers do indeed operate upon meaningful symbols is quite straightforward:

(1) Computers can store and operate upon things such as numerals, binary strings representing numbers, and so on.

(2) Things like numbers and binary strings representing numbers are symbols.

(3) Computers can store and operate upon symbols.

Of course, while one *could* design computers that operate (as Turing's fictional device did) upon things that are already symbols by independent conventions (i.e., letters and numerals), most of the "symbols" in production-model computers are not of this type, and so we need to tell a story about how we get from circuit states to markers, signifiers, and counters. I shall draw upon two examples here:

Example 1: The Adder Circuit

In most computers there is a circuit called an *adder*. Its function is to take representations of two addends and produce a representation of their sum. In most computers today, each of these representations is stored in a series of circuits called a *register*. Think of a register as a storage medium for a single representation. The register is made up of a series of "bistable circuits"—circuits with two stable states, which we may conventionally label 0 and 1, being careful to remember that the numerals are simply being used as the *labels* of states, and are not the states themselves. (Nor do they represent the numbers zero and one.) The states themselves are generally voltage levels across output leads, but any physical implementation that has the same on-off properties would function equivalently. The adder circuit is so designed that the pattern that is formed in the output register is a function of the patterns found in the two input registers. More specifically, the circuit is designed so that, under the right interpretive conventions, the pattern formed in the output register has an interpretation that corresponds to the sum of the numbers you get by interpreting the patterns in the input registers.

Example 2: Text in Computers

Most of us are by now familiar with word processors, and are used to thinking of our articles and other text as being "in the computer," whether "in memory" or "on the disk." But of course if you open up the machine you won't see little letters in there. What you will have are large numbers of bistable circuits (in memory) or magnetic flux density patterns (on a disk). But there are *conventions* for encoding graphemic characters as patterns of activity in circuits or on a disk. The most widely used such convention is the ASCII convention. By way of the ASCII convention, a series of voltage patterns or flux density patterns gets mapped onto a corresponding series of characters. And if that series of characters also happens to count as words and sentences and larger blocks of text in some language, it turns out that that text is "stored" in an encoded form in the computer.

Now to flesh these stories out, it is necessary to say a little bit about the various levels of analysis we need to employ in looking at the problem of symbols in computers and also say a bit about the connections between levels. At a very basic level, computers can be described in terms of a mixed bag of *physical properties* such as voltage levels at the output leads of particular circuits. Not all of these properties are related to the description of the machine as a computer. For example, bistable circuits are built in such a way that small transient variations in voltage level do not affect performance, as the circuit will gravitate towards one of its stable states very rapidly and its relations to other circuits are not affected by small differences in voltage. So we can *idealize* away from the properties that don't matter for the behavior of the machine and treat its components as *digital* —namely, as having an integral and finite number of possible states.[3] It so happens that most production-model computers have many components that are *binary* —they have *two* possible states—but digital circuits can, in principle, have any (finite, integral) number of possible states. Treating a machine that is in fact capable of some continuous variations as a digital machine involves some idealization, but then so do most descriptions relevant for science. The digital description of the machine picks out properties that are *real* (albeit idealized), *physical* (in the strong sense of being properties of the sort studied in physics, like charge and flux density), and *nonconventional*.

Next, we may note that a series of digital circuits will display some *pattern* of digital states. For example, if we take a binary circuit for simplicity and call its states 0 and 1, a series of such circuits will display some pattern of 0-states and 1-states. Call this a *digital pattern*. The important thing about a digital pattern is that it occupies a level of abstraction sufficiently removed from purely physical properties that the same digital pattern can be present in *any* suitable series of digital circuits independent of their physical nature. (Here "suitable series" means any series that has the right length and members that have the right number of possible states.) For example, the same binary pattern (i.e., digital pattern with two possible values at each place) is present in each of the following sequences:

$$
\begin{array}{cccc}
a & a & b & b \\
0 & 0 & 1 & 1 \\
\bullet & \bullet & | & |
\end{array}
$$

It is also present in the music produced by playing either of the following:

And it is present in the series of movements produced by following these instructions:

(1) Jump to the left, *then*

(2) jump to the left again, *then*

(3) pat your head, *then*

(4) pat your head again.

Or, in the case of storage media in computers, the same pattern can be present in any series of binary devices if the first two are in whatever counts as their 0-state and the second two are in whatever counts as their 1-state. (Indeed, there is no reason that the system instantiating a binary pattern need be physical in nature at all.)

Digital patterns are *real*. They are *abstract* as opposed to *physical* in character, although they are literally present in physical objects. And, more importantly, they are *nonconventional*. It is, to some extent, our conventions that will determine which abstract patterns are important for our purposes of description; but the abstract patterns themselves are all really there independent of the existence of any convention and independently of whether anyone notices them.

It is digital patterns that form the (real, nonconventional) basis for the tokening of symbols in computers. Since individual binary circuits have too few possible states to encode many interesting things such as characters and numbers, it is *series* of such circuits that are generally employed as units (sometimes called "bytes") and used as symbols and representations. The ASCII convention, for example, maps a set of graphemic characters to the set of seven-digit binary patterns. Integer conventions map binary patterns onto a subset of the integers, usually in a fashion closely related to the representation of those integers in base-2 notation.

Here we clearly have conventions for both markers and signifiers. The marker conventions establish kinds whose physical criterion is a binary pattern. The signifier conventions are of two types (see fig. 7). In cases like that of integer representation, we find what I shall call a *representation scheme,* which directly associates the marker type (typified by its binary pattern) with an interpretation (say, a number or a boolean value). In the case of ASCII characters, however, marker types typified by binary patterns are not given semantic interpretations. Rather, they *encode* graphemic characters that are employed in a preexisting language game that has conventions for signification; they no more have meanings individually than do the graphemes they encode. A string of binary digits in a computer is said to "store a sentence" because (*a*) it *encodes* a string of characters (by way of the ASCII convention), and (*b*) that string of characters is used in a natural language to express or represent a sentence. I call this kind of convention a *coding scheme.* Because binary strings in the computer encode characters and characters are used in text, the representations in the computer inherit the (natural-language) semantic and syntactic properties of the text they encode.

It is thus clear that computers can and do store things that are intepretable as markers, signifiers, and counters. On at least some occasions, things in computers are *intended* and *interpreted* to be of such types, though this is more likely to happen on the engineer's bench than on the end-user's desktop. It is worth noting, however, that in none of this does the computer's nature *as a computer* play any role in the story.

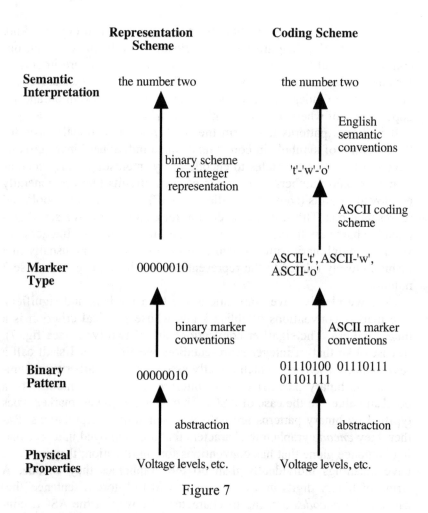

Figure 7

The architecture of the computer plays a role, of course, in determining what kinds of resources are available as storage locations (bistable circuits, disk locations, magnetic cores, etc.). But what makes something in a computer a *symbol* (i.e., a marker) and what makes it *meaningful* are precisely the same for symbols in computers as for symbols on paper: namely, the conventions and intentions of symbol users.

Now of course the difference between computers and paper is that computers can do things with the symbols they store and paper cannot. More precisely, computers can produce new symbol strings on the basis of existing ones, and they can do so in ways that are useful for enterprises like reasoning and mathematical calculation. The common story about this is that computers do so by being sensitive to the syntactic properties of the symbols. But strictly speaking this is false. Syntax, as we have seen and will argue further in the next chapter, involves more than functional description. It involves convention as well. And computers are no more privy to syntactic conventions than to semantic ones. For that matter, computers are not even sensitive to *marker* conventions. That is, while computers operate upon entities that happen to *be* symbols, the computer does not relate to them as symbols (i.e., as markers, signifiers, and counters). To do so, it would need to be privy to conventions.

There are really two quite separate descriptions of the computer. On the one hand, there is a functional-causal story; on the other a semiotic story. The art of the programmer is to find a way to make the functional-causal properties do what you want in transforming the symbols. The more interesting symbolic transformations you can get the functional properties of the computer to do for you, the more money you can make as a computer programmer. So for a computer to be *useful,* the symbolic features need to line up with the functional-causal properties. But they *need not in fact* line up, and when they do it is due to an excellence in design and not to any a priori relationship between functional description and semiotics.

5.2.2 A RIVAL VIEW REFUTED

Now while I think this last point is *true,* I can hardly pretend that it is *uncontroversial.* There is a rival view to the one that I have just presented, and this rival view has enjoyed quite a bit of popularity over the years. On this view, there is something about the functional nature of the computer that contributes to, and even explains, the symbolic character of what it operates upon. Due to the prevalence of this alternative theory, I think it is worth presenting it with some care and venturing a diagnosis of what has gone wrong in it.

Some writers claim that computer science has revealed important truths about the nature of symbols. Newell and Simon (1975), for example, claim that computer science has discovered (*discovered!*) that 'symbol' is an important natural kind, whose nature has been revealed through research in computer science and artificial intelligence. Their central concern is with what they call the "physical symbol system hypothesis." Newell and Simon describe a "physical symbol system" in the following way:

A physical symbol system consists of a set of entities, called symbols, which are physical patterns that can occur as components of another type of entity called an expression (or symbol structure).... Besides these structures, the system also contains a collection of processes that operate on expressions to produce other expressions.... A physical symbol system is a machine that produces through time an evolving collection of symbol structures. (Newell and Simon [1975] 1981: 40)

Their general thesis is that "a physical symbol system has the necessary and sufficient means for intelligent action" (ibid., 41). They define a physical symbol system as "an instance of a universal machine" (ibid., 42), but seem to regard this as a purely natural category defined in functional terms, not as a category involving the conventional component involved in markers, signifiers, and counters. Indeed, they claim that computer science has made an *empirical discovery* to the effect that symbol systems are an important *natural kind,* defined in physical, functional, and causal terms. It looks as though their "symbols" are supposed to be characterized precisely by "physical patterns" (ibid., 40), although perhaps the functional organization of the system plays some role in their individuation. Their characterizations of how symbols in such systems can "designate" objects and how the system can "interpret" the symbols are also quite peculiar:

Designation. An expression designates an object if, given the expression, the system can either affect the object itself or behave in ways depending on the object.

Interpretation. The system can interpret an expression if the expression designates a process and if, given the expression, the system can carry out the process. (ibid., 40)[4]

Newell and Simon regard the physical symbol system hypothesis as a "law of qualitative structure," comparable to the cell doctrine in biology, plate tectonics in geology, the germ theory of disease, and the doctrine of atomism (ibid., 38-39).

It is this kind of claim that has aroused the ire of critics such as Sayre (1986), Searle (1990), and Horst (1990), for whom such claims seem to involve gross liberties with the usage of words such as 'symbol' and 'interpretation'. In the eyes of these critics, Newell and Simon have in fact coined a new usage of words such as 'symbol' and 'interpretation' to suit their own purposes—a usage that arguably has a different extension from the ordinary usage and undoubtedly expresses different properties.

In one sense, I think this criticism still holds good. Here, however, I should like to draw a more constructive conclusion. For Newell and Simon are also in a sense correct, even if they might have been more circumspect about their use of language: computer science does indeed deal with an important class of systems, describable in functional terms, that form an empirically interesting domain. Their usage of the expressions 'symbol system' and 'symbol' do pick out important kinds relevant to the description of such systems. And the historical pathway to understanding such systems does in fact turn upon Turing's discussion of machines that do, in a perfectly uncontroversial sense, manipulate symbols (i.e., letters and numerals). But while it has proven convenient *within the theory of computation* to speak of functionally describable transformations as "symbol manipulations," this involves a subtle shift in the usage of the word 'symbol', and the ordinary notion of symbol is *not* a natural kind, nor are systems that manipulate symbols *per se* an empirically interesting class.

In order to illustrate this claim, it will prove convenient to tell a story about the history of the use of the semiotic vocabulary in connection with computers and computation. The story begins with Turing's article "On Computable Numbers" (1936)—the article in which he introduces the notion of a computing machine. The purpose of this article is to provide a general characterization of the class of computable functions, where 'computable' means "susceptible to evaluation by the application of a rote procedure or algorithm." Turing's strategy for doing this is first to describe the operations performed by a "human computer"— namely, a human mathematician implementing an algorithmic procedure (Turing *always* uses the word 'computer' to refer to a human in this article); second, to develop the notion of a machine that performs "computations" by executing steps described by Turing as being analogous to those performed by the human mathematician; and third, to characterize a general or "universal" machine that can perform any computations that can be performed by such a machine, or by anything that can perform the kinds of operations that are involved in computation.

It is worth looking at a few of the details of Turing's exposition. Turing likens

> a man in the process of computing a real number to a machine which is only capable of a finite number of conditions, q1, q2,..., qR, which will be called "*m*-configurations". The machine is supplied with a "tape" (the analogue of paper) running through it, and divided into sections (called "squares") each capable of bearing a "symbol". (Turing 1936: 231)

(Note the scare quotes around 'symbol' here. One plausible interpretation is that Turing is employing this word in a technical usage, not necessarily continuous with ordinary and existing usage.)

To continue the description: the machine has a head capable of scanning one square at a time, and is capable of performing operations that
move the head one square to the right or left along the tape and that
create or erase a symbol in a square. Among machines meeting this
description, Turing is concerned only with those for which "at each
stage the motion of the machine...is *completely* determined by the configuration" (Turing 1936: 232). The "complete configuration" of the
machine, moreover, is described by "the number of the scanned square,
the complete sequence of all symbols on the tape, and the *m* -
configuration" (ibid.). Changes between complete configurations are
called "moves" of the machine. What the machine will do in any complete configuration can be described by a table specifying each complete configuration (as a combination of *m*-configuration and symbol
scanned) and the resulting "behaviour" of the machine: that is, the operations it performs (e.g., movement from square to square, printing or
erasing a symbol) and the resulting *m*-configuration.

The symbols are of two types. Those of the first type are numerals:
0s and 1s. These are used in printing the binary decimal representations
of numbers being computed.[5] Those of the second type are used to represent *m* -configurations and operations; for these Turing employs Roman letters, with the semicolon used to indicate breaks between sequences. The symbols are typified by visible patterns,[6] and are meant to
be precisely the letters and numerals actually employed by humans.
Indeed, *the operations of the computing machine are intended to correspond to those of a human computer* (i.e., a human doing computation),
whose behavior "at any moment is determined by the symbols which he
is observing, and his 'state of mind' at that moment" (Turing 1936:
250). Again, Turing first describes the behavior of a human computer
(ibid., 249-251), and then proceeds to describe a machine equivalent of
what the computer (i.e., the human) does:

> We may now construct a machine to do the work of this [human/SH] com-
> puter. To each state of mind of the [human] computer corresponds an "*m*-
> configuration" of the machine. The machine scans *B* squares corresponding
> to the *B* squares observed by the [human] computer. (ibid., 251)[7]

To summarize, Turing's description of a computing machine is offered
as a model on which to understand the kind of computation done by

mathematicians, a model on which "a number is computable if its decimal can be written down by a machine" (ibid., 230).

Now there are two things worth noting here. First, if there is a similarity between what the machine does and what a human performing a computation does, this is entirely by design: the operations performed by the machine are envisioned quite explicitly as corresponding to the operations performed by the human computer (though Turing is not careful to say whether "correspondence" here is intended to mean "type identity" or "analogous role"). Second, while this machine is unproblematically susceptible to analysis *both* (*a*) in terms of *symbols* and (*b*) in the functional terms captured by the machine table, it is important to see that *the factors that render it susceptible to these two forms of analysis are quite distinct.*

On the one hand, it is perfectly correct to say that this machine is susceptible to a functional analysis in the sense of being characterizable in terms of a function (in the mathematical sense) from complete configurations to complete configurations. Indeed, that is what the machine table is all about. What renders the machine appropriate for such an analysis is simply that it behaves in a fashion whose regularities can be described by such a table, and any object whose regularities can be described by such a table is susceptible to the same sort of analysis, whether it deals with decimal numbers or not.

On the other hand, it is perfectly natural to say that Turing's machine operates upon symbols. By stipulation, it operates upon numerals and letters. Numerals and letters are symbols. Therefore it operates upon symbols. Plausibly, this may be construed as a fact quite distinct from the fact that it is functionally describable. Some functionally describable objects (e.g., calculators) operate on numbers and letters, while others (e.g., soda machines) do not. Likewise, some things that operate upon numbers and letters (e.g., calculators) are functionally describable, while others (e.g., erasers) are not (see fig. 8). Moreover, what makes something a numeral or a letter is not what the machine does with it, but the conventions and intentions of the symbol-using community. (Whatever one thinks about the typing of symbols *generally,* this is surely true for numerals and letters.)

Now how does one get from Turing's article to Newell and Simon's, forty years later? I suspect the process is something like the following. For the purposes of the theory of computation (as opposed to semiotics), the natural division to make is between the semantics of the symbols (say, the fact that one is evaluating a decimal series or an integral) and the formal

	Functional/Causal Description	No Functional/ Causal Description
Symbolic	Computers, Calcu-lators	Utterances, inscrip-tions
Non-symbolic	Most machines	Most continuous pro-cesses in nature

Figure 8

techniques employed for manipulating the symbols in the particular algorithmic strategy.[8] And from this standpoint, it does not matter a whit *what* we *use* as symbols—numerals, letters, beads on an abacus, or colored stones. And more to the point, it does not matter for the *functional* properties of the operations performed by the machine whether it operates on numerals and letters (as Turing's machine was supposed to) or upon equivalent sets of activation patterns across flip-flops or magnetic cores or flux densities on a disk. As far as the theory of computation goes, these can be treated as "notational variants," and from an engineering standpoint, the latter are far faster and easier to use than letters and numerals. And of course these circuit states (or whatever mode of representation one chooses) are at least sometimes "symbols" in the senses of being markers, signifiers, and counters: there *are* conventions like the ASCII convention and the decimal convention that group *n*-bit addresses as markers and map them onto a conventional interpretation, and there *are* straightforward mappings of text files in a computer onto ordinary text.

The occupants of computer memory thus live a kind of double life. On the one hand, they fall into *one* set of types by virtue of playing a certain kind of role in the operation of the machine—a role defined in functional-causal terms and described by the machine table. On the other hand, they fall into an *independent* set of types by dint of (possible) subsumption under conventionally based semiotic conventions. Both of these roles are necessary in order for the machine to plausibly be said to be "computing a function"—for example, evaluating a differential equation—*but they are separate roles*. If we have functional organization without the semiotics, what the machine does cannot count as being, say, the solution of a differential equation. This is the difference between calculators and soda machines. More radically, however, addresses in computer memory only count as storing *markers* ("symbols" in the most basic sense) by virtue of how they are interpreted and used. (We *could* interpret inner states of soda machines as symbols—that is, invoke conventions analogous to the ASCII convention for thus construing them—but why bother?) On the other hand, we also do not get computation if we have semiotics without any functional-causal organization (writing on paper) or the wrong functional-causal organization (a broken calculator).

Now I think that what Newell and Simon have done is this: they have recognized that computer science has uncovered an important domain of objects, objects defined by a particular kind of functional organization that operates on things that correspond to symbols. And because they are interested more in the theory of computation than in semiotics (or the description of natural language), they have taken it that the important usage of the word 'symbol' is to designate *things picked out by a certain kind of functional-causal role* in systems that are describable by a machine table. What they have not realized is that this usage is critically different from an equally important, but distinct, usage necessary for talking about semiotics. Nor, as Searle and Sayre have noted, do writers who make this move seem adequately sensitive to the dangers of paralogistic argument that emerge from this oversight.

5.3 A NEW INTERPRETATION OF 'SYMBOL' AND 'SYNTAX'

It thus appears that there is good reason to think that some writers in computer science have at least implicitly employed the words 'symbol' and 'syntax' in a fashion that has proven quite fruitful in their investigations, and yet which bears marked discontinuities with the ordinary uses of those words. In rough terms, in the context of discussion of computers, the words 'symbol' and 'syntax' are sometimes used to designate entities and their properties that play particular roles in a functionally describable system. This is of interest for our purposes in evaluating CTM for the simple reason that it may be *this* usage of the words 'symbol' and 'syntax' that CTM's advocates have in mind, and not the ordinary usage explicated in the Semiotic Analysis in chapter 4. However, in order to clarify how this would affect CTM, it is necessary to make this implicit usage of 'symbol' and 'syntax' more exact by supplying some technical terminology and an analysis. We may develop this category in the following way. The role that is played by the things Newell and Simon call "symbols," quite simply, is one of picking out items in functionally determined categories. The technical use of 'symbol' serves to pick out entities that fall into types based upon the role they play in a functionally described system. A corresponding technical use of 'program' or 'formal rule' picks out causal regularities between functionally described structures in such a system.

We may even state a formal definition for this use of 'symbol,' which may be replaced with the technical term 'machine-counter':

> A tokening of a machine-counter of type T may be said to exist in C at time t iff
>
> (1) C is a digital component of a functionally describable system F,
>
> (2) C has a finite number of determinable states $S: \{s_1,..., s_n\}$ such that C's causal contribution to the functioning of F is determined by which member of S digital component C is in,
>
> (3) the presence of a machine-counter of type T at C is constituted by C's being in state s_i, where $s_i \in S$, and
>
> (4) C is in state s_i at t.

The notion of a machine-counter is defined wholly in nonconventional terms. It can also do an important part of the work that is to be done by the legitimately syntactic categories employed to describe computers: machine-counter types correspond to the counter types that could be used to describe a computer in syntactic terms. To put it differently, for every machine-counter type T of a system S, there is a syntactic description of S available in principle that contains a counter type T^* such that T and T^* provide functionally equivalent characterizations of S. Any X that is a machine-counter of type T is thus interpretable-in-principle as a counter of type T^*. Notice, however, *that the notion of a machine-counter is not built out of a simpler notion corresponding to a marker*. Since functional role is constitutive of machine-counter type, the typing of machine-counters is not dependent upon some prior categorization. In this respect, the functional description of computers differs from semiotic description, which depends on a fundamental level of marker typing.

5.4 IMPLICATIONS OF A SEPARATE
USAGE OF 'SYMBOL'

It may well be, then, that the usage of the words 'symbol' and 'syntax' in computer science poses a challenge to the Semiotic Analysis presented in chapter 4, but the challenge it seems to pose is *not* that that analysis is the wrong analysis of the ordinary usage of 'symbol', but that there is a *new* and *technical* usage of 'symbol' which needs to be considered as well. The implications of this for the assessment of CTM seem to be fairly straightforward. It looks as though there are at least two sorts of things the advocate of CTM might mean in talking about "symbols" in computers: (1) she might mean that they are markers and counters, or (2) she might mean that they are machine-counters. And hence when she speaks of "mental representations" being "symbols," she might mean that they are markers and counters, or she may mean that they are machine-counters. These are very different kinds of claims and need to be assessed separately.

Arguably the case is similar with respect to semantics. It is not clear that there is any coherent nonconventional notion of "meaning" forthcoming *from computer science* that is analogous to the notion of a machine-counter. But it could be that CTM's advocates are using the semantic vocabulary in some fashion distinct from that used to express semiotic-semantic properties. Or, even if they do not yet see the problem and hence do not explicitly mean to use the semantic vocabulary in a new way, they might most charitably be viewed as doing so. That is, it is possible that the best way to read advocates of CTM when they speak of mental representations as having "semantic properties" is to read them as attributing not semiotic-semantic properties, but some distinct class of properties peculiar to mental representations. These properties, presumably nonconventional in character, we might designate *MR-semantic properties,* in contrast with the semiotic-semantic properties of symbols and the mental-semantic properties of mental states. It is unclear what such properties might be, but we may nonetheless signal with our terminology that some distinct, nonconventional set of properties is intended in this way.

It has turned out, then, that computers can be said to "manipulate symbols" both in the ordinary sense of doing things with objects that in fact have conventional interpretations and in a distinct technical sense having only historical connections to the ordinary usage explicated by the Semiotic Analysis. This, however, by no means undercuts the Semiotic

Analysis as an analysis of what it is to be a symbol in the ordinary sense. On the one hand, that analysis applies perfectly well to many things in computers. On the other hand, the technical usage—that is, the notion of a machine-counter—expresses distinct properties from those expressed by the ordinary usages of the word 'symbol'. It is necessary to see whether *either* family of usages of the semiotic vocabulary will provide a viable version of CTM, and it is necessary to explore both.

The analysis of these alternative interpretations of CTM will be undertaken in chapters 7 through 9. Chapter 7 will assess the merits of CTM if it is interpreted as attributing the semiotic properties to mental representations. Chapters 8 and 9 will look at two ways of interpreting CTM in a way other than that suggested by the semiotic categories.

CHAPTER SIX

Rejecting Nonconventional Syntax and Semantics for Symbols

Chapter 4 presented the Semiotic Analysis of the nature of symbols, syntax, and symbolic meaning. According to this analysis, linguistic symbolism is thoroughly dependent upon conventions and intentions of language users. Indeed, this is not merely some contingent fact about symbols in public languages that accrue to them because of their public character, but a feature that is built into the very logical form of predicates in the semiotic vocabulary: to attribute semantic or syntactic properties to a symbol—or indeed, even to call a thing a symbol—just is to relate it to certain conventions and intentions. We saw in the last chapter that this has the consequence that the semantic vocabulary is in fact ambiguous, expressing different properties when applied (*a*) to symbols and (*b*) to mental states. In particular, attributions of meaning to symbols are conceptually dependent upon attributions of meaningful mental states. We saw as well that symbols in computers are symbols in the ordinary semiotic sense, although the paths by which interpretive conventions apply to them may be more circuitous than in the case of discursive symbols such as utterances and inscriptions. But we also saw that recent discussions of symbols in computers by writers such as Newell and Simon seemed sometimes to imply a distinct and technical usage of the word 'symbol' that picked out not the semiotic properties of the symbols, but their functionally defined type. I argued there that functional typing and symbol typing in the ordinary sense are conceptually (and ontologically) distinct, and that bringing them together in computers is a highly contingent matter that in fact makes up much of the programmer's art. The relationship between functional analysis and semiotics is one of craft and not of definition or dependence.

This will lay the basis for the analysis of CTM that will take place in the three chapters that follow this one. Chapter 7 will evaluate the prospects of CTM on the assumption that the "symbolic" and "semantic" properties imputed to mental representations are semiotic-semantic properties. Chapters 8 and 9 will address the possibility that CTM's "symbols" are *not* "symbols" in the ordinary semiotic sense, but simply functionally typed entities. These two chapters will explore two avenues for interpreting the "semantic" properties imputed to mental representations in a fashion that does not impute convention- and intention-based semiotic-semantic properties.

Before proceeding to this analysis, however, it is perhaps prudent to address a possible objection to the Semiotic Analysis. Thus the present chapter will address the objection that we can separate "pure syntactic" and "pure semantic" components of our analysis of symbols from a conventional component that accrues to them solely because they are used for public languages. In particular, we shall examine the claim that Tarskian semantics provides us with such a "purely semantic analysis," as seems often to be assumed by philosophers of mind.

6.1 A CRITICISM OF THE SEMIOTIC ANALYSIS

While the analysis of symbols and symbolic representation presented in chapter 4 is in certain ways novel and no doubt will be regarded as controversial in some respects, one general thrust of the analysis—the idea that the nature of utterances and inscriptions depends upon the conventions and intentions of speakers and writers—may plausibly be regarded as a "mainstream" view. It is a view widely held by writers both within cognitive science and outside it,[1] and is indeed endorsed in some form by CTM's most important advocates (see Fodor 1981). There are some particular twists to my articulation of this view—notably, the distinction between the technical sortal terms 'marker', 'signifier', and 'counter', the claim that there are four separate "modalities" of conventional being, and the claim that not only semantics but also syntax and symbolhood are conventional in nature. But most objections to these features of my account *as an account of utterances and inscriptions* would probably take the form of an intramural debate between writers who embrace a semiotics based on convention and intention.

When this account is offered as a *general* account of symbols and symbolic meaning, however, it sometimes meets with greater resistance. For it is often claimed that what this analysis really gives us is an account of the nature of specifically *communicative* or *discursive symbols* —or perhaps of *symbols-as-used-communicatively*—and not an account of symbols, syntax, and symbolic meaning generally, much less a general account of representation.

Now to this latter claim—that the account in chapter 4 is not a general account of *representation* —I gladly defer. It was not my intent there to supply a general account of representation or an analysis of the uses of the word 'representation', nor does doing so fall within the rhetorical scope of this book. What we are discussing here, after all, is not the general claim that thought involves representation, but the more specific claim that it involves *symbolic* representation.[2] It is necessary, however, to address the claim that the Semiotic Analysis presented here is somehow specifically an analysis of discursive symbols—of *symbols-used-communicatively,* and not *symbols per se.*[3] For this claim will be of direct relevance to the analysis of CTM, as the symbols postulated by CTM's advocates are not used communicatively. For purposes of brevity, I shall put this objection in the critic's voice:

CRITIC : Look here, Horst. The analysis you give may be very well and good as an analysis of discursive symbols such as utterances and inscriptions, but you have been far too hasty in drawing the conclusion that *all* symbols are conventional in nature on the basis of these examples. The decision to confine yourself to conventional examples seems to be a matter more of *fiat* than of principle; and as a consequence, the analysis is question begging if it is presented as a general account of symbols and symbolic meaning. What you really have here is a *hybrid* analysis: what it describes is not precisely *what it is* for symbols to *have* semantic properties, but *also* how they *come by them* in a fashion that is conducive to *communicative use* of the symbols. *Other* symbols (e.g., mental representations and some representations in computers) also have semantic properties, but are not used communicatively. Arguably, the only reason that discursive symbols are conventional in nature is that this is necessary for their communicative role in natural languages. And so there is no reason to suppose that the semantic properties of mental representations would share this feature.

Moreover, as to your contention that there is no such thing as a symbol simply "being" of a particular type or "meaning something" apart from how it is interpretable under conventions, intended, interpreted, or interpretable-in-principle, you have really shown less than you think. You are right that there is no question of a *discursive* symbol "meaning something" apart from how it is interpretable, intended, and so forth—at least if you mean by this (*a*) that you can't have discursive symbols that are meaningful without getting their meanings in these ways, and (*b*) that telling the story about how symbols are conventionally interpretable and how they were intended, and so on, already says all there is to say about what they mean. That is, your conditions are both necessary and sufficient for the attribution of symbolic meaning in the case of discursive symbols. But this is quite compatible with the possibility that symbolic meaning is a *distinct property* from conventional interpretability, authoring intentions, and the like. Consider the following analogy: suppose that someone wanted an analysis of redness, and you were to give an analysis in terms of the reflectance properties that a solid object would have to have in order for it to be red. It would be true that solid objects could not be red without having these reflectances and, arguably, that once you had said that an object had these reflectances, there was nothing more to *its* being red to be told. But it does not follow that *being red* is *in general* simply a matter of reflectances or that nothing can be red without having these reflectance gradients. There is, for example, red *light*. The property of being red is accounted for in one way in solids and in another way in light. It is thus a mistake to identify the property of being red with the properties *solids* must have to be red *solids,* because things other than solids can be red. Similarly, you can have physical triangles and abstract triangles, and the latter do not have all of the properties one would expect of the former. Give an account of triangularity that builds physical properties into the picture and you leave out abstract triangles. Similarly, temperature is mean kinetic energy of molecules for gases, but not for plasma. Give an account of temperature *simpliciter* as mean kinetic energy of molecules and you've arbitrarily ruled out the possibility of plasma having a temperature.

Now look at your examples: you've restricted your domain to communicative signs. You may be right that, for this domain, an account of semantic properties (and symbolhood and syntax) has to advert to conventions and intentions. But all that means is that things in *this* domain cannot realize such properties except by way of conventions and intentions. In some sense, temperature *is* mean kinetic energy for gases, but not for plasma. Redness *is* a matter of reflectance curves for solids, but not for light. And similarly, meaning *is* a matter of conventional and actual interpretation for communicative signs (and likewise syntax and symbolhood), but (as you point out) not for mental states, *and arguably not for symbols that represent but are not used communicatively*. You have confused the analysis of the property of meaningfulness *simpliciter* with an account of how it gets realized in particular kinds of objects—namely, objects whose function is to communicate meaning. If you want to analyze *meaning,* the analysis had better work for non-communicative symbols like mental representations as well. And if you do that, arguably you will end up with an analysis that is applicable to mental states, too, thus circumventing the conclusion that the semantic vocabulary is paronymous.

I cease to speak in the critic's voice.

6.2 INITIAL RESPONSE

Now I think that this is in some ways a very difficult criticism to properly come to grips with.

First, I am in some ways uneasy about the examples. (And let me hasten to point out that they are *my* examples—any problems with them are not the responsibility of other parties.) For it is not clear to me that we really ought to say that there *is* a single property called "temperature" or one called "redness," given that they require completely different accounts in different media. (I am more compelled by "triangle," though arguably there simply are not any concrete triangles.) That is, I am not fully persuaded that these terms have not been proven to be ambiguous, or at least ill defined. Or, insofar as there *do* seem to be properties of temperature and redness, they seem in some way to be observer-related: temperature in terms of kinds of measurements and redness in terms of the propensity to produce particular sensations.

Beyond this, however, there seem to be two things that one would need in order for the critic's objection to be made to carry much impact. First, the critic would have to justify the criticism of my choice of paradigms by pointing to things that were said to be "symbols" and to have "syntax" and "meaning" in the same sense in which these things are said about utterances and inscriptions, yet which were susceptible to non-conventional analysis. Second, she would need to show how to provide an alternative analysis of symbols and symbolic meaning that could "factor out" the alleged "purely syntactic" and "purely semantic" components from the "merely communicative" aspects. I shall attempt to show that these enterprises are not viable in the remainder of this chapter. In some places I will try to argue directly against the critic's analysis; in others, I shall try to show where the critic's story seems to have gone wrong.

6.3 THE CHOICE OF PARADIGM EXAMPLES

First, let us consider whether I have been arbitrary in my choice of utterances and inscriptions as my paradigm examples of symbolhood and symbolic meaning. In particular, are there in fact other paradigm examples available such that (a) the words 'symbol' and '(symbolic) meaning' are predicated of those examples in the same sense in which they are predicated of utterances and inscriptions, and (b) there is no covert reference to conventionality or mental states when these words are used of the alternative paradigms?

Now in a certain way, I find this a very odd objection. It is not as though we were overrun with things we call "symbols" that jump out as alternative paradigms. It is true that the *words* 'symbol' and '(symbolic) meaning' are *said of* other sorts of objects. In what follows, however, I shall argue that all of these usages are either (a) *homonymous* and express different properties from those expressed by the same words when they are applied to utterances and inscriptions, or (b) *contentious* in ways that render illicit their use as alternative paradigms in the present context.

6.3.1 SOME EXISTING USES OF 'SYMBOL' AND 'MEANING'

First, there are clearly some alternative uses of the words 'symbol' and 'meaning' in ordinary English and existing usage in the sciences. Jung, for example, wrote a book entitled *Man and His Symbols,* in which the word 'symbol' is applied to things other than utterances and inscriptions. There seems to be a similar and related usage in cultural anthropology, which is interested, among other things, in the "symbols" employed by a culture—meaning not their linguistic tokens, but the way they express themes and mythic forms. However, it seems very unlikely that the fact that linguistic tokens and Jungian archetypes might both be called "symbols" indicates that there is a property (being-a-symbol) that is common to both sets of objects. It seems more likely that the word is homonymous and expresses different properties in the two cases, the existence of a common word being a function of family resemblance or analogy rather than property sharing.

Similarly, the word 'meaning' and its variants has some different ordinary uses. We say, for example, that dark clouds "mean rain," and that what some human beings long for most is a "meaningful relationship" with another human being. But here again it seems wrongheaded to assume that the word 'meaning' expresses the same property when applied to utterances, clouds, and relationships. Even in the case of theorists who speak of "natural meaning" or "natural signs" (e.g., Grice 1957)—and it is almost never "natural *symbols* "—it seems clear that the words 'meaning' and 'sign' (or, at a stretch, 'symbol') are used here precisely to express the relation that is sometimes called "indication" (see Dretske 1981, 1988), and not to express the same property that is predicated of utterances and inscriptions. To be sure, some writers (notably Dretske 1981, 1988) have tried to make a case that the kind of "meaning" that accrues to language (i.e., the usage of the word 'meaning' that is applied to linguistic tokens) can ultimately be *explained* in a fashion that depends heavily upon indication. But their point is not to give an account of *what property* is *expressed by* the ordinary usage of 'meaning', but to give an account of how this property arises. Indeed, Dretske (1988: 55-56) explicitly embraces Grice's distinction between two uses ("natural" and "nonnatural") of the word 'meaning'. If an indicator theory should prove adequate as an explanation of linguistic meaning, the status of that theory would be that of an empirical account that explains the presence of the property P expressed by the "nonnatural" sense of 'meaning' in terms of a distinct property Q that is expressed by the "natural" sense, and not an analysis of what property that word is used to express.

Thus these examples are of no help to the critic. First, the Semiotic Analysis in no way claims that the words in the semiotic vocabulary may not be ambiguous in additional ways, or that they cannot be used to express properties other than those mentioned in the analysis. (Who on Earth would want to claim *that?*) Second, if the usage of the semiotic vocabulary in CTM is related to the convention-bound usage explicated in the Semiotic Analysis only by way of analogy or family resemblance, this seriously undercuts much of the appeal of CTM. For one thing, if these words express different properties when applied to mental representations, we are entitled to some explanation of what these properties are. For another, what the computer paradigm shows us how to do is to link up causal powers with the semiotic-semantic properties of the symbols. If the salient properties expressed by the words 'symbol' and 'meaning' in CTM express not semiotic-semantic properties but a distinct set of properties, we will need a new assessment of how *these* properties can be linked up with causal role, and hence we will require a new assessment of the status of the vindication of intentional psychology, which depended so heavily upon the computer paradigm.[4]

6.3.2 MENTAL REPRESENTATIONS AS A PARADIGM

Perhaps, however, one might say that one has an alternative paradigm of nonconventional symbols in mental representation itself. After all, people in cognitive science have been talking about mental representations in their theories for years, and most of them seem pretty clear that they do not mean to be talking about convention-dependent symbols. *Ergo* we have an alternative paradigm.

This approach, however, is *either* a case of homonymous usage *or else* it is question begging. For talk of symbolic representations in the mind is either (*a*) an attempt to apply existing usage, fixed by the older paradigms, in a new domain, or else (*b*) its relationship to existing usage is merely one of analogy or family resemblance. Fodor at one point seems to recognize this issue, but is rather cavalier in dismissing it. He writes,

> It remains an open question whether internal representation... is sufficiently like natural language representation so that both can be called representation 'in the same sense'. But I find it hard to care much how this question should be answered. There is an analogy between the two kinds of representation. Since public languages are conventional and the language of thought is not, there is unlikely to be *more* than an analogy. If you are impressed by the analogy, you will want to say that the inner code is a language. If you are unimpressed by the analogy, you will want to say that the inner code is in some sense a representational system but that it is not a language. (Fodor 1975: 78)

It seems to me that Fodor *ought* to worry a bit more about his options here. For the notions of "symbol" and "meaning" play an absolutely central role in CTM, and so one should wish to know just what properties these words are supposed to express when applied to Fodor's hypothesized mental representations. If these words express the same properties they express when applied to linguistic tokens, they would seem to require whatever analysis is given to linguistic tokens generally. These turn out to be convention- and thought-dependent, and Searle and Sayre have suggested that this kind of dependence renders these notions unfit for explaining the intentionality of mental states. (This view will be argued in detail in chapter 7.) But if the use of words like 'symbol' and 'meaning' signifies only an *analogy* with language, one needs to hear what properties these words *do* express when applied to mental representations, in order to see if these properties are even candidate explainers for the intentionality of mental states. In the first case, the critic's use of mental representation is contentious and question begging, as the question at hand is one of whether the ordinary usage of 'symbol' and 'meaning' *can* be applied to some internal states in a fashion that will do what CTM's advocates claim. In the second case, what we have is not an objection to the Semiotic Analysis, but a claim that it is not an analysis of the usage of the semiotic vocabulary employed by CTM's advocates.

6.3.3 SYMBOLS IN COMPUTERS

It also might be suggested by some that symbols in computers present a counterexample to the Semiotic Analysis. In light of the discussion in the previous chapter, however, this is clearly a confusion. On the one hand, things in computers that are normally thought of as symbols—for example, representations of numbers or text encoded in the ASCII format—are clearly convention-dependent in exactly the same senses as are utterances and inscriptions. On the other hand, the implicit usage of the words 'symbol' and 'syntax' in writers like Newell and Simon (1975) to denote functionally typed kinds is clearly a *distinct* (i.e., homonymous) usage of the words 'symbol' and 'syntax'. Things in the computer do not fall into semiotic kinds *because* they fall out of the functional description of the computer; nor do they fall under the functional description provided by the machine table *because* they are markers and counters. Rather, the great accomplishment of successful program design is to get the semiotic types to line up with the functional types so that the computer will perform operations that happen to be of interest when interpreted as symbol manipulations. So, far from providing a counterexample to the Semiotic Analysis, computers are only properly understood when that analysis is employed.

6.4 FURTHER OBJECTIONS

In spite of the lack of clear alternative paradigms for the usage of the semiotic vocabulary, it nevertheless might be argued that even the ordinary usage of words such as 'symbol' in fact involves two distinct elements: a *non* conventional element that defines the essence of symbolhood, syntax, or semantics, and a conventional element that is required in the case of utterances and inscriptions only because they are symbols-used-communicatively, and conventions are needed for communication. Indeed, one might suggest that the notion of a machine-counter provides an analysis of a "pure" notion of symbolhood and syntax, while some other kind of analysis might do the same for semantics. The notions of "symbol" and "syntax" might thus be adequately and perspicuously developed along functional or functional-causal lines, while semantics might be given a nonconventional analysis in terms of the kind of semantic theory proposed by Tarski, which depends on an (arguably nonconventional) notion of *satisfaction*.

6.5 THE ESSENTIAL CONVENTIONALITY OF MARKERS

First, let us consider the bare notion of "symbol" captured by the term 'marker'. Is it possible to factor out the conventionality of the analysis of markers by attributing that conventionality to the fact that letters and numerals are symbols-used-for-communication? Or, alternatively, is it simply part of the essence of markers that they be conventional?

I think that it is not possible to factor out the conventional aspect of markers. To begin with, it seems quite clear that categories like "rho" and "0" and their genera "letter" and "numeral" are legitimate categories that form an important domain for characterizing some aspects of human life. The issue, then, is not one of legitimating these categories but of providing a proper analysis of them. I wish to argue that these categories cannot be adequately cashed out either (*a*) in terms of physical properties, including abstract physical properties such as patterns, or (*b*) in functional terms like those defining the notion of a machine-counter.

First, consider physical pattern. The problem here is that physical pattern is not a rich enough condition to distinguish between marker types. For two distinct marker types (e.g., *P* and rho) may share criteria such that the same set of physical patterns is employed for both types— that is, anything that is rho-shaped is *P*-shaped, and vice versa. Thus physical pattern is not sufficient for the explanation of marker types themselves, even though its presence is a sufficient condition for a physical particular to count as being interpretable as a token of such a type *given the existence of the conventions associating the type with particular physical patterns*. This, however, presumes the conventional type and does not explain it. Intuitively, what seems to be required is the additional fact that rhos and *P*s play distinct roles in the language games of distinct linguistic communities—and hence marker types are defined in part by the role they occupy in the linguistic lives of communities of language users.[5]

Yet the critic might very well seize upon this very characterization to make her point in another way. She might reason: if markers are determined by the *role* they play in a system of interactions between persons, then they are *functionally defined*. And hence they would appear to be a subspecies of machine-counters. Now perhaps the "system" needed for defining markers as a species of machine-counters would have to be a very complex one, involving entire linguistic communities, rules for coining new symbols, revising practices, and so on, but it is nonetheless a functionally describable system. Hence markers are machine-counters. It is just that the "machine" here is something on the order of a human society. The reason that markers have to be conventional is that the makeup of this particular system requires it for communication and decoding between individuals. (I cease to speak for the critic.)

This is admittedly a very seductive characterization. However, I believe that it suffers from several weaknesses. First, for anyone even a little bit taken with the work of Ryle, the late Wittgenstein, or *Lebenswelt-philosophie*, it is *contentious at best* to claim that the notion of a role within a language game or a form of life *can* be cashed out as a functional relationship in the bare mathematical sense of function required for a machine-counter. I shall not belabor this point here, but it seems that really what the critic ought to say is not that the role of a marker in a language game is *fully explicated by* something appearing in a machine-table description of a language, but rather that we can *abstract away from* a language in such a way that what we end up with is a machine table, and that we can do so in such a way that some of the machine-counters appearing on the table correspond to markers in the ordinary practice being described. This, however, raises two questions: (1) Can one in fact do this for the practices involved in marker usage? And (2) even if we can, does this amount to factoring out the notion of marker-hood into the notion of a machine-counter plus conventions needed for communicative usage?

I think that the only really honest answer to the first question at the moment is *we don't really know*. There are notorious problems with characterizing and simulating linguistic practices in situ, such as those described in Dreyfus (1972, 1992), Weizenbaum (1976), and Winograd and Flores (1986). Arguably, some such problems could be developed to apply not only to semantic and pragmatic competence but to the ways that marker-related practices are embedded in a larger web of practices as well. In brief, we are not really entitled to assume that these problems can be overcome, given the fact that the remarkable amalgam of brainpower, person-hours, and research dollars represented in the artificial intelligence community has not managed to overcome them in the space of several decades.

But even if one could produce a machine table for linguistic groups that isolated markers in the desired way, it is not so clear that this would accomplish all that the critic desires. For while functional describability—even of the special sort required for machine-counters—may pick out kinds that are of interest for the purposes of computer scientists or others, functional kinds (in this *mathematical* sense of "functional") are notoriously cheap. As Block (1978) and others have noted, any object or system of objects one likes has some functional description—or, better, a very large number of such descriptions. But surely *even if* one were to factor out a conventional component of markers, what one should *wish* to have left as a "pure notion of symbolhood" is something more robust than mere functional describability. What the critic wants is something that is plausibly a nonconventional characterization of *symbols per se* —something that should be *common* to things that one might plausibly think are or involve symbols (say, computer memory states, brain states, and inscriptions) but *not* predicable of things that are functionally describable, yet not plausibly construed as symbols (say, molecules of water in a bucket). But the functional properties distinctive of machine-counters are not robust enough to do this. At best, they do half of what is needed—namely, unite computer memory states, brain states, and (perhaps) inscriptions—but they fail to distinguish these as a kind from the rest of creation. Thus the kind of "definition by role" one finds for machine-counters does not appear to be rich enough to explain the kind of "definition by role" needed for markers.

6.6 SYNTAX, FUNCTIONAL ROLE, AND COMPOSITIONALITY

It is likewise tempting to see the notion of a machine-counter as a way of factoring out a "purely syntactic" element from the notion of a counter, thus allowing us to treat the conventional aspects of the syntax of natural languages as features that accrue to them only because natural languages are used for communication. Some writers might well assert that the notion of a machine-counter cashes out what we *do* mean in talking about symbols and syntax—namely, that syntax really is nonconventional in character, and is rightly understood purely in terms of combinatorial properties of the units manipulated—in short, that syntactic properties are purely a matter of functional role in the mathematical sense of 'function'.

Again, I wish to be very careful here. It is certainly possible to use the word 'syntactic' for any nonsemantic properties, or for any such properties that have to do with ordering or combination. However, I think that this does not do justice *either* to the *ordinary* usage of the word 'syntax' or to the categories that *linguists* call "syntactic." It does seem right, of course, to say that linguists are interested, among other things, in the formal, combinatorial properties of syntactic categories, much as physicists are interested in the formal properties of bodies *qua massive* (e.g., in Newton's laws) or as chemists are interested in the combinatorial properties of classes of elements. But the formal and combinatorial nature of the objects in these other sciences does not exhaust what it is to be, say, a gravitational body or a halogen; and likewise the formal characterization of syntax does not exhaust what it is to be of a particular syntactic category.

There are, I think, several good reasons for rejecting a formal or functional-role view of the nature of syntactic categories.

(1) The word 'syntax' has a natural domain: namely, linguistic entities. But Turing tables are applicable to any object that has a functional description. It has been argued by numerous writers (e.g., Block and Putnam) that all sorts of strange entities might have functional descriptions and be truly describable in terms of a Turing table. Syntax cannot simply be ordering, because there are plenty of things that are concatenated (e.g., cars in a traffic jam) that are not ordered *syntactically*. And even when there is a whole functionally describable system, it is not *eo ipso* a syntactic system. Were this so, the whole world would turn out to be syntactically structured, as everything has at least some trivial functional description. Of course, one *could* use the word 'syntax' in this way, but to do so would be to abuse a word that already has perfectly good uses, and would certainly underdetermine the kinds of distinctions envisioned by the linguist.

(2) At least some of the linguist's primitive categories tend to have *semantic* and *pragmatic* overtones: notions like "noun," "plural formative," "connective," and "pragmatic formative" all seem to be typed not only according to combinatorial properties but in terms of what kinds of things they are used to express or to do. Perhaps, however, one might say that these categories are *identified* in part on semantic and pragmatic grounds, and the task of the student of syntax is to find purely syntactic (i.e., formal) characterizations of the categories.

(3) More problematically, *we seem to need a semantically pregnant notion of syntax if we are to use it to explain compositionality*. For here is a common story about compositionality: a complex symbol string means what it means because of (*a*) the meanings of the primitives, and (*b*) the function of its syntactic structure. But the "function" implied here cannot simply be "function" in the mathematical sense, and it cannot amount merely to a description of the formal properties of the string of symbols. For lexical semantics plus formal syntax only tells us how we can concatenate meaningful lexical items in legal strings; it does not tell us how to interpret them. That is, it may tell us that "Borin bit the bear" and "The bear bit Borin" are both grammatical sentences in English, but it does not tell us who bit whom in each case. For there is a consistent interpretation of English (or any language) that reverses assignments of agent and patient (with corresponding changes for transformations into the passive voice), and there is nothing about the formal properties of the language that tells us which is operative. Indeed, there is nothing about the formal properties of the language that distinguishes meaningful sentences from grammatically well-formed nonsense (this despite a tendency of writers such as Tarski and Davidson to use the word 'meaningful' to mean "well formed"). So if we want to tell a more or less familiar story about compositionality, we need syntactic categories that are partially defined in terms of their contribution to compositional semantics, and such categories are not purely formal. Likewise, we need rules for composition that are semantically pregnant.

Nor is this a feature only of natural languages. The same observations could be made with respect to, say, predicate calculus or Hilbert's geometry. It is quite possible for a person to understand (*a*) the uses of the individual symbols, and (*b*) formal rules for symbol manipulation, without understanding what is asserted by a given equation. (I expect, for example, that this is the case for many students in college differential equations classes.)

(4) Two languages can have the same syntactic categories while having different functional descriptions. Intuitively, one wishes to say that different languages (e.g., English and French) share some of the same syntactic categories (e.g., "count noun," "plural affix"). But these categories enter into different combinatorial relations with other categories in the different languages, and hence differ with respect to their formal properties. We might, of course, conclude that they are thus, contrary to appearances, distinct syntactic categories. But this seems quite arbitrary and does violence to the natural construal of what the linguist is

up to. A simpler solution is to conclude that syntactic categories are not typed precisely by their formal properties.

(5) Two speakers of the same language may have different dialects that, say, permit different collocations and allow different replacements, but employ the same syntactic categories. But if the functional-role interpretation of syntax were correct, this could not be the case: differences in what are taken to be legal sentences and legal transformations would require differences in syntactic category as well. But surely this consequence is intolerable.

(6) A given speaker may revise her way of speaking (a functional change) without thereby replacing her syntactic categories. It may be, of course, that children learning a language at some points entertain incorrect hypotheses about how their native language works, and so when they learn the correct rules they are in fact trading in old categories for new. But adult speakers can also change their grammatical competence in ways that do not seem to require this kind of interpretation. (They can learn, for example, that 'as' becomes 'so' after 'not'.) It is surely too radical an interpretation to say that they are learning new syntactic categories just on the basis of the fact that the functional description of their syntactic competence changes. Better to say that syntactic categories are not defined in functional terms, although they may receive a functional description, and to say that the proper functional description of a category may differ across languages, dialects, and even changes in a given speaker's competence.

Thus it seems wrong to say that the notion of a machine-counter is a good explication of the *ordinary* sense of 'syntax' or of the *linguist's* sense of that word (and this even if many linguists *think* that syntactic categories are precisely functional-role categories). And it seems that convention is needed to get from combinatorial "syntax" to full-blooded syntax. In particular, it is needed to explain compositionality: two languages could have the same interpretation scheme for the lexical primitives and combinatorially equivalent syntaxes and yet have different assignments of meanings to complex expressions. Why? Because full-blooded syntax provides, among other things, a mapping from the ordered pair [meanings of primitives, syntactic form] to the meaning potential of the complex expression, and combinatorial properties do not provide such a function.[6] What does provide such a function? In natural languages, it is surely a matter of convention. It is, of course, worth considering whether the combinatorial properties to which machines can be sensitive can be combined with something nonconventional to provide the same results as combinatorial syntax, but that is quite another question (one that will be addressed in a later chapter).

6.7 WHAT FUNCTIONAL DESCRIPTION CAN'T DO

Given this analysis of the relationship of the functional description of computers to symbols and syntax, we might do well to ask what might have made the opposing view seem attractive in the first place. I believe that answer is to be found in a certain misunderstanding of what is going on in functional description. It has been noted by writers like Cummins (1975) and Block (1980) that there are several different uses of the word 'functional' and its variants. Cummins distinguishes the *mathematical* notion of function, which is employed in machine-table functionalism, from what he calls "functional analysis," which describes an object in terms of its role in a larger system (e.g., the function of the heart is to pump blood). And Block distinguishes functionalism as a thesis about the nature of mental states from "psychofunctionalism," which is a thesis about the meanings of mental-state *terms*.

I believe that the dangers of conflating mathematical functionalism and functional analysis are very great. CTM is concerned with mathematical functionalism insofar as mental processes are said to be describable by something like a machine table. It is sometimes claimed additionally that asserting that the mind has a math-functional description is tantamount to asserting that the *nature* of mental states is given by the completed machine table, and hence given in nonintentional terms—namely, that a math-functional description of the mind yields a functional *analysis* of mental states as well, or that it is the functional description of things in computers that confers upon them the status of symbols with syntactic properties. I wish to argue that this is false, and rests upon a basic misunderstanding of what goes on in functional analysis.

What *does* go on in math-functional description is simply a special case of mathematical abstraction employed in the modeling of real-world phenomena. Functional description, whether in computer science, in psychology, or in linguistics is completely parallel with the formation of mathematical models in mechanics and thermodynamics. (I do not mean to imply, of course, that the machine-table paradigm in psychology is as mature or as well confirmed as are our models of mechanics or thermodynamics—merely that they are the same kind of enterprise.) And from the standpoint of scientific research in psychology, this is one of the cardinal virtues of the computer paradigm: it purports to provide the resources for a mathematization of psychology, much as, say, analytic geometry and the calculus provided the resources for the mathematization of classical physics.

But what is involved in mathematical modeling? First, such models involve *abstraction*. Laws of gravitation abstract away from other forces that are almost always operative in vivo: mechanical force, electromagnetism, strong and weak force. And likewise for laws governing the other fundamental forces. Additionally, all macroscopic laws abstract away from the statistical possibility of freak quantum events, and so on. Mathematical models are not universally quantified propositions about individual objects or events. If they were, most of them would be false. Rather, they are propositions about how certain forces in nature contribute to events in abstraction from other processes. (An interesting corollary: Why is psychology so hard to make into a rigorous science? Hint: abstract characterization becomes exponentially more difficult as the number of mutually dependent variables increases. How many mutually dependent modules are there in the brain?)

Second, you get an *exact* and *rigorous* model only when you can express it in some mathematical form. Many of the most famous such models are expressed in the form of equations. (Laws are models expressed by equations.) But there are other kinds of mathematical structure that need not involve equations—for example, alternative geometries can serve as models of space-time, and geometric models are not equations. A model can be mathematically exact without relying heavily on laws. And even when laws are central to a theory, they are only a part of a larger model. For example, classical mechanics involves a model of space and time whose structure is Euclidean, while relativistic mechanics involves a model whose structure is non-Euclidean. Thus grasping a theory such as general relativity involves more than being able to manipulate the equations as algebraic entities. One needs to understand what relationships they express against the background of the larger model.

This leads to a third point, which is intended to be the real emphasis of this section. There are really two different ways of looking at a mathematical model, which correspond to two different levels of abstraction away from real-world phenomena in vivo. The ultimate goal of modeling, of course, is to describe and explain real-world phenomena. But real-world phenomena are messy, and scientific description aims at capturing such order as is to be found in their behavior. This involves separating different factors that are at work in vivo (gravity from mechanical force in physics, rationality from emotion in psychology) by way of abstrac-

162   Symbols, Computation, and Thoughts

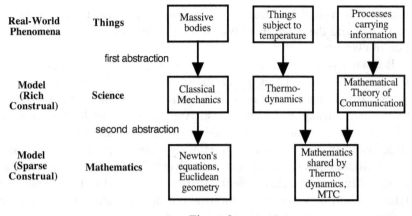

Figure 9

tion. We know we have a sufficient degree of orderliness when we can provide a mathematical model. This is the first level of abstraction: the use of a mathematical model to describe real-world phenomena (for example, the use of Newton's equations to describe gravitational attraction). At this level of abstraction, which I shall call the "rich construal" of the model, the model is by definition a model *of* some particular real-world phenomenon. Newton's equations are not just equations (i.e., algebraic entities); they are equations that express relationships between the real-world phenomena of gravitational force, mass, and distance.

But we may also perform a second act of abstraction and look at Newton's model purely in mathematical terms. We can perform algebraic operations upon his equations, for example, or examine the Euclidean assumptions of classical physics purely as geometric assumptions. The physicist and the mathematician often operate upon the same models, but do so under different constraints. The mathematician is concerned with the model as a purely mathematical entity. The physicist is concerned with it as a description of real-world phenomena. If we say that the physicist is concerned with a *rich* construal of the model, let us say that the mathematician is concerned with a *sparse* construal (see fig. 9).

It is important to see that any scientific theory is always more than the mathematics that sums it up. The formula describes the relevant *form* of a process or relation, but it does not itself determine what it is whose form it describes. For example, there are always an indefinite number of purely abstract objects that are described by the same mathematics, and usually a multitude of uninteresting concrete objects (for example, those formed by meriological relations) that are described by it as well. Moreover, there are sometimes multiple *non*trivial natural systems described by the same mathematics. The most notable case is probably that of thermodynamics and mathematical theory of communication, which treat of distinct domains but happen to share a substantial portion of their mathematical descriptions. The equations employed in these domains, treated as equations (that is, sparsely construed), do not tell you that they are *about* heat or information. The mathematics of thermodynamics and information theory does not provide a complete analysis of the *nature* of heat or information. What it provides is an exact description of relationships between the kinds of entities that are relevant to the domain in question. If you want to know about heat or information in detail, you will need the mathematics. But if *all* you have is the mathematics, you will not be able to derive a full-blooded description of the real-world phenomena merely from their mathematical characterization.

And so, in general, mathematical modeling does not provide an analysis of the full nature of a phenomenon, though it tends to specify that nature in more exact detail. At a rich level of description, we know that we are talking about, say, gravitation simply because we embarked upon the enterprise with the intention of talking about gravitation. Newton's equations do not tell us what gravitation is, they merely specify its form. At a sparse level of description, we are no longer talking about gravitation at all; we are merely talking about equations as equations. They are no longer being treated as a model of anything, and there is nothing about them that has the conceptual riches needed to explain real-world phenomena like gravitation or heat.

6.7.1 FUNCTIONALIST THEORIES OF THE MIND

I believe that this is precisely the case with functional description of the mind and of language as well. Let us begin with math-functionalist theories of the mind. The founding hypothesis here is that math-functional description of the sort provided by machine tables or general-purpose programming languages provides mathematical tools adequate to the task of describing the form of mental states and processes. Mental states and processes are real-world phenomena, and describing them is bound to involve some abstraction. We treat as irrelevant things like mechanical force (though people do get banged on the head, often to the detriment of their thinking) and gravitation (though it has been claimed that some individuals are affected by the full moon and NASA does psychological experiments on the effects of free-fall), and so on. Likewise we treat some physiological factors like blood sugar and hormone levels as constant much the way we treat voltage levels in a computer as constant, abstracting away from the fact that variations in these things affect real-world performance in ways that are of considerable concern to doctors and systems operators, respectively. Perhaps this strategy for mathematizing psychology will pan out in the long run. Perhaps it is fundamentally flawed, as claimed by Dreyfus (1972) and Winograd and Flores (1986). Perhaps it would work in principle, but the number of mutually dependent variables makes it impossible to carry out in practice. My concern is not with the prospects of this strategy but with what it *would* provide if carried out in detail. And what it would provide is precisely analogous to what, say, Newton's equations provided for classical mechanics: a mathematically exact model of mental states and processes.

Given the preceding discussion of mathematical models in general, it should be clear why a psychological machine table would not be an analysis of the *nature* of mental states and processes. At a rich level of description, the model is indeed a model *of mental processes*. But we know it is a model of those processes and not something else for the very pedestrian reason that we knew it all along: the model is a model of the mind because producing a model of the mind was our goal from the outset. Such models can be better or worse insofar as they involve better or worse approximations of the form that is really present in processes in vivo (that is, in the sense that Einstein's model is better than Newton's, and Newton's is better than Descartes's). So at the rich level of abstraction, the *content* of the model is not a consequence of its mathematical form alone. At this *rich* level we do indeed have a description in which we can identify mental states as such and characterize them in terms of their location in a network of other mental states, inputs, and outputs. However, we can do this only by *assuming* the individual mental states as mental states and assuming the network, and then characterizing the relations precisely in terms of the machine table. At best, we can analyze one mental state in terms of its relation to the others, holding their existence and relatedness as a kind of background assumption. But in doing this we never break out of the web of the intentional, unless it should prove possible to define *all* of the inner states in terms of a neutrally characterized set of inputs and outputs. But arguably this "best case" itself involves a misunderstanding. For in such a case what we are doing is picking out a set of states by dint of an abstract description of their causal interrelations. But this by no means assures that those causal relations are the essential properties of the states involved, nor that there could not be a variety of distinct state types that could occupy isomorphic causal roles.

On the other hand, at the sparse level the "model" is now just a mathematical entity. This level of abstraction is indeed useful even for the scientist (as opposed to the mathematician) at times, such as when one is interested in seeing whether, say, classical mechanics is a special case of relativistic mechanics, and does so purely by mathematical manipulations. But it cannot tell us what the interpretations of the mathematical symbols used to express the theory might be. For example, the formulas used for information theory do not tell us whether they are being used to express a model of information or of heat—or indeed that they are being used to express any real-world properties at all. In the case of the mathematization of psychology, here all we have is the machine table, which is a representation of a function in the mathematical sense. There is nothing about the table that tells us what the domain of the table is. Indeed, it could serve equally well as a functional description of all kinds of things: some abstract objects, some interesting real-world phenomena, some monstrous meriological contrivances. If we do not start out knowing that we are talking about the mind, there is nothing about the math-functional description that will tell us that we are doing so.

In short, math-functional description cannot provide us with an analysis of the *nature* of mental states and processes any more than equations for entropy can teach us the difference between heat and information. What it would do is no more and no less than what other mathematical models do in the other sciences: namely, to specify exactly the mathematical form of real-world phenomena of whose existence and nature we have some kind of independent knowledge.

6.7.2 FUNCTIONALIST THEORIES OF LANGUAGE

The functional analysis of language runs a parallel course. It is, of course, true that linguists are interested in formal descriptions of things like rules for legal formation of expressions in a language, transformation rules, and so on. And some might say that even conventions can usefully be examined as a kind of rule-governed activity which is subject to a more precise description. If linguistic conventions are established by practices of language users, perhaps this very network of practices can be mathematized into a machine table or something of the sort, in which case language will have been given a functional characterization in terms of a system as large as a human society.

I do not wish to debate the long-range prospects of such an analysis here. My point is simply this: if such a characterization were to be given, its status would be the same as that of any other mathematically precise model. On a rich construal, it is a description *of a language,* but *only* because that was the intent of the modelers from the outset. On a sparse construal, it is simply some abstract mathematical structure without an interpretation. The nature of the mathematical structure— that is, the math-functional properties—does not explain the nature of language as language, though of course differences in such properties are important for, say, differentiating two natural languages (which may have different morphemic categories and different grammars) or differentiating natural languages from other language games like first-order predicate logic. Looking at the formal properties of a language or comparing those of different language games is a very useful enterprise, and reveals a great deal about particular language games. What it does not do is reveal the nature of symbolhood and syntax in their own right. Either one knows what those are beforehand and asks about *specific* systems of symbols and syntax (rich construal of the model) or else one has an uninterpreted mathematical model that does not essentially describe languages (sparse construal).

6.7.3 THE "FALLACY OF REDUCTION"

This error of mistaking the properties of the mathematically reduced model of a phenomenon for the essential properties of the phenomenon itself is sufficiently important to merit a name: I shall call it the "Fallacy of Reduction." The Fallacy of Reduction is committed when you abstract away from features of a real-world process to give a more rigorous characterization of some of its features, and then assume that it is only the properties that survive the abstraction, and not those that are abstracted away from, that are relevant to the nature of the real-world phenomenon. Thus it is an instance of the Fallacy of Reduction to conclude that heat and information are "the same thing" because they share a mathematics. Likewise it is an instance of this fallacy to conclude that the functional *describability* of the mind would license the conclusion that mental states are functionally *defined.*

6.8 THE POSSIBILITY OF PURE SEMANTICS

It remains to consider the question of whether the Semiotic Analysis does justice to *semantic* notions—in particular, whether the analysis is a hybrid of a nonconventional "purely semantic" element and a conventional element that accrues only to symbols-used-communicatively. What I propose to argue is as follows: (1) The Semiotic Analysis provides a proper analysis of the properties expressed by the semantic vocabulary as they are applied to the paradigm examples of symbols such as utterances and inscriptions. (2) This analysis cannot be factored into a nonconventional "purely semantic" component combined with a conventional component that is needed only for communicative symbols. This has the consequence (3) that people who wish to speak of "meaningful mental representations" must either be (*a*) attributing to those representations the convention-dependent semiotic-semantic properties, or else (*b*) attributing to those representations a set of properties distinct from those normally expressed by the semantic vocabulary—properties that we may for convenience dub "MR-semantic properties."

In order for the critic to succeed in his objection to my analysis of symbols and symbolic meaning, it must be possible to show that the semiotic analysis I have presented blurs the distinctions between two different kinds of properties that accrue to discursive symbols: (1) their nonconventional semantic properties, and (2) the conventional properties that accrue to them specifically because they are symbols *used in communication*. The critic need not show that symbols such as utterances and inscriptions can in fact possess the first sort of property without the second, but merely that there is in fact an element of such symbols that is indeed semantic but in no way conventional, and that this element is what is shared by discursive symbols, the mental representations posited by CTM, and perhaps mental states as well. In effect, he must show that the Semiotic Analysis really shows only a causal dependence of semantic properties upon convention, not a conceptual dependence. And to do this, it must be possible to present an analysis of semantics that can apply to the paradigm examples of "meaningful symbols" which "factors out" a nonconventional semantic component from a convention-laden communicative element.

The basic issue that faces us in any attempt to factor out a "purely semantic" component from the Semiotic Analysis is this: one thing that does seem clear about the Semiotic Analysis is that there are separate claims to be made about conventional interpretability, authoring intention, actual interpretation, and interpretability-in-principle. How are we to isolate a "purely semantic" component that is common to these four modalities in such a fashion that we do justice to the distinction between the modalities? That is, it seems that we have important and coherent notions of "being interpretable under convention C as signifying Y " and "being intended by S as signifying Y " and so on. The critic's claim is that these notions are in fact amalgams of a "purely semantic" component plus factors that are not essential to semantics but necessary for communication. So one important constraint upon an analysis developed along these lines is that it ought to do justice to the notions it is intended to analyze.

We may schematize the approach in the following way. For each of the modalities, the critic needs to articulate an alternative analysis that separates two components: a "purely semantic" component and a convention-bound communicative component. We may indicate our predicates as follows:

CONV($X,T,C_{i,L},L,Y$): X is of a type T that is interpretable under convention $C_{i,L}$ of L as signifying Y

AUTH*(S,X,Y)*: *X* was intended by *S* as signifying *Y*

INT*(H,X,Y)*: *X* was interpreted by *H* as signifying *Y*

PPLE*(X,Y)*: *X* is, in principle, interpretable as signifying *Y*

Let us further note the putative "purely semantic" meaning as a relation as follows:

M*(X,Y)*: *X* means *Y* (in the "purely semantic" sense)

The critic's strategy has to be one of formulating certain biconditionals for each of these modalities that reduce each of them to a claim that M*(X,Y)* plus some residual claim about communicative use. The form of such claims might be indicated thus:

CONV($X,T,C_{i,L},L,Y$) $\Leftrightarrow$ M(X,Y) & CONV*

where CONV* denotes the aspects of CONV that remain once M*(X,Y)* has been factored out. (The exact logical form of CONV* would require such an analysis to be performed in detail.) And similarly for the other modalities:

AUTH(S,X,Y) ⇔ M(X,Y) & AUTH*

INT(H,X,Y) ⇔ M(X,Y) & INT*

PPLE(X,Y) ⇔ M(X,Y) & PPLE*

What the critic wants is a notion of "pure meaning" that is *necessary* for each of the semiotic modalities, but is not *sufficient* for any of them (except of course for interpretability-in-principle, which is trivially satisfied with or without a "pure" notion of meaning). For if "pure meaning" is a sufficient condition for one or more of the other modalities, then it somehow secures the conventional element of semiotic-semantics as well, which is precisely what the notion of "pure meaning" is an attempt to avoid.

Although I have never seen an explicit attempt to analyze the conventional semantics of natural and technical languages in terms of a purely semantic component plus something else, such an analysis seems implicit in the way many people regard the kind of analysis of language associated with Tarski and Davidson. On this view, a "language" is viewed not as a historical entity distributed over a community situated in time and space, but as an abstract object that associates expressions with interpretations and may be adopted by one or many individuals—or, for that matter, by none at all. A language thus conceived is an abstract object, and hence presumably its existence is not dependent upon conventions. Where convention enters the picture is when a community or an individual *uses* such an object as a public language or an idiolect.[7] It is here, in the *adoption* of a preexisting abstract object, that convention and intention enter the picture. There just *are* these abstract entities $L_1,...,L_n$ called "languages," and each of them essentially involves a semantic interpretation that is a mapping from expressions to their interpretations. Convention enters the picture when a community or an individual decides to use some particular language L_i as its language of representation and communication.

This view of languages points to a possible analysis of the semiotic modalities: The *conventional interpretability* of an expression E in community L amounts to (1) the fact that E means Y in language L_i (i.e., the interpretation mapping for L_i maps E onto Y), and (2) the fact that community L has adopted L_i. Likewise, authoring intention and actual interpretation are to be analyzed in terms of the employment of an abstract language. If S intends E to mean Y, then S is adopting language L_i and E means Y in L_i. If H interprets E as meaning Y, then H is adopting the interpretive conventions of L_i and E means Y in L_i. The basic notions here are (1) meaning-Y-in-L_i and (2) *adopting* a (preexisting) language as a language of communication and representation.

It is perhaps clear why this view should be popular among people who wish to provide a semantics for mental representation—or, more generally, to "naturalize" semantics. For this view separates the purely semantic element (a mapping from expressions to objects and states of affairs) from a particular way that that semantic component gets hooked onto communicative languages. And this leaves open the door to the possibility that the *same* semantic properties might get connected with other things (mental representations, thoughts) in other ways. If one takes this view, the question for the cognitive scientist is to find a different relationship that plays the same meaning-conferring role for mental representations that conventional and intentional adoption of languages plays for natural languages.

This "pure semanticist" view has gained a great deal of currency. It also enjoys a great deal of intuitive plausibility, and it is attractive to many in cognitive science precisely because it seems to relieve cognitive science of the problems of conventionality by treating conventionality as a feature of the adoption of language *for communicative use*. It is a worthy opponent for the Semiotic Analysis. I happen to think that this view is wrong in some very fundamental ways, but to see why this is so we shall need to examine a concrete example of a project in "pure semantics" of "abstract languages," and do so in some detail. Those who have brought up this kind of view as an objection to the Semiotic Analysis tend to refer to it as "Tarskian semantics" or "Tarski-Davidson semantics," due to the influence of Tarski's work in semantics for formalized languages. Given Tarski's insistence that his analysis did *not* apply to natural languages, the connection here is not completely solid. But the basic moves of postulating "abstract languages" and treating their semantic analysis extensionally are indeed to be found in Tarski, and both the strengths and weaknesses of the view are to be found clearly in his work. Thus it seems in order to take a careful look at Tarski's work to see whether it really provides a viable "pure semantics."

6.9 TARSKI'S SEMANTICS

Tarski's work on truth, and Davidson's claim that Tarski's technique can also yield an account of meaning, have garnered a great deal of attention, and have been met with sharply polarized reactions. A great deal of the literature discussing Tarski's work has been devoted to the problem of deciding just what Tarski's theory *does* —what *kind* of theory it is, and what it is a theory *of*. It is thus probably wise to begin by presenting some of the essentials of Tarski's account. Tarski wrote three papers that are of central importance to his work on truth: "The Concept of Truth in Formalized Languages" (1956b), "The Establishment of Scientific Semantics" (1956c), and "The Semantic Conception of Truth and the Foundations of Semantics" (1944). The first of these papers begins with the following synopsis: "The present article is almost wholly devoted to a single problem—*the definition of truth*. The task is to construct—with reference to a given language—*a materially adequate and formally correct definition of the term 'true sentence'*" (Tarski 1956b: 152, italics in original).

Tarski's project is thus one of providing a "definition" of truth that is "materially adequate" and "formally correct." This terminology requires some comment. It is somewhat controversial what Tarski meant by 'definition' and what he achieved in this regard. For the term 'definition' has acquired a specialized usage in metamathematics that implies something at once weaker and stronger than some more ordinary uses of the word. If one constructs a set theoretic model $M(D)$ of a mathematical domain D, then a concept C in D is said to be "defined by" the set-theoretic construction that corresponds to it in the model.[8] Thus, for example, in the *Principia Mathematica,* numbers are said to be "defined" by sets; but it is quite controversial whether this result really has any consequences with respect to the nature of numbers. So a logician's expressed desire to provide a "definition" of truth may easily turn out to be merely a desire to provide a model-theoretic construction characterizing truth in a language. However, Tarski does say things that indicate that he may be interested in more than this. He says, for example, that it is his wish to provide a "definition" that corresponds as closely as possible to familiar uses of the word. And what he explicitly cites as the familiar use of the word is the "classical" conception of truth in which truth is understood informally to consist in correspondence to reality. This will have direct consequences for the conditions Tarski believes to be relevant to the "material adequacy" of a truth definition.

Before moving on to these conditions, however, it is important to note that Tarski does not aim to supply any *general* definition of truth-in-any-language-whatsoever. Rather, truth-definitions are relativized to languages. The stated reason for this is that the "same sentence" can appear in different languages, and may be true in one, false in another, and meaningless in a third. Tarski is thus taking "sentences" or "expressions" to be defined in terms of concatenations of graphemes or phonemes, an assumption that arguably is not completely unproblematic. So the search for a "definition of truth" is really a search for the conditions that must be met by a definition of truth relative to any language L. One such condition is "material adequacy," by which Tarski means, informally, that the truth-theory for L, $T(L)$, should have the consequence that, for any sentence S in L, $T(L)$ assigns S the value TRUE iff what is asserted by S is true. Tarski suggests that the constraint that truth theories have such biconditionals as consequences be formulated in terms of a schema, which he calls convention T:

(T) X is true, if and only if, p,

where p is a schematic letter to be replaced by a sentence of L and X is a schematic letter to be replaced by a ("structure revealing") name of the sentence that replaces p. He writes that "we shall call a definition of truth 'adequate' if all these equivalences follow from it" (Tarski [1956b] 1985: 50). Tarski refers to this conception of truth as "the semantic conception of truth" (ibid., 51), the point being that truth is defined in terms of relationships between expressions and states of affairs in the world (hence a semantic relationship), rather than being defined *syntactically* in terms of derivability from formally specified axioms. (This point, often glossed over today, was perhaps the most significant feature of Tarski's approach in the climate in which it was first propounded.)

The issue of "formal correctness" is driven by several concerns Tarski raises with respect to classes of languages that are *not* subject to the kind of definition he desires. First, he believes that languages can be characterized in the desired fashion only if they are "exactly specified," in the sense that in "specifying the structure of a language we refer exclusively to the form of the expressions involved" (ibid., 52). This excludes languages that involve lexical ambiguity and elements that are dependent upon pragmatics or context, such as demonstratives and indexicals. Second, he points out that certain classes of languages—languages that he calls "semantically closed"—are inconsistent because they are prone to the generation of paradoxes such as the antinomy of the liar. Languages are said to be "semantically closed" if they contain the resources for naming expressions occurring within the language, for applying the term 'true' to sentences in the language, and for stating the truth conditions of the language (ibid., 53). The concept of truth, Tarski claims, is not definable for semantically closed languages.

It is perhaps obvious that these observations lead to the conclusion that truth is not definable for natural languages, since these are lexically ambiguous, employ demonstratives and indexicals, and have resources for referring to their own elements and making truth-assertions about them. Tarski embraces this conclusion, though other writers have since attempted to treat these features in a way that avoids Tarski's negative result. More easily overlooked is the fact that the linguistic features that interest Tarski include things like axioms and theorems, which play a large role in logic and mathematics, and are strongly connected with the notion of truth in those domains, yet are notably absent (not to mention irrelevant to empirical truth) in natural languages. This would be highly problematic if Tarski's stated aim was to provide a general "definition" of truth, but is perhaps innocuous so long as one is carefully attentive to the fact that what he is about is providing a model-theoretic characterization of truth for those languages for which this might be done.[9]

The definition of truth is constructed out of a more basic notion of *satisfaction*. Satisfaction is a relation that obtains between any objects and a special class of expressions called "sentential functions," which are expressions such as "X is white" or "X is greater than Y." (Sentential functions are differentiated from sentences in that they may contain free variables.) Intuitively, an object O satisfies a sentential function F if replacing the variable in F with the name of O results in a true sentence. This, however, will not serve as a definition of "satisfaction" for Tarski's purposes, as his aim is to define "truth." And so he employs another strategy—that of "defining" satisfaction for a language L in an extensional fashion. In order to accurately represent Tarski here, I shall cite his own text:

> To obtain a definition of satisfaction we have rather to apply again a recursive procedure. We *indicate* which objects satisfy the simplest sentential functions; and then we state the conditions under which given objects satisfy a compound function—assuming that we know which objects satisfy the simpler functions from which the compound one has been constructed. (Tarski [1956b] 1985: 56, emphasis added)

From this definition of satisfaction for sentential functions, one follows for sentences (functions in which there are no unbound variables). Sentences are either satisfied by all objects (in which case they are true) or else they are satisfied by no objects (in which case they are false). This, indeed, provides a definition of truth: "Hence we arrive at a definition of truth and falsehood simply by saying that *a sentence is true if it is satisfied by all objects, and false otherwise"* (ibid.).

Here we have a general schema for talking about truth in a language *L*, given that *L* falls within the specified class of languages. It is a *schema* for truth-definitions rather than a *general* truth-definition because the two more basic semantic notions of *naming* (or, to employ Field's [1972] useful paraphrase "primitive denotation") and *satisfaction,* are defined for expressions only relative to a language.

Given Tarski's desire for the introduction of all semantic terms only by definition, it is important to be attentive to the way in which satisfaction and primitive denotation are treated in Tarski's articles. For the "definition" of 'satisfaction' for a language *L* consists merely in (*a*) providing a mapping from simple functions to the sets of objects that satisfy them, and (*b*) providing a recursive rule for producing such a mapping for complex functions, given the values of the simple functions. And similarly, one may assume that the "definition" that would be given for the relation of primitive denotation would simply be a mapping from a class of expressions to a set of objects. These are "definitions" in the mathematician's sense of exactly specifying the function performed in set-theoretic terms. But they are surely not "definitions" in the sense of explaining what satisfaction or designation *consist in.* This has led to some criticisms of the scope of Tarski's accomplishment, some of which (Field 1972 and Blackburn 1984) I shall allude to in developing a more general analysis of the problems with the notion of pure semantics. These and other concerns cast some doubt upon whether Tarskian semantics in fact provides the "pure semantics" desired by the critic of the Semiotic Analysis.

6.9.1 A NONCONVENTIONAL ANALYSIS?

First, it is by no means clear that the analysis presented by Tarski renders the semantics of languages essentially nonconventional. Tarski says that we *"indicate* which objects satisfy the simplest sentential functions" (Tarski [1956b] 1985: 56, emphasis added). But in the context in which he is speaking, this "indication" can be interpreted in either of two ways, *both of which are plausibly interpreted in conventional terms.* On the one hand, one might wish to supply a semantic analysis of an *existing* formal language (say, Hilbert's geometry). In this case, one is approaching an existing *public* language game that is conventionally established. The ability to "indicate" the objects that satisfy the sentential functions in such a language game by no means shows that the relationship of satisfaction is essentially nonconventional. Making the mapping from expressions to interpretations explicit in no way implies that the preexisting system is nonconventional. And indeed the way in which the mapping is indicated in the formal model is itself conventional.

Alternatively, one may be defining a new language game de novo, and hence *stipulating* its semantic assignments. Here there is no preexisting convention-laden public language game. But in doing this one is necessarily defining a *convention* for semantic interpretation. Doing so by no means shows there is an independent stratum of meaning or even satisfaction that obtains apart from the conventions established by the theorist. At best, *if* there is a preexisting set of *markers,* there are infinite numbers of *mappings* between that set and sets of objects. And mappings do, indeed, exist independent of mapping conventions. But a mapping, per se, is not a semantic relationship. Nor does the existence of *mappings* that are independent of conventions establish the existence of *semantic relations* that are independent of conventions. Semantic assignments are *represented by* mappings and involve mappings, but mappings are not themselves semantic.

6.9.2 THE CONVENTIONALITY OF THE MARKERS

Tarski has also made an illicit move in assuming that "sentences" and "expressions" that constitute the domain of the mapping can be defined in terms of concatenations of graphemes or phonemes, and the pure semanticist would be wrong in concluding that this amounts to a nonconventional definition. There are at least three problems here. First, as argued above, markers and counters are conventional in character. Thus, while it may be right to say that the same physical patterns may get concatenated in more than one language, it does not follow that the same complex markers are employed, nor that identical strings of markers in two languages are the same sentence. Second, the marker kinds themselves are underdetermined by physical pattern and are essentially conventional. Third, if sentences are defined in terms specific to their mode of representation, it is not clear how one is to account for the fact that the same sentence can be both spoken and written, and can potentially be represented in other modalities (e.g., Morse code, ASCII coding, etc.) as well. As an idealization, Tarski's move is permissible within certain bounds; as a real definition, it seems inadmissible. This seems to undercut the pure semanticist's claim that Tarski's semantics is free from conventional taint. Even if we agree that the mapping from expressions to objects is nonconventional, the overall language is still conventional because the domain of expressions is conventionally established.

6.9.3 FIELD'S ARGUMENT

In a justly famous article, Hartry Field (1972) undertakes an extensive examination of Tarski's theory of truth. Field argues that Tarski succeeded in reducing truth to what Field calls "primitive denotation," but failed to define primitive denotation in nonsemantic terms. And thus, in Field's view, the remaining project in semantics for naturalists such as himself is to provide a nonsemantic account of primitive denotation. The crux of Field's argument is that merely extensional characterization of semantic notions such as denotation or satisfaction, while adequate for model-theoretic purposes, does not constitute a genuine reduction of semantic terms, any more than we may produce a genuine reduction of the notion of valence that proceeded by saying

$(\forall E)$ $(\forall N)$ (E has valence $n \equiv E$ is potassium and n is +1, or ... or E is sulphur and n is -2.) (Field 1972: 363)

There seem to be at least two problems with merely extensional characterizations, on Field's view. First, they do not reduce semantic properties to nonsemantic properties in the sense of "reduction" employed in the sciences and relevant to the incorporation of semantics within the project of physicalism. Second, they seem to license unfortunate would-be "reductions": "By similar standards of reduction, one might prove that witchcraft is compatible with physicalism, as long as witches cast only a finite number of spells: for then 'cast a spell' can be defined without use of any of the terms of witchcraft theory, merely by listing all the witch-and-victim pairs" (ibid., 369).

Field seems right in his claim that Tarski's extensionally based account of his primitive semantic properties fails to yield any robust account of their nature. What Field directly argues is that Tarski's characterizations do not yield a reduction of these properties in terms that demonstrate compatibility with physicalism, but we shall see below in Blackburn's criticisms that this point can be generalized beyond Field's physicalistic agenda as well.

6.9.4 BLACKBURN'S ARGUMENT

Simon Blackburn (1984) argues that Tarski in fact gives *no* definition of any semantic notions, but merely describes a "neutral core" that "connects together truth, reference, and satisfaction" but "gives us no theory of how to break into this circle; that is, of how to describe what it is about a population which makes it true that any of their words or sentences deserve such semantic descriptions" (Blackburn 1984: 270). Blackburn's chapter on Tarski and truth presents a number of insights that are not easily separated. But one important observation he makes is that the specific character of Tarski's characterizations of the semantic notions renders them ill suited to serve as definitions. In particular, there are two problematic features of these characterizations: their extensional character and their relativization to a language. First, in giving a list-description of, say, names in language L and their denotations, one does nothing to explain what the property is that is being characterized. A list-description tells you what objects are named by what terms, *given that you know* that the property characterized by the mapping is supposed to be *naming* in a particular language, but it tells you nothing about naming per se. One can make use of these lists only if one also knows that they are descriptions of how L -speakers use this set of expressions *as names,* and hence we have no real definition here (see ibid., 268-269). Second, the definition of, say, 'satisfies' for $L1$ is completely different from the definition of that same word (or a corresponding word) relative to $L2$. The satisfaction relation is provided merely in terms of extensional characterization *for particular languages*. It is defined differently for each language individually, because there is a different mapping of expressions onto objects in each language, and there is no overarching notion of satisfaction apart from those relativized to particular languages. If satisfaction were really defined extensionally (indeed, even if it were fully *accounted for* in extensional terms), it would seem to be the case that there is no property or function called "satisfaction" common to $L1$ and $L2$, but rather it would be more accurate to speak of separate notions of satisfaction-for-$L1$ and satisfaction-for-$L2$. This, Blackburn observes, is a problem for Tarski's account. For although Tarski is surely right in relativizing truth to a language,

> it does not follow that there is *nothing in common* to... truth as expressed in English sentences, and as expressed in those of any other language whatsoever. Reflection upon the application of an abstract semantic system to any actual population shows that there must be. (ibid., 270)

In other words, there is clearly something in common to notions such as truth or satisfaction across languages. But list-accounts for individual languages do not provide any indication of this common feature. Hence Tarski's analysis does not do an adequate job of "defining" the semantic properties.

I believe that this part of Blackburn's analysis is quite right. For our purposes, however, there is a certain aspect of Blackburn's approach that cannot be simply accepted without some justification. For when Blackburn says that Tarski does not tell us how to "break into the circle" of truth, reference, and satisfaction, he glosses this by saying that it gives us no theory "of how to describe what it is about a population which makes it true that any of their words or sentences deserve such semantic descriptions" (ibid., 270). Blackburn explicitly rejects the idea that one can separate a purely semantic account from a pragmatic account that ties a purely abstract language to the actual practices of a community (ibid., 269). This is, of course, very much in accord with what I wish to argue in this chapter. But by the same token, it is the very point which the fictional critic of this chapter wishes to contest. So the most we are really permitted to take from Blackburn here is the conclusion that Tarski's analysis does not provide a definition of the semantic terminology in nonsemantic terms (except perhaps in the model-theoretic sense of "definition"). What we are not licensed to conclude from Blackburn's arguments is the more robust thesis that the notions of satisfaction and primitive denotation presented by Tarski do not constitute notions that are legitimately semantical, yet do not have conventional elements.

At best, we might be able to make the following argument towards that conclusion on the basis of Blackburn's considerations. We might regard Tarski's "definitions" in one of the following two ways: (1) as attempts to give accounts of familiar semantical notions in nonsemantic terms, or (2) as stipulative definitions of how he is going to use those terms. If we interpret the definitions as stipulative in character, Blackburn's observations are enough to show that "denotation" and "satisfaction" thus defined are not really semantical notions at all, but merely model-theoretic counterparts of semantical notions. If we interpret Tarski in the first way, Blackburn's arguments show that Tarski has not successfully reduced the familiar semantical notions, but Blackburn has not shown that these notions are not "pure" in the sense of containing no conventional (or "pragmatic") element. This will require a further original consideration of the import of Tarski's work.

6.10 "PURE SEMANTICS" AND
"ABSTRACT LANGUAGES"

The suggestion at hand, then, is that Tarski has employed notions of denotation and satisfaction and has characterized them for model-theoretic purposes in purely extensional (and nonconventional) terms—and, while his list-accounts do not provide *any* account of the nature of denotation or satisfaction (conventional or otherwise), the relations of denotation and satisfaction may yet be nonconventional in nature. And, moreover, the critic claims that the extensional characterization provided by Tarski is sufficient to show that we have notions here that can be applied indifferently to discursive symbols, thoughts, and mental representations.

I believe that this is the wrong moral to draw from Tarski's work. I further believe that the plausibility this thesis may enjoy derives from a common misunderstanding of what is going on in the formal (modeltheoretic) characterization of a language. Tarski himself differentiates between what he calls "descriptive semantics," which is concerned with describing how an actual group of people employs words, and what he calls "pure semantics," in which a language is considered in the abstract. Blackburn calls the domain of pure semantics "abstract languages," and *this kind of locution, I submit, is the crux of the difficulty.* For speaking of "abstract languages," as opposed to "languages considered in the abstract" suggests that there are these purely abstract entities called "languages," and it is to these that semantics applies, and the only job left for the descriptive theorist is to link a concrete community of speakers with the right abstract language. Thus many writers seem to see the problem of meaning as being identical to the problem of figuring out which abstract language a given community or individual speaks. Partitioning the problems in this way leads one to think that issues of semantics are all handled on the side of abstract languages which are, from the theorist's standpoint, stipulative in their semantic assignments. (I suppose from the metaphysician's viewpoint they are necessary and eternal.) Issues of conventionality, on the other hand, lie on the side of descriptive semantics. And if you view descriptive semantics as a matter of hooking up an abstract language, complete with semantics already intact, to a community of speakers, then it is natural to view the semantics of language per se as something outside of the web of convention and in the pristine world of abstract objects.

This story is alluring, but it is wrong. To see why it is wrong, it is necessary to tell a better story. The general moral is this: it is every bit as misleading to confuse *languages-considered-in-the-abstract* with "abstract languages" as it is to confuse *material bodies-considered-in-the-abstract* (e.g., in terms of mechanical laws) with "abstract bodies" (e.g., point-masses). Strictly speaking, there are neither abstract bodies nor abstract languages, and features that are bracketed for purposes of abstract analysis are not thereby proven to be inessential. In short, the belief that the domain of semantics is a kind of abstract object called an "abstract language" is to fall prey to another instance of the Fallacy of Reduction discussed earlier in this chapter.

The Fallacy of Reduction, you will recall, consists in giving an abstract description of a phenomenon as a model and then treating the properties that are clarified by the "reduced" model (e.g., the mathematical description) as precisely those properties that are constitutive of the original phenomenon. There are some cases, no doubt, in which the properties retained in the model *are* precisely those constitutive of the original domain, but such is not generally the case. The mathematics of thermodynamics, for example, does not tell you that the subject matter is *heat*. (Indeed, the same mathematics applies to information.) And in the case of abstractions such as point-masses, one is indeed faced with fictional entities that do not exist in nature. So long as one bears in mind that one is involved in a theoretical activity that involves abstraction, speaking of point-masses is completely benign. But if we forget the act of abstraction and treat point-masses as the real domain of mechanics—or even as a real part of the objects of mechanics—we have been deceived by our own use of language.

So what is one doing in giving a formal model of semantics for a language? What one is doing here is really just a special case of what one does in giving a model generally—for example, in mechanics or thermodynamics. (Tarski himself is really quite explicit about this, likening the relationship between metamathematics and particular mathematical domains to that between one of the natural sciences and the objects it studies.) In any of these cases, one begins with an intuitively characterized domain consisting of a set of objects one wishes to characterize (say, bodies or sentences) and a set of properties or relationships to rigorously specify (say, gravitational attraction or truth-functionality). One then abstracts or idealizes the objects in one's domain in a fashion that brackets those properties the objects have in vivo that are irrelevant to the problem at hand. One brackets features of bodies such as color, magnetism, and even size when one is doing a theory of gravitation, treating bodies as point-masses. And one brackets features of languages such as pronunciation, dialectical variation, nonassertoric sentences, linguistic change, and the conventionality of the symbols people really use when one is doing a theory of deduction. This kind of idealization is perfectly legitimate so long as the properties that are bracketed are truly irrelevant to the features one wishes to rigorously characterize. But of course the question of what may safely be bracketed depends entirely upon what aspects of the intuitively characterized domains one wishes to specify: a formal model of particle collisions *should* be sensitive to size and shape even if a model of gravitation is not, and a formal model of phonetics or pragmatics should be sensitive to features that are irrelevant to truth-functionality.

A formal model of a language (or of anything else) is thus a characterization of a language, viewed under a certain aspect and screening out other aspects of the language in vivo. It is, indeed, possible in some cases to construct artificial languages that actually lack some of the features that one idealizes away from in natural languages—for example, languages that lack lexical ambiguity, ambiguity in surface structure, notational variation, nonassertoric aspects, and change in usage. And indeed one usually constructs one's languages for mathematics and other deductive systems (Tarski's main interest) in a fashion that avoids these features. However, in the description of natural languages, one merely idealizes away from these features. Moreover, even with specialized languages, formal modeling idealizes away from other features—notably, those tied to the way the language is employed by its users. For example, what is called "denotation" in the model is bound up in what the language user does in referring in vivo, "satisfaction" is bound up with what the language user does in predicating in vivo, and so on.

As argued earlier in this chapter, there are two importantly distinct ways of looking at a model, corresponding to two different levels of abstraction one may adopt with respect to the intuitively characterized domain. At the first and milder level of abstraction, one views the model precisely *as* a model *of* the specified domain. One views Newton's equations as a model of gravitational interaction between bodies, or a Tarskian truth-definition as a model of truth in a language L. Here one is in fact looking at the initial domain, but viewing it abstractly through the lens of the model. One is making assertions about bodies, albeit bodies-considered-as-point-masses, or assertions about truth in a language L, but truth-characterized-extensionally. This is the "rich" characterization of the model. Yet one may also perform a second act of abstraction and look at the model itself in abstraction from what it is a model *of*. One may, for example, look at Newton's laws simply as equations that can be satisfied under particular constraints and can be evaluated using particular techniques, or one may look at a Tarskian model simply in terms of the set-theoretic relations it employs and the valid deductions one may make on the basis of those. Here one has ceased to look at the mathematical construction that started out as a model *as a model* (for a model is a model *of* something), and treats it as an independent entity. This is the "sparse" interpretation of the model.

Now this does indeed have the consequence that formal modeling distills a purely abstract object—the construction that is the model sparsely characterized. However, it is incorrect to view this as an "abstract language," for it is not a *language* at all, but merely an object consisting of a set of expressions, a set of objects, and some mapping relationships between them. One applies the *names* 'denotation' and 'satisfaction' to some of these relationships, but that is simply an artifact of the process through which we got to the model sparsely characterized. There is nothing about the model sparsely characterized qua set-theoretic construction that makes particular mappings count as denotation or satisfaction. Indeed, there is nothing about the model sparsely characterized that makes them count as *anything* but arbitrary mappings. (This, I think, is the essence of Blackburn's point.)

Now indeed in the model richly characterized, we are entitled to speak of these functions as "denotation" or "satisfaction"—or, perhaps more correctly, as the extensional characterization of the denotations and satisfaction conditions of particular languages. But the reason for this is that we started out talking about such relations as the features of the intuitively characterized domain that we wished to speak about, and have merely constructed a model that gives a rigorous specification of these properties in a fashion that is "materially adequate and formally correct."

Compare the analogy with a theory of gravitation. If we look at Newton's laws just as equations—as a model sparsely characterized—they tell us nothing about what relationships they are supposed to describe. We may call the variables by names like 'mass' and 'distance', but they are no longer variables signifying mass and distance. Of course they do signify those properties in the model richly characterized, but again that is only because the model richly characterized is obtained by starting from an intuitively characterized domain, performing certain idealizations, and applying a rigorous description to what is left. A mathematization tells us only the relationships between the features we wish to describe—be they mass or denotation. It specifies only the form of the relations and not the nature of the relata.

As a consequence of this, it is important to see that a formal model of a language no more implies the existence of something called an "abstract language" than a formal model of gravitation implies the existence of things called "point-masses." The model richly characterized is precisely a description of a familiar intuitively characterized domain that uses an abstract object to describe certain properties of that domain. The "language" here is the full-blooded language we set out to describe, not some formal subset of it, and it is fraught with conventionality. The model sparsely characterized is not a language at all, even if we misleadingly use words like "denotation" for a mapping function it employs. All it is is a construction consisting of sets of expressions and objects and a set of mappings between them. Mappings in themselves no more add up to denotation than equations employed in mechanics or thermodynamics or the Mathematical Theory of Communication add up to mass or heat or information.

In short, there is no level at which we find what the critic needs: an "abstract language" that has genuinely semantic relationships but no conventionality. The model richly characterized has semantic relationships, but they are the conventional ones of full-blooded languages. The model sparsely characterized does not suffer from semantic conventionality (though it still presupposes the conventionally sanctioned symbol types that constitute its domain); but it does not involve genuine semantic relationships either, but merely the mathematical-logical form that those relationships in real languages share with many other nonlinguistic systems with which they are isomorphic. Tarskian semantics deals with (real, full-blooded) languages in abstraction from many features found in vivo, including their conventionality. But it does not succeed in uncovering "abstract languages" that can provide the domain for a "pure semantics."

6.11 CONCLUSION

It would seem, then, that it is not true that semantics is properly concerned with a set of abstract entities called "abstract languages." It is true that we can begin with full-blooded languages and abstract away from their real-world features in order to be left with an object that is more suitable to rigorous study, much as we may do so in, say, physics. Indeed, in both cases there are *two* levels of abstraction: a *richly* construed model that treats the real-world processes in terms of their mathematical relations, and a *sparsely* construed model that is a purely abstract mathematical entity. Neither of these, however, has the features needed to count as an "abstract language." The rich model indeed has the features needed to count as a *language,* but is not truly *abstract:* the linguistic categories it works with are the convention-laden ones of the full-blooded *language.* The sparse model is indeed abstract, but there is nothing about the model, as such, that would make it count as a language. This is equally true for the semantic and the syntactic aspects of language. And hence the criticism that the Semiotic Analysis is really a hybrid of a nonconventional pure semantics (and pure syntax) *plus* a conventional element required only for symbols used in communication fails.

PART III

The Critique of CTM

CHAPTER SEVEN

Semiotic-Semantic Properties, Intentionality, Vindication

The preceding chapters have brought us to a point from which it is possible to return to the issues that were raised in the discussion of Searle's and Sayre's objections to CTM in chapter 3. There it was suggested that, if it were to turn out to be the case that words used in the attribution of intentionality and semantic properties are systematically homonymous, this might pose problems for CTM's account of the intentionality and semantic properties of mental states. The reason for this concern was straightforward: CTM attempts to account for the semantic and intentional properties of mental states by saying that these are "inherited" from those of the mental representations they contain. The general schema for explaining the semantic properties of a mental state M would appear to be something like this:

Mental state M has semantic property P because

(1) M involves a relationship to a mental representation MR, and

(2) MR has semantic property P.

But if it should turn out to be the case that the semantic properties predicated of mental states *are not the same properties* as those predicated of symbols, then this schema is at best in need of refinement and at worst betrays a deep confusion about semantic properties, because the expression "semantic property P" cannot be said univocally of symbols and of mental states, and hence one cannot sensibly speak of "inheritance."

Now the results of chapters 4 and 5 have borne out the suspicion that the terms used in attributions of semantic properties are systematically homonymous. The kinds of "semantic properties" attributed to symbols are both different from and conceptually dependent upon the kinds of "semantic properties" attributed to mental states. It was suggested that we can mark this distinction by adding prefixes to words such as 'semantic' and 'intentional' so as to disambiguate these crucial terms. The kinds of semantic properties attributed to mental states we may designate *mental-semantic* properties, and similarly the intentionality of the mental we may designate *mental intentionality*. In contrast, we may refer to the kinds of semantic properties attributed to symbols as *semiotic-semantic* properties, and the kind of intentionality attributed to symbols as *semiotic intentionality*.

In order to determine whether this analysis will have any consequences for CTM, it is necessary first to revise CTM's schema for explaining intentionality and semantic properties in light of these new distinctions. It seems clear that the kinds of properties of *cognitive states* that are to be explained by CTM are their mental-semantic properties. What is less clear is just what kinds of "semantic" properties *mental representations* are supposed to possess, in virtue of which they can provide the basis for an account of the mental-semantic properties of mental states. There seem to be three basic possibilities: (1) they are mental-semantic properties, (2) they are semiotic-semantic properties, or (3) they are neither mental-semantic properties nor semiotic-semantic properties but some other sort of properties that have not yet been clearly identified or distinguished from mental- and semiotic-semantic properties. This third possibility must be considered, since it could be that references to the "semantic properties of mental representations" are best construed as attributions of some kinds of properties that are *particular to* mental representations. It is not clear what these properties are supposed to be, but if someone were to advance the claim that there are such properties, the properties might be distinguished from mental- and semiotic-semantic properties by calling them *MR-semantic properties,* where "MR" is short for "mental representation."

I should like to separate the task of exegesis of texts by Fodor and other proponents of CTM from the task of analyzing variations on the account of intentionality. I do not wish to place too much emphasis upon the exegetical task. That task may well be pointless: it does not look as though Fodor recognizes the ambiguity of the semantic vocabulary, and if this is so, there is no point in asking which leg of the ambiguity he intended. In spite of this, however, it makes perfect sense to ask what various construals of CTM amount to, what their prospects are, and what advocates of the theory might need to provide in order to lend further support to their account. So here are three variations upon CTM's account of the intentionality of mental states:

Mental-Semantic Version

Mental state *M* has mental-semantic property *P* because

(1) *M* involves a relationship to a mental representation *MR*, and

(2) *MR* has mental-semantic property *P*.

Semiotic-Semantic Version

Mental state *M* has mental-semantic property *P* because

(1) *M* involves a relationship to a mental representation *MR*, and

(2) *MR* has semiotic-semantic property *X*.

MR-Semantic Version

Mental state *M* has mental-semantic property *P* because

(1) *M* involves a relationship to a mental representation *MR*, and

(2) *MR* has MR-semantic property *Y*.

7.1 A BRIEF DISCUSSION OF THE THREE VERSIONS

These three versions of CTM's account of semantics and intentionality are not all of equal interest. The first version, the mental-semantic version, seems plainly to be of little merit. While it may be that this version best reflects the fact that CTM's advocates fail to distinguish between different kinds of "semantic properties," it is also quite hard to see what it would mean to attribute mental-semantic properties to mental symbols—or indeed to anything other than a mental state. To embrace the mental-semantic version would be to say that mental representations are symbols that do not "have semantic properties" in the normal sense in which symbols are said to have semantic properties, but instead (and unlike other symbols) have the very same kind of semantic properties that one attributes to mental states. I am hard pressed to see what such a claim could really mean, and am fairly confident that none of CTM's advocates would wish to offer it as a clarification of his theory.

The semiotic-semantic version of the account would clearly seem to be the best candidate for an interpretation of Fodor's account of intentionality. After all, Fodor repeatedly characterizes mental representations as "meaningful symbols," and one would certainly be justified in assuming that these representations are supposed to be symbols that are "meaningful" in the sense that *symbols* (as opposed to, say, mental states or discussions with one's therapist) are said to be "meaningful." If Fodor meant something *else* by 'meaningful' ('semantic', etc.), one would certainly expect that he would have said so. For this reason alone, the semiotic-semantic version should count as the default reading of CTM's account of semantics. In addition, CTM is supposed to be an application of the paradigm of machine computation; and in the case of symbols in computers, when we speak of their "semantic properties" it is their semiotic-semantic properties with which we are concerned. Therefore it seems reasonable to assume that it is semiotic-semantic properties that are attributed to the mental representations over which mental computation is supposed to take place. Moreover, it seems the only option that is really on the table. The only senses of semantic terms that we have become acquainted with are those that denote semiotic-semantic properties and those that denote mental-semantic properties, and it clearly will not do to attribute mental-semantic properties to mental representations. It may be that someone could develop another usage of words such as 'semantic' and 'intentionality' that could be used in denoting some other class of properties, perhaps properties particular to such mental representations as there might be; but to the best of my knowledge no one *has* clearly stated such an alternative usage, nor made clear what it might be used to denote. The problem with analyzing the MR-semantic version of the account of intentionality is not so much that an explanation of intentionality based on such a peculiar usage of semantic terminology is either impossible or that it would not prove fruitful. The problem is, rather, that it is difficult to criticize an account that has not yet been articulated. There is, however, a trend towards causal explanations of the semantic properties of mental representations that could, in principle, be taken as pointing towards a usage of semantic terminology that would be peculiar to mental representations, and this is worthy of some investigation.

What I propose to do, therefore, is to examine in this chapter the prospects of CTM for explaining intentionality and vindicating intentional psychology, on the assumption that the "semantic properties" of mental representations are semiotic-semantic properties. In the two chapters that follow, I shall explore two ways of developing an account of "semantic properties" for representations that diverges from semiotic-semantics.

7.2 SEMIOTIC-SEMANTIC PROPERTIES AND CTM'S ACCOUNT OF INTENTIONALITY

The first order of business, then, is to consider the prospects of CTM's account of intentionality on the assumption that the "semantic" properties imputed to mental representations by CTM are the same kinds of semantic properties normally imputed to symbols—that is, that they are what I have called semiotic-semantic properties. In order to proceed here, it might be helpful to return to Fodor's own characterization of cognitive states in *Psychosemantics:*

Claim 1 (the nature of propositional attitudes)

For any organism O, and any attitude A toward the proposition P, there is a ('computational'-'functional') relation R and a mental representation MP such that

MP means that P, and

O has A iff O bears R to MP. (Fodor 1987: 17)

On the current interpretation, the condition "MP means that P " may be interpreted as "MP semiotically means that P." But this does not yet leave us at a point at which we can evaluate this claim, for the simple reason that claims about semiotic-meaning are ambiguous: they might be claims about interpretability, about intended interpretation, about actual interpretation, or about interpretability-in-principle. So even if we confine ourselves to semiotic-meaning of mental representations, there are really four distinct accounts of cognitive states that might be seen in Fodor's characterization:

Authoring Intention Version

For any organism O and any cognitive attitude A towards a proposition P, there is a relation R and a mental marker MP such that

MP was intended as signifying (that) P, and

O has A iff O bears R to MP.

Actual Interpretation Version

For any organism O and any cognitive attitude A towards a proposition P, there is a relation R and a mental marker MP such that

MP was actually interpreted as signifying (that) P, and

O has A iff O bears R to MP.

Interpretability Version

For any organism O and any cognitive attitude A towards a proposition P, there is a relation R and a mental marker MP such that

MP is <u>interpretable under convention C as</u> signifying (that) P, and

O has A iff O bears R to MP.

Interpretability-in-Principle Version

For any organism O and any cognitive attitude A towards a proposition P, there is a relation R and a mental marker MP such that

MP is <u>interpretable-in-principle as</u> signifying (that) P, and

O has A iff O bears R to MP.

Our task thus becomes one of examining each of these four versions of the account and determining whether any of them can succeed in providing an explanation of the intentionality and semantic properties of mental states.

In the following sections, I intend to address each of these versions of Fodor's representational theory of mental states and to argue that none of them can provide an account of the semantic and intentional properties of such states. The arguments against three of the versions of the theory—those based upon interpretability, authoring intentions, and actual interpretation—are roughly cognate with one another, and hence these three versions will be addressed together. The case against the version based on interpretability-in-principle is quite different, and will be addressed separately.

7.3 INTENTIONS, CONVENTIONS, AND THE REPRESENTATIONAL ACCOUNT

The first three versions of Fodor's representational account of cognitive states share an important feature: all three involve (covert) appeals to intentions and conventions. The Authoring Intention Version involves the claim that cognitive states involve mental representations that are *intended as* signifiers. But the logical form of the locutional schema 'is intended as signifying (that) P' requires a specification of some *author* of the marker token *whose intention it was* that the token signify (that) P. Likewise, the Actual Interpretation Version involves the claim that cognitive states involve mental representations that are *interpreted as* signifiers. But for something to be interpreted as a signifier, there must be some symbol *user* who does the interpreting. The Interpretability Version involves the claim that cognitive states involve representations that are *interpretable* as signifiers. But for a marker to be interpretable as a signifier, there must be a convention licensing the interpretation; and for there to be such a convention, there must be a community of symbol users who share a common understanding that such an interpretation is licensed.

In each of these cases, the resulting account of intentional states itself contains further reference to intentional states. In the case of authoring intentions and actual interpretation, it will involve reference to the intentional states involved in intending the marker to bear an interpretation or in construing it as bearing an interpretation, respectively. In the case of interpretability under a convention, the situation is only slightly more complex: conventions themselves are not intentional states, but the presence of a shared set of beliefs about how marker types may be used is a necessary (if not quite sufficient) condition for the presence of a semantic convention.

It thus turns out that versions of CTM based on interpretability, authoring intention, and actual interpretation are infected with exactly the kind of covert reference to cognitive states that was discussed in the development of the Conceptual Dependence Objection in chapter 3: the logical forms of attributions of intentional and semantic properties to symbols contain references to cognitive states. What remains to be seen is whether this fact imperils these versions of the account. I wish to claim that such accounts face four serious problems. First, they are empirically implausible. Second, they do not provide an explanation of the intentional and semantic properties of cognitive states. Third, they undercut one of the fundamental tenets of representational accounts of mind: namely, the intuition that access to extramental reality is mediated by mental representations. Finally, they lead to circularity and regress.

7.4 THE EMPIRICAL IMPLAUSIBILITY OF THE ACCOUNT

The first problem with the versions of CTM based on convention and intention is that they are highly implausible as empirical theories. Indeed, they are *so* empirically implausible that it would be difficult to find any stranger theories in the history of science. For suppose that the semiotic-semantic version of CTM is true. If this is so, then whenever you have a belief that (for example) Lincoln was president, you have a mental representation *MP* that means (that) Lincoln was president. And if one of the three versions of CTM based on conventions or intentions is correct, 'means (that)' has to be cashed out in terms of conventions or intentions. According to the Interpretability Version, you can only have a belief that Lincoln was president if there is some convention C that licenses the interpretation of *MP* as meaning that Lincoln was president. According to the Authoring Intention Version, you can only have such a belief if someone "authored" *MP* and intended it to mean that Lincoln was president. And according to the Actual Interpretation Version, you can only have such a belief if someone apprehends *MP* and takes it to mean that Lincoln was president.

All of these possibilities seem very unlikely, to say the least! *Who is it*, after all, whose intentions, interpretations and conventions are supposed to explain the meaningfulness of *MP?* One possibility would be that it is the thinker's own intentions, interpretations, or conventions. But there are two problems here, both of which may be familiar from criticisms of Hume offered by Thomas Reid and Edmund Husserl. First, there is certainly no *experience* of authoring or interpreting a symbol in ordinary cognition. (And it is not clear what it would mean to interpret or author a symbol one does not and cannot apprehend.) Second, in order to intend or interpret a symbol token as being about something else, one must have access both to the symbol and to the thing it is to represent. As we shall see below, this runs afoul of some basic motivations for representational theories of mind.

But perhaps the relevant conventions and intentions are not those of the organism itself, but of some other being(s). It is, perhaps, conceivable that there are some supernatural beings, or perhaps some very sophisticated Martian psychologists, who have subtle enough access to human brain states to view them as computers—for example, by constructing Turing machine descriptions for each human being. But it really seems quite unlikely. And according to these convention- and intention-based versions of CTM, humans could only be said to be in cognitive states if there *were* such beings. A theory that appeals to the unlikely to explain the matter-of-fact surely has to be regarded as highly suspect.

7.5 THE IRRELEVANCE OF CONVENTIONS AND INTENTIONS

In addition to being highly unlikely, the presence of beings who do in fact interpret human psychological states *is quite irrelevant to* our ascriptions of intentional states to humans, and to ascriptions of semantic and intentional properties to those states. For suppose that there are two possible worlds that are indiscernible with respect to all features accessible to human observers. In one world—all it the Demon World—there are beings called demons, indetectable to humans, who have a kind of access to and understanding of human mental processes that is simply uncanny. Among other things, they can instantly see how a particular human being's mind is describable as a Turing machine, and can assign interpretations to the operations and the symbols picked out by this Turing machine description in such a fashion that the person has a mental state of type A with content P when, and only when (a) the human, described as a Turing machine, has a tokening of a symbol of type MP in a particular functional relationship R with the rest of the "machine," and (b) the demon's interpretation scheme associates MP - tokens with P and associates the propositional attitude A with functional relationship R. Let us assume, moreover, that these demons do "read off" humans' mental states, and that they can even effect tokenings of intentional states by causing tokenings of symbols in humans. In the Demon World, humans *do* have states for which there are conventional interpretations, there *are* acts of interpretation of these symbols, and there *are* authoring acts in which these symbols are intended to have particular meanings.

Consider now a second world. It is indiscernible from the Demon World in all aspects accessible to human observers. But this world—all it the Demon-Free World—contains no beings who have the peculiar kind of access to human psychological states and processes that the demons in the Demon World have. Humans in the Demon-Free World have exactly the same experiences as humans in the Demon World. And ideally completed empirical psychologies in the two worlds would come to precisely the same conclusions. The two worlds are, by stipulation, indiscernible with respect to all features accessible to human observers.

Now let us pose the following question: would there be any differences in what mental states we should ascribe to humans in the Demon World and humans in the Demon-Free World? Would they have different beliefs, desires, and hopes? I think that the answer is, clearly, *no*. If the two worlds are indiscernible both with respect to the experiences of individual human beings and with respect to everything an empirical psychology might discover, it is hard to see how there could be any grounds for attributing different intentional states in the two worlds. Moreover, it is impossible for *us* to know beyond Cartesian doubt which sort of world *we* live in. It is epistemically possible that there are, in fact, such demons; it is similarly possible that there are not. But this realization does not (and should not) put us into any kind of doubt about whether we have particular beliefs or desires.

But if the intentionality of our mental states were a matter of our being in relationships with mental representations that were bound to meanings by conventions or intentions, then the existence of beings who employ such conventions or have such intentions would be a necessary condition for our being in intentional states. Since the existence of such beings is patently irrelevant to our attributions of intentional states, it follows that the intentionality of mental states is not dependent upon the association of symbols with meanings via conventions or the acts of symbol users.

Consider, in addition, the following concerns. Suppose that the demons in the Demon World suddenly decide to *change* their interpretive conventions, and they then start interpreting human psychological states in new ways. There is no change in what people *experience* when they are in particular psychological states, but the demons shuffle their assignments of interpretations to marker types. Should we say that there is a corresponding change in what intentional states we should assign to humans in the Demon World? Surely not. Questions about what intentional states people are in are surely not dependent upon anything so contingent as externally imposed interpretations. If this is the consequence of the versions of Fodor's account based upon conventions and intentions, those accounts fail to provide conditions that are relevant to proper ascription of cognitive states and of the kind of semantic and intentional properties normally ascribed to cognitive states.

7.6 CONFLICTS IN THE NOTION OF REPRESENTATION

In the previous section it was argued that *external* impositions of interpretations to mental representations (through authoring intentions, interpretive acts, or conventions) would be irrelevant to the ascription of mental states to an organism, and to the ascription of semantic and intentional properties to an organism's mental states. *Internal* impositions of interpretations are likewise problematic, albeit for a different reason. The reason to be developed here is suggested by Thomas Reid (1983) and Edmund Husserl ([1913] 1931). Reid and Husserl both offer arguments against theories of mind that postulate representations as objects that mediate the mind's access to extramental objects. Both philosophers' arguments are directed primarily against Hume, but their objection to representational theories can, as Keith Lehrer (1989) has recently suggested, be marshaled against contemporary theories as well, including CTM.

Reid and Husserl both claim that representational theories postulate "immanent objects" that mediate perception and cognition in order to account for the intentionality of perceptual and cognitive states. They also claim that, for a theory to be truly *representational,* this "immanent object" must be *interpreted or taken as standing for* the extramental object. But in order for such an act of interpretation to be possible, Reid and Husserl argue, the subject must have some kind of access *both* to the representation *and* to the thing represented: If Jones uses a symbol *S* to stand for some object *X* —if he *judges* "*S* stands for *X*" or *decides* "*S* shall stand for *X*," he must be cognizant of *S*, and he must also be cognizant of *X*. And his access to *X* must be *independent* of his access to *S*, since his acquaintance with *X* must precede the interpretive act that associates *S* with *X*. But this, argue Reid and Husserl, has the consequence that mental representation is possible only where there can be independent access to extramental objects. And this consequence undercuts the whole motivation for the postulation of mental representations, since these were introduced to explain how access to extramental reality is possible.

The Reid-Husserl objection is straightforwardly applicable to the variations on Fodor's account of cognitive states that we are presently considering. As formulated, it can stand as an attack upon the Actual Interpretation Version, since it addresses theories in which someone must interpret a representation as standing for an extramental object. According to such a theory, an organism *O* can be in a cognitive state *A* about some object *X* only if (1) *O* is in a functional relation *R* to a mental representation *MP*, and (2) *O* interprets *MP* as being about *X*. As Reid and Husserl point out, the appeal of representational theories lies largely in what is to be gained by saying that access to extramental reality is mediated by mental representations. But if the representations are only "meaningful" in the sense of being interpreted as being about extramental objects, then this motivation for a representational theory is undercut. In order to interpret *MP* as being about *X, O* must have access to *MP* and have *independent* access to *X*. If *O* is to apprehend *MP* and decide, "Aha! This is about *X*," *O* must have some kind of access to *X* that is not mediated by *MP*. His access to *X* may be very distant and dim, and might well be mediated by something *other* than his access to *MP*. (For example, *X* might be a number with a very large decimal, and *O* might know of X only under the aspect of being the limit of a particular series of rational numbers.) But in order for O to interpret a particular symbol MP as being about X, he must have some idea of X that is not mediated by his apprehension of *that particular* symbol. Otherwise interpreting MP as being about X amounts to forming the judgment "MP is about what MP is about."

Now the appeal of representational theories lies in large measure in what is to be gained by saying that access to extramental reality is mediated by representations. But if representation can take place only if someone actually interprets the representation as standing for a particular object, and this requires access to the object that is not mediated by the representation, then it turns out that representational accounts of intentionality are self-defeating. For if we postulate a representation MP in order to explain an organism O 's access to an object X, but the very definition of representation ensures that using MP to represent X presupposes having access to X that is *not* mediated by MP, then it is simply fruitless to explain access to objects by postulating mental representations. If there *can* be access to objects that is not mediated by representations, it is unnecessary to postulate such representations. If there *cannot* be access to objects that is not mediated by representations, there cannot be representations either, because one can only interpret a symbol as being about X if one has some independent idea of what X is.

Neither Reid nor Husserl develops the objection specifically against *symbolic* representations, and neither of them seems to realize that there are several senses in which an object can be said to be a symbolic representation in addition to the sense of actually being interpreted as referring to something else. But the objection can easily be adapted so as to be applicable to representational accounts based on conventions or authoring intentions as well. First, suppose that an organism O has a belief about Lincoln just in case (1) O is in a particular functional relationship R to a mental representation MP, and (2) O has a convention C whereby the representation MP is interpretable as being about Lincoln. To have such a convention, O must know about Lincoln in a way that is not mediated by MP. Similarly, suppose that O has a belief about Lincoln just in case (1) O is in a particular functional relationship R to MP, (2) O authored MP, and (3) O intended that MP be about Lincoln. In order for O to intend that MP be about Lincoln, O must know about Lincoln in a way that is not mediated by MP. In any of these cases, it is impossible to make sense of the notion of mental representation without supposing that the organism also has access to the thing represented in a fashion that is *not* mediated by such a representation. But if this is the case, postulating that there is a mental representation through which *O* apprehends Lincoln is pointless. Thus any such representational account of intentionality is bound to be self-defeating.

7.7 CIRCULARITY AND REGRESS

Finally, an account of the intentionality of mental states based upon interpretability, authoring intentions, or actual interpretations of mental representations would be circular and regressive. Consider first what would be involved in a claim that *O*'s mental state *A* means (that) *P* because it involves a representation *MP* that is either (*a*) intended (by some agent *A*) to mean (that) *P* or (*b*) interpreted (by some symbol user *H*) as meaning (that) *P*. If either account were correct, *O* could only be said to be in a mental state *A* that means (that) *P* if some organism *O** (possibly, but not necessarily, distinct from O) were in some particular intentional states—namely, those involved in (*a*) intending that *MP* mean (that) *P* or in (*b*) interpreting *MP* as meaning (that) *P*.

But if this is the case, the strategy for explaining the intentionality of mental states has serious problems. First, it is circular: the intentionality and meaningfulness of mental states is accounted for by appealing to the meaningfulness of symbols, while the meaningfulness of the symbols is accounted for by appealing to the mental states involved in bestowing meaning upon those symbols. Second, the account is regressive: each time we account for the intentionality of a mental state *A* of an organism *O*, we allude to the "meaningfulness" of a representation *MP*. But the kind of "meaningfulness" we invoke involves covert reference to the intentional states *A** of some organism *O**. But since we are looking for a general account of the intentionality of mental states—not just an account of *O* 's mental states—we must account for the intentionality of *O**'s mental states as well. Presumably, to account for *O**'s mental state *A**, we would have to posit a meaningful representation *MP**, whose meaningfulness would in turn have to be cashed out in terms of the interpretive acts of some organism *O***, and so on. The resulting account would not explain the intentionality of mental states in nonintentional terms; it could account for the intentionality of a *given* mental state only in terms of *another* mental state.

A very similar argument can be given against accounts where the "meaningfulness" of mental representations is to be understood in terms of interpretability under a convention. For while linguistic conventions are not themselves mental states, they only obtain by virtue of several beings having a shared understanding of how certain symbols may be used. (Or, if one wishes to refer to the meaning assignments of idiolects as conventions, these obtain because one being has an understanding of how certain symbols may be used, and this understanding could, in principle, be shared by other language users as well.) And it is surely a necessary (if not a sufficient) condition for this shared understanding that the beings who share in it be in mental states that are similar in relevant ways. This, I take it, would have to be a part of the analysis of what it is for a group of language users to share a linguistic convention. But if this is the case, then a convention-based account of meaningfulness of mental representations is no better than an intention-based account, since it too ultimately depends upon allusions to intentional states and hence ends in the same kind of circularity and regress.

7.8 THE INTERPRETABILITY-IN-PRINCIPLE VERSION

There is, however, a fourth modality under which marker tokens can be said to be signifiers: namely, interpretability-in-principle. The Interpretability-in-Principle Version of CTM explained the semantic and intentional properties of an organism O's cognitive state A —say, its meaning (that) P —by positing a mental representation MP and a functional relation R such that (1) MP is interpretable-in-principle as meaning (that) P, and (2) O is in relation R to MP. Two coextensive definitions for semantic interpretability-in-principle were offered in chapter 4. One definition was framed in terms of counterfactuals about conventions, the other in terms of the availability of a mapping from marker types to interpretations. Since the former definition seems clearly to risk running afoul of the same problems about convention that have already been discussed, we may assume that the second definition holds more promise for CTM. This definition was formulated as follows:

(S4*): An object X may be said to be *interpretable-in-principle as signifying Y* iff

(1) X is interpretable-in-principle as a token of some marker type T,

(2) there is a mapping M from a set of marker types including T to a set of interpretations including Y, and

(3) $M(T) = Y$.

Now semantic interpretability-in-principle is a very permissive notion. *Every* object is interpretable-in-principle as a token of a marker
type (i.e., can, in principle, be used as a marker if someone comes up
with a suitable marker convention); and every marker type can be
mapped onto whatever interpretation one likes. Therefore, for every
object X and every interpretation P, X is interpretable-in-principle as
meaning (that) P.

One thing that should be noted about the notion of interpretability-in-
principle is that the connection it makes between marker types and interpretations is *not* dependent *either* upon actually existing semantic
conventions *or* upon acts of authoring or interpretation. And this has
the significant consequence that the Interpretability-in-Principle Version of Fodor's account of cognitive states is immune to the criticisms
raised against the Interpretability, Authoring Intention, and Actual Interpretation Versions. To put it differently, the logical form of attributions of semantic interpretability-in-principle does *not* involve references to semantic conventions or meaning-bestowing acts, with the
consequence that the preceding arguments do not show that the Interpretability-in-Principle Version suffers from the pernicious kind of
conceptual dependence that threatened the other versions.

I shall argue, however, that the Interpretability-in-Principle Version is
also incapable of supplying a viable account of the semantic and intentional properties of cognitive states. In particular, there are four distinct
problems. First, such an account would impute to mental states semantic
and intentional properties which they clearly do not have. Second, it
would impute the kinds of semantic and intentional properties that we
ascribe to mental states to objects that clearly do not have them. Third, it
would not provide an *explanation* of the intentionality and "semanticity"
of mental states. And, finally, the definition of being interpretable-in-
principle as a signifier token presupposes being interpretable-in-principle
as a marker token—and that *does* involve conventions in a way that leads
to circularity and regress, albeit not at the semantic level.

7.8.1 SPURIOUS PROPERTIES

The first problem with the Interpretability-in-Principle Version is that it
would impute to mental states intentional and semantic properties that
they clearly do not have. According to the Interpretability-in-Principle
Version, for example, an organism O can have a belief about Lincoln just
in case (1) O is in the right functional relationship R to a mental representation MP, and (2) MP is interpretable-in-principle as being about

Lincoln. Now if there are mental representations, it is surely the case that any mental representation *MP* is interpretable-in-principle as being about Lincoln—the definition of interpretability-in-principle is so permissive as to assure that. But by the same token, the definition is also so broad as to assure that *MP* is interpretable-in-principle as being about the number two, the Crimean War, or anything else. Indeed, for *every* interpretation *P, MP* is interpretable-in-principle as being about *P*.

Now suppose that (as the Interpretability-in-Principle Version suggests) *O* 's being in relation *R* to *MP*, in conjunction with *MP* 's being interpretable-in-principle as being about *P*, are conditions jointly sufficient for ascribing to *O* a belief about *P*. If this is the case, then *O* has beliefs about *everything,* since each marker token *MP* is interpretable-in-principle as being about everything. Indeed, *each* of *O*'s beliefs is about everything, since each belief involves a marker token that is interpretable-in-principle as being about everything.

Surely this consequence of the Interpretability-in-Principle Version is intolerable. There may sometimes be some unclarity, vagueness, and ambiguity as to just what our beliefs are about, but not to the extent that each of our beliefs is about everything! And as this is a consequence of the Interpretability-in-Principle Version, so much the worse for that account.

7.8.2 STRANGE COGNIZERS

Depending upon how one takes the words 'organism' and 'functional relation' in Fodor's characterization of cognitive states, there may be a second problem for this version of the account as well. For one might well think that Fodor does not really mean to restrict his characterization of cognitive states to *organisms*. To do so in the context of a *computational* theory of mind would be very odd indeed! Perhaps the word 'system' could usefully replace the word 'organism'. And one might well think that the word 'functional' is used in the sense that it is used when one classifies digital machines according to their machine tables—that is, according to functional relationships between current states and succeeding states.

But if one does interpret Fodor in this way, it would seem that all kinds of things turn out to be cognizers. For, according to Fodor's account, it would seem that if (*a*) two systems are functionally equivalent, and (*b*) their symbols have the same semantic and intentional properties, and (*c*) they are in equivalent functional relations to their symbols, then it should be the case that they are in the same cognitive states. But consider the following problem. If a cognizer is describable in purely formal terms, it *must* be the case that there is an abstract formal system that is functionally equivalent to the cognizer. And if interpretability-in-principle is all that is needed to give a symbol system the kind of intentional and semantic properties that mental states enjoy, then it would seem to be the case that abstract symbol systems have intentional and semantic properties in just the same senses that mental states do. Presumably this would be enough to include such systems in the class of cognizers. But surely such a conclusion would be absurd.

7.8.3 LACK OF EXPLANATORY FORCE

Even if we could avoid these problems, it is difficult to see how the interpretability-in-principle of a marker token could supply anything in the way of an *explanation* of the meaningfulness or intentionality of a mental state. Suppose that I wish to know why a particular state of Jones's is about Lincoln and someone tells me that it is because Jones is in a particular functional relationship to a mental representation, and that representation is about Lincoln. I then ask, "Why is that mental representation about Lincoln?" If the reply is merely, "Because there is a mapping from that representation's marker type to Lincoln," then I have not received an explanation. Even if I believe everything that I have been told, I still don't know why Jones's cognitive state is about Lincoln. Pointing to the availability of a mapping just does not supply the kind of information that would answer my question. (It is not clear just what *would* supply the right kind of information, but it is clear that *this* reply does not.)

7.8.4 THE REAPPEARANCE OF CONVENTIONALITY AT THE MARKER LEVEL

Finally, upon closer inspection, it turns out that the notion of semantic interpretability-in-principle is not so free of convention as at first it seemed. The connection between marker types and interpretations is, indeed, not conventional. But for an object to be interpretable-in-principle as a signifier, it must first be interpretable-in-principle as a *marker,* and the expression 'interpretable-in-principle as a marker' *does* have a conventional aspect. For remember how this notion was defined:

(M4): An object X is said to be *interpretable-in-principle as a token of a marker type T* iff

(1) a linguistic community could, in principle, employ conventions governing a marker type T such that any object having any pattern $Pi \in P : \{P_1,..., Pn\}$ would be suitable to count as a token of type T,

(2) X has a pattern p_j, and

(3) $p_j \in P$.

An object's being interpretable-in-principle as a marker is *not* just a matter of there being a mapping from one object to another, because marker types are necessarily conventional. The very *notion* of a marker is convention-dependent.

This has the consequence that the Interpretability-in-Principle Version *does* involve conceptual dependence upon cognitive notions. For while attributions of semantic interpretability-in-principle do not involve tacit ascriptions of *semantic* conventions or intentions, they do involve tacit reference to *marker* conventions. Any explanation of marker conventions, like semantic conventions, would have to involve reference to a community of symbol users who share a certain understanding about marker types and tokening. And this shared understanding must surely consist in large measure in the members of the community being in relevantly similar mental states. But if this is so, the Interpretability-in-Principle Version is bound to end in the same kind of circularity and regress as the other versions.

7.9 APPLICABILITY OF THESE CRITICISMS

Now one might wish to pause at this point and consider how directly these criticisms affect CTM. For one might be tempted to think that in developing my terminology I have set up a straw man that my arguments are suited to knocking down. Fodor and other proponents of CTM acknowledge, after all, that there are differences between the fashions in which discursive symbols, mental states, and mental representations have semantic properties. In particular, they acknowledge that discursive symbols get their semantic properties from those of the mental states they are used to express. They simply deny that the same is true of those symbols that serve as mental representations. It might therefore seem that, in likening mental representations to discursive symbols, I am arguing against a position that Fodor and others have explicitly rejected.

But this is not the case. What Fodor claims is that discursive symbols, mental states, and mental representations all have the same kind of semantic properties, but come by them in different ways. I have shown that mental states and discursive symbols do *not* have the same kind of semantic properties, and that it is not clear what sort of "semantic properties" mental representations are claimed to have. Here I have been concerned with examining what happens if you suppose that mental representations have the same kinds of semantic properties—namely, semiotic-semantic properties—that symbols may uncontroversially be said to have. All of the problems that have arisen here arise purely from saying that the kinds of semantic properties representations have are semiotic-semantic properties. The problems do not arise because of some *additional* feature having to do with how they *came* by those properties; the problems arise because of the kinds of properties that are being attributed, and what they are used to explain. The position may be easily knocked down, but it is not the one that Fodor clearly rejects, and is in fact the most plausible interpretation of the ambiguous characterization that he offers.

7.10 TWO POSSIBLE RESPONSES

Now there are two kinds of objections that one might expect to hear at this point, each based on key differences between computers and paper or other passive media for the storage of symbols. First, computers do not just store individual symbols. The computer's sensitivity to the syntactic features of the symbols and its ability to generate new representations in accordance with formal rules allow the overall system to encode the semantic relationships between the symbols as well. If we ask how a symbol-manipulation process in a computer counts as, say, addition, we must not talk merely about the interpretations sanctioned by programmers and users, we must say something about the process that goes on in the computer as well. It looks as though the computer has its own contribution to make towards the symbols it stores having semantic values. If there is more to tell about the meaningfulness of symbols in computers than can be told in terms of the conventions and intentions of language users, the objections offered here may not undercut CTM's account of semantics and intentionality entirely.

Second, computers can be equipped with transducers that allow them to be sensitive to features of their environments. As a consequence, it is possible for the tokening of symbols in a computer to covary in regular ways with the presence of particular kinds of objects and circumstances in their environments. If a computer is able to detect when a light has been turned on, and inscribes "The light has been turned on" whenever it detects the light being turned on, one might be inclined to think that such an inscription is *about* the light being turned on in a way that a random inscription of the same symbol string would not be about the light being turned on. One might well think that the computer paradigm suggests more than a semiotic explanation of the intentionality of mental states: if the mind is a computer, and computers can support causal covariations between objects in the environment and the tokening of symbols of particular types, this kind of causal covariation might well form an important part of the explanation of the intentionality of mental states as well.

In the following sections, I propose to argue that neither of these lines of argument can rescue CTM's account of intentionality. The first line of argument fails because one can talk about the systematicity of meaning relationships in a symbol system only if one can first talk about assignments of interpretations; systematicity contributes nothing to the assignment of interpretations. The second line of argument may present an interesting theory, but that theory is simply not CTM's representational account of the semantics and intentionality of cognitive states. Moreover, as we shall see shortly, there are additional problems for CTM that arise from the fact that *syntax,* as well as *semantics,* is conventional in character.

7.11 SYSTEMATIC SYMBOL MANIPULATION

Computers do not merely store isolated, inert symbols. Indeed, much of what seems special about the computer paradigm is to be found in the way things in a computer are interrelated in the right ways—the way semantic relationships are mirrored by syntactic relationships, for example, and the way that derivations of symbol structures are truth-preserving under the right interpretation. Moreover, the systematic nature of the computer places constraints on how the symbols may sensibly be interpreted: the larger and more complex the representational system, the fewer reasonable interpretations are available. Haugeland, for example, writes that "an interpretation that renders a system's theorems as truths is a rare and special discovery," and that there is a sense in which "random interpretation schemes don't really give *interpretations* at all" (Haugeland 1981: 27).

And this is, in large measure, correct: interpretation schemes *do* take on special interest when they have certain properties: notably, (*a*) when they map marker strings onto *true* propositions, (*b*) when the interpretation of a marker coincides with something that is *causally related* to the tokening of that marker, and (*c*) when the interpretation scheme "makes sense" of the overall performance of the system—that is, when it gives it an interpretation that makes it seem as though it is acting rationally.[1] It is important, however, to distinguish the question of how a symbol system is *suitable for bearing* a particular interpretation from the question of *how the symbols may be said to bear any interpretation in the first place.* In the case of a computer, the answer to the first question has two parts: (*a*) a specification of how all of the semantic relationships necessary for a given interpretation scheme are reflected in syntactic relationships, and (*b*) an account of how the formal rules that allow truth-and sense-preserving derivations are linked to causal regularities through the functional architecture of the machine. The answer to the second question—the question of how the symbols may be said to bear any interpretation at all—has little or nothing to do with computers per se. The question of how it is that symbols may be said to have meanings is a question about semiotics, and the answer would have to be given in terms of the interpretability of the symbols under conventions, the intentions and interpretations of the symbol users (programmers, designers, and users of computers), and the fact that symbols are interpretable-in-principle as bearing any interpretation whatever.

As we saw in chapter 5, the functional analysis of the computer and its semiotic interpretation are two distinct issues: getting them to coincide is a virtue of good programming and not a fundamental axiom of semiotics. What we saw in that chapter was that semiotic analysis and math-functional analysis were distinct enterprises. That result still holds good here. But one may also argue a stronger point: namely, that the functional organization of the computer as a symbol manipulator cannot uniquely determine a single privileged semiotic-semantic interpretation scheme for symbols in a computer, either—and hence *even the combination of semiotic-semantics with functionally describable symbol manipulation* cannot explain the unique mental-semantic properties of mental states.

Let us reconsider the example of a computer programmed to perform operations corresponding to addition. The computer has three storage locations—A, B, and C—each of which bears a string of binary digits representing an integer under some interpretation scheme *I*. The computer proceeds by sampling the symbols present at A and B and causing the tokening of a symbol at C. The computer is so designed and programmed that (1) the syntactic patterns of the symbols present at A and B will determine what symbol is tokened at C, and (2) the symbol that is tokened at C will be mapped by interpretation scheme I onto the number that is the sum of the numbers represented (under I) by the symbols stored at A and B. Now if we ask, "What makes this system such that one might sensibly refer to what it does as addition?" part of our answer will have to make reference to the features of the system as a *system*. We might express what is needed in algebraic terms: if we take the set B of binary strings that can be present at A, B, and C, and the function F that maps pairs of strings found at A and B onto the string that they would cause to be tokened at C, we may speak of a group G which is defined in terms of the elements of B and the function F.[2] Now for this system to be suitable for supporting "computer addition" of some subset of the integers, there must be some subset of the integers S such that the group consisting of the elements of S under the operation of addition is isomorphic to G. That is, there must be a one-to-one mapping M between binary strings in B and integers in S such that for any three binary strings b_1, b_2, and b_3 in B, and any three integers i_1, i_2, and i_3 in S, if $M(b_1) = i_1$, $M(b_2) = i_2$, and $M(b_3) = i_3$, then $F(b_1, b_2) = b_3$ iff $i_1 + i_2 = i_3$. Or, to put things quite informally, what is needed to render a computer system suitable for supporting an interpretation is that it have a functional description that has the right formal properties for supporting that interpretation.

Now when a computer has a particular functional description—say, that described by group G described above—this renders it suitable for supporting *any number* of interpretations. If, for example, its operations are suitable to be interpreted as addition over the first n natural numbers, they are equally suitable to be interpreted as addition over the first n even natural numbers, or indeed as addition over any set of numbers generated by taking the first n natural numbers and multiplying each by the same real number r. And it is suitable for bearing any number of other interpretations as well—some purely mathematical, some referring to systems involving concrete objects.

But the suitability of a system as a whole for bearing an interpretation scheme that interprets both the symbols and the operations does not fully determine what the symbols or the operations may be said to mean. A system's formal properties render it interpretable-in-principle under a number of interpretation schemes, but confer pride of place upon none of them. This does CTM no good: if the formal properties of cognitive processes render the overall system interpretable-in-principle under several different interpretation schemes, but do not uniquely pick out the right interpretations for each representation, then semantic interpretability-in-principle of mental representations, even when applied to a whole system of such representations, is not sufficient to account for the semantic properties of cognitive states. And it is definitely the case that the formal properties of cognitive processes *would* leave them interpretable-in-principle in more than one way, because it can be shown that any formal system has more than one consistent interpretation. In particular, each has an interpretation in number theory. If semantic interpretability-in-principle of mental representations *were* a sufficient condition for the meaningfulness of mental states, it would turn out that all of our thoughts are about numbers, since any system of computations over mental representations would have a consistent number-theoretic interpretation. But it is clearly not the case that all of our thoughts are about numbers; therefore there must be more involved in the meaningfulness and intentionality of cognitive states than the availability of a consistent systematic interpretation of mental representations. And hence the fact that computers manipulate symbols does not save CTM's account of intentionality if the "semantic" properties attributed to mental representations are semiotic-semantic properties.

7.12 CAUSALITY AND COMPUTERS

There is, however, a second avenue of response to the arguments offered in this chapter. This response starts from the observation that computers may be equipped with transducers in such a fashion as to render them sensitive to environmental stimuli. What it is for a computer to be "sensitive to environmental stimuli" is for it to be so configured that it will dependably produce particular symbol tokens when particular conditions are present or when particular events take place in its environment. That is, a computer is sensitive to environmental stimuli to the extent that there are regular, causal covariations between conditions or events in the computer's environment and tokenings of particular symbol types in the computer: for example, that it writes out "Hello, Professor Pembrooke" whenever Professor Pembrooke enters the room, or tokens "My, but it's dark in here!" whenever the lights go out.

Now it is very tempting to assume that an inscription of "My, but it's dark in here!" that is produced *whenever* the lights go out and *because* the lights have gone out is *about* the lights going out, and in a fashion

that an inscription of the same symbol string that was not causally connected to the lights going out would not be about the lights going out.
Regular, causal covariations with objects and conditions in the environment, moreover, are plausibly a factor relevant to the intentionality
of mental states as well: it seems plausible that part of what it is for my
thoughts to be about Lincoln is for there to be a causal chain stretching
back to Lincoln and including my thoughts.

It is little wonder, then, that advocates of the computer metaphor in
philosophy of mind have often gravitated towards an application of the
computer paradigm that involves a causal component—in particular,
towards accounts that explain the intentionality and semantics of mental representations in terms of regular, causal covariations between objects or conditions in the environment and the tokening of symbols of
particular types. Fodor, for example, has placed increasing emphasis on
causality. In *The Language of Thought,* published in 1975, his emphasis was completely upon the "internal code" of intrinsically meaningful
representations in a language of thought. In "Methodological Solipsism," published in 1980, the emphasis was still upon meaningful representations, but Fodor hinted at the possibility of a naturalistic theory
of reference (though he argues that this possibility is dubious, and says
nothing about a naturalistic theory of meaning). *Psychosemantics*
(1987) and *A Theory of Content* (1990) include the articulation of a
sketch of a semantic theory that still accounts for the intentionality and
semantics of cognitive states in terms of the intentionality and semantics of mental representations, but also tries to ground semantics and
intentionality in causal relationships with objects in the environment.

But how is this supposed to rescue the account of intentionality?
The answer, it seems, depends upon the relationship of the causal
component of the theory to the representational component. I see
four basic possibilities for such a relationship. First, causal regularity just adds an *additional* condition for intentionality over and above
what is supplied by the representational account. Second, the semi-
otic-intentional properties of mental representations are still supposed to provide an adequate account of the mental-semantic properties of cognitive states, but causal regularities, in turn, are supposed to provide an account of the semiotic-semantic properties of
mental representations. Third, the causal account is supposed to
provide an alternative *definition* for semantic terms as applied to
mental representations. Fourth, semantic terms are applied in some
undisclosed way to mental representations, and these "semantic"
properties are still supposed to explain the mental-semantic properties

of cognitive states, while causal regularities are supposed to explain these "semantic" properties. These four possibilities will now be examined in more detail.

7.12.1 REPRESENTATION PLUS CAUSATION

The first possibility is what John Searle (1980) has called "the Robot Reply" to his arguments against computational theories of mind. According to the Robot Reply, computation over symbols does not, indeed, provide a sufficient condition for the ascription of cognitive states, meaning, or intentionality. But if the computer were, additionally, connected to the external world in the right ways by means of transducers, then it would provide a model for understanding cognition. On this account, the semiotic-semantic properties of mental representations would not be sufficient to account for the intentionality and semantics of cognitive states, because part of what is involved in a belief being about Lincoln is that it be part of a causal chain involving Lincoln. But if one were to provide an account of cognitive states that alluded both to the meaningfulness of mental representations and to the causal chains involved in the formation of beliefs (and other cognitive states), this problem could be remedied.

Now it might well be possible to formulate a useful theory along these lines. As Searle has pointed out, however, this is no longer the same theory that was originally offered as part of CTM. The original claim was that "the objects of propositional attitudes are symbols (specifically, mental representations) and that this fact accounts for their intensionality and semanticity" (Fodor 1981: 24). But if one must, additionally, appeal to causal factors to explain the "intensionality and semanticity" of cognitive states, then one cannot account for it merely by saying that the objects of the attitudes are symbols. If an account of the intentionality and semantics of cognitive states needs to appeal to mental representations and needs, *additionally,* to appeal to causality, then CTM's account of the intentionality and semantics of cognitive states is not viable.

7.12.2 CAUSALITY EXPLAINS SEMANTICS

Now while some writers certainly endorse the Robot Reply, it is not clear that this is Fodor's strategy when he appeals to causality in explaining semantics. In *Psychosemantics,* for example, Fodor invokes causality at the level of explaining the semantic properties of mental representations. In so doing, he appears to be taking up a project at the point at which he left it off at the end of the introduction to *RePresentations*. In that introduction, Fodor gives what is perhaps his best articulation of CTM and how it emerged. He also give a clear indication of what it is intended to accomplish: "It does seem clear what we want from a philosophical account of the propositional attitudes. *At a minimum, we want to explain how it is that propositional attitudes have semantic properties*" (Fodor 1981: 18, emphasis added). Yet if CTM is supposed to provide an explanation of "how it is that propositional attitudes have semantic properties," it is curious that Fodor writes on the last page of that introduction, "What we need now is a semantic theory for mental representations; a theory of how mental representations represent. Such a theory I do not have" (ibid., 31). Now one way of reading this passage would be as an admission that CTM has thus far failed miserably at meeting Fodor's own standards for a theory of cognitive states. Such, however, is hardly the tone of the chapter in which it occurs. A better way of making sense of this passage, and of Fodor's subsequent treatment of the semantics of representations in *Psychosemantics* would be as follows: Fodor believes that CTM's representational account of the semantic and intentional properties of cognitive states *is* successful. Saying that cognitive states involve meaningful representations is enough to explain the meaningfulness of cognitive states: for example, saying that Jones is in a particular functional relation to a mental representation that means "Lo! a horse!" is all that needs to be said to provide an explanation of why Jones believes that there is a horse before him. But this still leaves an additional problem: how do we account for the semantic and intentional properties of the *representations?* Why does the mental representation mean "Lo! a horse!"? And it is here that Fodor wishes to give a causal answer—to the question of why mental representations that mean "horse" do, in fact, refer to horses. Fodor's initial, "crude" formulation of such a theory is that "a plausible sufficient condition for 'A's to express A is that it's nomologically necessary that (1) every instance of A causes a token of 'A'; and (2) only instances of A cause tokens of 'A'" (Fodor 1987: 126).

So it sounds as though Fodor wishes to make two separate claims: the first is just the representational account of the semantics and intentionality of cognitive states: namely, that cognitive states "inherit" their semantic and intentional properties from the representations they involve. The second claim is a causal theory of the semantic properties of mental representations. (Fodor gives only a sketch of such a theory, and repeatedly voices doubts that a full-fledged semantic theory can be developed.)

In order to assess these claims, it is absolutely crucial at this point to determine (1) just what Fodor *means* when he uses words like 'intentional' and 'meaningful' of mental representations, and (2) how the way Fodor picks out semantic properties is related to his causal account of semantics. The first and most obvious possibility is that Fodor is applying semantic terms to symbols in the ordinary way: that is, using them to attribute semiotic-semantic properties. This should, I think, be the default reading of expressions like 'meaningful symbol'. After all, if someone says he is bringing you "healthy food" and produces a live fish in a bowl, you might well think that he is using language in a peculiar manner—a reaction that will not be changed if he explains, "Well, he is food, after all, and you've never seen a fish that was in better health!" Similarly, if someone says that cognitive states are meaningful (referential, intentional, etc.) because they involve "meaningful symbols," you may reasonably expect that he is using 'meaningful' in the way it is usually used when it modifies 'symbol'—and that, if he is *not* using it in *that* way, he should specify just how he *is* using it. Fodor and other advocates of CTM give no warning that they are using semantic terminology in nonstandard ways, so it is reasonable to begin by assuming that the standard (i.e., semiotic) usage is in force.

If the standard usage is in force, however, CTM's representational account of semantics and intentionality for cognitive states fails, for reasons described earlier in this chapter. And if the causal account of the semantics of mental representations is supposed to be independent from the representational account of the semantics of cognitive states, it can do nothing to bolster it. If the semiotic-semantic properties of representations cannot explain the mental-semantic properties of cognitive states, it does not matter, for purposes of an account of the intentionality of cognitive states, how the representations get their semiotic-semantic properties. *Whatever* the answer to that question might be, it does the representational account of the intentionality of cognitive states no good.

7.12.3 CAUSAL AND OTHER DEFINITIONS OF SEMANTIC TERMS

The final two candidates for the relationship between representation and causal covariation do not really fall within the purview of this chapter. One candidate was the view that the usages of terms such as 'meaningful' and 'semantic' should simply be *defined* in causal terms. The other was the view that the usages of semantic terms as applied to mental representations do not denote semiotic- or mental-semantic properties, and are not to be defined in causal terms, but that they denote properties that can be explained through causal covariations. In either case, the theory offered does not explain the mental-semantic properties of cognitive states in terms of the semiotic-semantic properties of representations. Whether such variations on CTM can provide any solace will be examined in the next two chapters.

7.13 COMPOSITIONALITY AND THE CONVENTIONALITY OF SYNTAX

Thus far we have shown systematic disregard for one feature of CTM which is in some ways quite important—namely, that it is supposed to support semantic compositionality. The representations envisioned by Fodor and other advocates of CTM, after all, are not all lexical primitives; the vast majority of them are made up of a large number of primitives combined with the help of syntactically based compositional rules. The semantic properties of complex representations are explained by (*a*) the semantic properties of their atomic constituents, in combination with (*b*) the compositional rules by which those constituents are combined.

Now this feature of CTM leaves the theory no better off with respect to the objections already raised: if the "semantic properties" are of the conventional or intentional kind, the fact that compositionality is thrown in does not rescue the theory from circularity or regress. *Any* taint of semantic convention or intention is enough to scuttle the whole project. But the appeal to compositionality does introduce a further problem: according to the analysis of symbols in chapter 4, syntax, as well as semantics, is conventional in nature, and hence there is a second kind of conventionality involved in CTM for the complex representations, assuming that CTM's advocates mean by "syntax" what one normally means when speaking of the syntactic properties of symbols.

The problem might be looked at in the following way. In order for there to be compositionality, it is not enough to have assignments of interpretations to primitive elements and rules governing legal concatenations of symbols. That is, it is not enough to assign "Lincoln" to A and "Douglas" to B and say that there is a legal schema for expressions 'x-&-y' into which A and B may be substituted. There must, additionally, be a rule that will further determine that 'A & B ' counts as meaning "Lincoln and Douglas" and not, say, "Lincoln, Douglas" (a list), "Lincoln is greater than Douglas," or "Lincoln or Douglas." Semantic compositionality requires a notion of syntax that consists in more than rules for legal concatenation—it requires a notion of syntax that delivers complex semantic values. (It is worth noting that *most* of the time when we speak of syntactic categories we speak of them in ways that have some semantic overtones: for example, "count noun," "dependent clause," "conjunction symbol," or even "Boolean operator.")

But this is quite problematic if we try to move from natural languages (where conventions are a commonplace) to an inner language of thought, where conventions are an embarrassment. For the only way we have of generating complex *symbolic* meanings from atomic meanings is through syntactically based combinatorial rules, and the only such rules we have are conventional rules. But if the meanings of mental states are dependent upon syntactic convention, the old problems about semantic conventions reassert themselves at a different level: in brief, (1) the actual existence of such conventions is extremely dubious, (2) their existence is in fact irrelevant to the meanings of our mental states, and (3) positing such conventions would lead to a regress of mental states.

This problem with the conventionality of syntax, moreover, in some ways poses a problem for CTM more fundamental than that posed by the conventionality of semiotic-semantic properties. As we shall see in the next chapter, one might try to rescue CTM by developing a notion of "semantic properties" for representations that is not convention- or intention-dependent. Some would say we already have such notions. It is far less clear, however, that we *do* have or *could* have an account of compositionality that was not ultimately based upon conventions, and hence this objection will recur for the versions of CTM to be explored in chapters 8 and 9.

7.14 SEMIOTIC-SEMANTICS AND THE VINDICATION OF INTENTIONAL PSYCHOLOGY

Up to this point, this chapter has been directed towards showing that CTM cannot provide an account of the intentionality and semantics of mental states based upon the semiotic-semantic properties of mental representations. What about CTM's other claim—the claim to provide a vindication of intentional psychology? There is a fairly straightforward case that, because the semiotic-semantic version of Fodor's account cannot explain the mental-semantic properties of mental states, it proves unable to vindicate intentional psychology as well. The reason this is so is that the vindication of intentional psychology turns out to be dependent upon an account of intentionality in ways that may thus far have been unforeseen. To see how this is so, consider how CTM was supposed to provide a vindication of intentional psychology.

What the computer paradigm was supposed to show was that the semantic properties of symbols in computers can be coordinated with their causal powers because semantics can be coordinated with syntax, and in a computer a symbol's syntactic type determines its causal role. If we assume that the mind is a computer, and that the semantic properties of mental states are inherited from the symbols which it uses in its computations, then explanations cast in intentional vocabulary can (in principle) pick out psychological categories in a fashion that gets the causal regularities right.

This line of reasoning, however, is compromised by the analysis of symbols and semantics in chapters 4 and 5. For what we need for a vindication of intentional psychology is an account of how the mental-semantic properties of mental states can be coordinated with causal properties, and the most that a computational theory of mind can give us, it seems, is an account of how the semiotic-semantic properties of mental representations can be coordinated with causal powers. Of course, if one could account for the mental-semantic properties of mental states in terms of the semiotic-semantic properties of mental representations, the vindication of intentional psychology could proceed intact. But what we have seen in this chapter is that one cannot account for mental-semantic properties in this fashion. So even if there are mental representations with semiotic-semantic properties, and even if the semiotic-semantic properties of these are coordinated with causal roles, this does intentional psychology little good, because it does not explain how the kind of semantic properties ascribed to beliefs and desires can link up with causal regularities.

7.15 SUMMARY

CTM claims that the mind is a computer that operates upon mental representations that are symbols having semantic properties. But we have seen that the expression 'semantic properties' is ambiguous. In order to see just what CTM might be claiming, and how this might or might not support the claims to explaining intentionality and to vindicating intentionality psychology, it was necessary to substitute different senses of 'semantics' into CTM's account. Here we have seen that neither the account of intentionality nor the vindication of intentional psychology can proceed upon the assumptions that the "semantic properties" ascribed to mental representations are mental-semantic or semiotic-semantic properties. (That is, they cannot proceed upon the assumption that they are the kinds of semantic properties ascribed to mental states or to symbols, respectively.) In the following chapter we shall examine whether substituting some other sense of the expression 'semantic property' might produce more hopeful results.

CHAPTER EIGHT

Causal and Stipulative Definitions of Semantic Terms

In the last chapter we began a project of assessing CTM's claims (1) that the intentionality of mental states can be explained in terms of the semantic properties of mental representations, and (2) that this will also provide a vindication of intentional psychology. The basic claim about the intentionality and semantics of mental states that we set out to examine was this:

Mental state M has semantic property P because

(1) M involves a relationship to a mental representation MR, and

(2) MR has semantic property P.

In light of the distinction between mental- and semiotic-semantic properties, however, it was necessary to revise this schema for explaining intentionality in the following fashion:

Mental state M has mental-semantic property P because

(1) M involves a relationship to a mental representation MR, and

(2) MR has _____-semantic property X.

The lacuna in clause (2) is to be filled by some more specific kind of "semantic property." What was shown in the last chapter is that filling the lacuna in a way that offers an account in terms of mental-semantic properties or semiotic-semantic properties will not provide an explanation of the intentionality of the mental. And indeed, the problems arise not only at the level of the conventionality of semiotic *meaning,* but also involve problems with the conventionality of *syntax* and even of mere *marker-hood.* If 'symbol' means *marker,* then it will not do to speak of the mind as a manipulator of symbols, since that would again involve us in a regress of conventions.

However, we saw in chapter 5 that it is possible to develop the notion of a machine-counter in a fashion that seems to provide everything CTM should require when it speaks of "symbols" and "syntax," yet in a way that avoids commitments to conventions or intentions. It is therefore necessary to consider whether CTM might provide a viable account of the mind if we interpret the talk about "symbols" not as talk about *markers and counters,* but as talk about *machine-counters.* In order to do this, however, we will require more than the notion of a machine-counter. That notion might be sufficient for an articulation of the kind of "syntactic theory of mind" advocated by Stich (1983), but an interpretation of CTM will also require an interpretation of talk about the "semantic properties of the symbols" that supplements the notion of a machine-counter with a nonconventional notion of semantics. In this chapter, therefore, I shall present a way of interpreting CTM that avoids problems of the conventionality of symbols and syntax by interpreting CTM as dealing with machine-counters. Additionally, I shall explore two ways of interpreting CTM's use of semantic vocabulary as expressing some set of properties distinct from semiotic-semantic properties. First, I shall explore the possibility of treating Fodor's causal covariance theory of content as a stipulative definition of his use of semantic terms as applied to mental representations. Then, I shall explore the possibility of treating the semantic vocabulary in CTM as a truly theoretical vocabulary, whose meaning is determined by its use in the theory.

8.1 THE VOCABULARY OF COMPUTATION IN CTM

In order to reinterpret CTM's claims so as to avoid the taint of convention and intention, we must find alternative interpretations for its talk about "symbols," "syntax," and "semantics." Chapter 5 already gives us a plausible alternative construal of talk about "symbols" and "syntax." For there we saw that some writers in computer science, like Newell and Simon (1975), seemed implicitly to use the word 'symbol' to denote not the convention-based semiotic typing, but a typing tied directly to the functional analysis of the machine. There we suggested the technical notion of a machine-counter in an effort to make this usage more precise. The notion of a machine-counter was developed as follows:

A tokening of a machine-counter of type T may be said to exist in C at time t iff

(1) C is a digital component of a functionally describable system F,

(2) C has a finite number of determinable states $S: \{s_1, ..., s_n\}$ such that C's causal contribution to the functioning of F is determined by which member of S digital component C is in,

(3) the presence of a machine-counter of type T at C is constituted by C's being in state s_i, where $s_i \in S$, and

(4) C is in state s_i at t.

I argued in that chapter that this functional typing is quite distinct from the semiotic typing and can serve neither as an analysis nor as an explanation of it. But at the same time, this kind of functional typing may provide just what CTM needs to escape from the conventionality of markers and counters. It is thus only natural to try to reconstruct CTM in a way that substitutes an unobjectionable notion like that of a machine-counter for the problematic convention-laden notions of "symbol" and "syntax." Intuitively, the idea is that the mind has a functional analysis in terms of a machine table, and there are things in the mind or brain that (*a*) appear as machine-counters in such an analysis and (*b*) covary with content. We are thus ready to reconstruct CTM in a way that avoids the problems of conventionality explored in earlier chapters.

8.2 A BOWDLERIZED VERSION OF CTM

In Victorian England, there was a practice of producing editions of books that had been expurgated of all objectionable material (references to ankles and other such scandalous license). Such books were said to have been "bowdlerized," the word deriving from the name of one of the notable practitioners of such editing. What I propose to do here is to describe a bowdlerized version of CTM—BCTM—which avoids objectionable suggestions that MR-semantic properties are mental- or semiotic-semantic properties by characterizing MR-semantics in terms of the work that the semantic vocabulary seems to do in CTM. Note that it is CTM in particular that is under discussion, and not cognitive theories generally: the operative meaning of semantic terminology might turn out quite differently if one were discussing other philosophical theories (e.g., those of Dennett, Searle, or Dretske) or if one were discussing particular empirical work (say, that of Colby, Newell and Simon, Marr, or Grossberg).

So, without troublesome references to symbols and semantics, it seems to me that what CTM wishes to claim is the following:

Bowdlerized Computational Theory of Mind (BCTM)

(B1) The mind's cognitive aspects are functionally describable in the form of something like a machine table.

(B2) This functional description is such that

(*a*) attitudes are described by functions, and

(*b*) contents are associated with local machine states. Call these *cognitive counters*.

(B3) These cognitive counters are physically instantiable.

(B4) Intentional states are realized through relationships between the cognizer and cognitive counters. In particular, for every attitude A and every content C of an organism O, there is a functional relation R and a cognitive counter type T such that O takes attitude $A[C]$ just in case O is in relation R to a tokening of T.

BCTM may be regarded as a special form of machine functionalism. It is stronger than *mere* machine functionalism in several respects. Condition (B1) asserts that machine functionalism is applicable to minds. Condition (B2) goes beyond this to make special claims about how the attitude-content distinction will be cashed out in functional terms. Machine functionalism, in and of itself, does not make such a claim and indeed does not even assure that the attitude-content distinction *will* be reflected in a psychological machine table. Nor does machine functionalism claim, as (B4) does, that things that are picked out by functional description will also play a role in determining content.

If we interpret computational psychology in the way suggested by BCTM, the notion of rule-governed symbol manipulation becomes more of a guiding metaphor for psychology than the literal sense of the theory. Cognitive counters are not "symbols" in the ordinary semiotic sense, but machine-counters—specifically, they are the things that occupy the slots of machine-counters in the functional analysis of *thought,* as opposed to other functionally describable systems. On this view, the mind shares with computing machines the fact that the salient description of their causal regularities is math-functional in character, but differs in that what is described by the function table is not a set of entities with conventional semiotic interpretations but—well, something else whose true nature is not yet known. If the theory is right, we presently know cognitive counters and their MR-semantic properties only through the role they play in contributing to something we know more immediately: namely, intentional states and mental processes.

I should stress that I view this as a reconstruction of CTM and not as an attempt to guess at what its advocates had in mind. I think it seems clear that Fodor and others have generally assumed the univocity of the semantic vocabulary, and likewise assumed that there was a perfectly ordinary usage of terms like 'semantics' and 'meaning' that could be extended to mental representations. In light of the problems that have already been shown to exist for that assumption, I am now trying to see whether there is an alternative interpretation of computational psychology that can avoid the problems already raised. (I am trying to pull CTM's chestnuts out of the fire, if you will.) In the end, I think there are two very different questions here: one about the viability of computational psychology as an *empirical research programme,* and another about the distinctively *philosophical claims* CTM's advocates have made about explaining intentionality and vindicating intentional psychology. In the remainder of this chapter, I shall try to argue that BCTM does not allow the computationalist to make good on these philosophical claims. In the final section of the book I shall explore an alternative approach to computational psychology that liberates its empirical research agenda from unnecessary philosophical baggage.

8.3 THE PROBLEM OF SEMANTICS

If successful, the analysis of semantic properties in chapters 4 through 6 has shown several important things about the task of explaining the intentionality of mental states. First, what we call "meaning" and "intentionality" with respect to mental states are not exactly the same properties we ascribe to symbols when we use those words. Second, the properties we ascribe to symbols are conceptually dependent upon those we ascribe to mental states. And hence, as shown in chapter 7, we cannot use semiotic-semantics to explain mental-semantics. Most articulations of CTM have seemed to assume, on the contrary, that the semantic vocabulary can be predicated univocally to mental states, overt symbols, and mental representations, and that the semantic properties of representations could be used to explain those of mental states through "property inheritance" because they are, after all, the very same properties and need only be passed up the explanatory chain.

In light of the previous chapters, this direct explanation of intentionality by way of "property-inheritance" seems to be closed off. If the "semantic properties" of mental representations are semiotic-semantic properties, they cannot explain mental semantics. And if they are *not* semiotic-semantic properties, it remains to be seen what kind of properties they *are* supposed to be. However, it is possible that waiting in the wings there is a way to finesse this problem the way we were able to finesse problems of syntax and symbolhood by way of the notion of a machine-counter. That is, perhaps the semantic vocabulary expresses some *distinct* kind of property when applied to mental representations, and this kind of property gives us what we need to explain the intentionality of mental states. Of course, we do not have a *theory* until we spell out what these properties are supposed to be. But we may for the meantime indicate the fact that they are supposed to be distinct from mental-semantic properties and semiotic-semantic properties by indicating them as "MR-semantic" properties. (That is, the kind of properties expressed by the semantic vocabulary when applied to mental representations.)

Presumably, what is common to mental-, semiotic-, and MR-semantic properties is that in each case there is a relationship between the typing of the theory (i.e., types of intentional content, types of symbol, types of representation) and a set of objects or states of affairs. Indeed, presumably the mathematically reduced abstractions of the three sets of properties are in very close correspondence: since words are expressions of thoughts, words-to-world mappings will closely track thought-to-world mappings. And if there are indeed mental representations, presumably representation-to-world mappings will closely parallel thought-to-world mappings. (In the ideal case, they will be isomorphic. But it is likely that the relationship falls short of isomorphism due to factors such as two words expressing the same concept or one word ambiguously expressing multiple concepts.) As we have seen, this does not add up to a "common" notion of semantics, because the nature of the relations expressed by the mappings is different in each case. (For example, in the semiotic case it is essentially conventional, while in the case of intentional states it is not.)

The problem for a computational-representational semantics is to articulate a theory of MR-semantics that can meet the following desiderata: (1) the MR-semantic typing of representations must correspond to their machine-counter typing; (2) the relation that establishes a mapping between representation types and their MR-meanings must be such as to be able to explain the presence of the mental-semantic properties of mental states; and (3) the mapping so established for *representations* must have a proper degree of correspondence to that of the semantics of *mental states*.

In this chapter I shall explore two possible ways of developing such a semantics for mental representations. First I shall examine the possibility of using Fodor's Causal Covariation Theory of Intentionality (CCTI) as a *stipulative definition* of the properties expressed by the semantic vocabulary when it is applied to representations. Later, I shall turn to the possibility that the semantic vocabulary, as applied to representations, is a true *theoretical* vocabulary, where the meanings of the terms are determined by the explanatory role they play in the theories in which they are introduced.

8.4 A STIPULATIVE RECONSTRUCTION OF THE SEMANTIC VOCABULARY

If, then, the semantic vocabulary is being used in some novel way when applied to mental representations, how is it being used? One reasonable hypothesis would be to suppose that it is being used to supply precisely the properties that Fodor ascribes to representations in his own theory of representational semantics—the so-called "causal covariation account." To repeat, I do not think that Fodor was in fact offering his semantic theory as a stipulative definition of the semantic vocabulary. But if the theory works, and the semantic vocabulary is in need of definition for mental representations, it seems a viable candidate. And if, as a stipulative definition, it is incapable of meeting the desiderata listed above, it will fail as a nondefinitional account as well, and so time spent critiquing it will not be ill spent.

Consider, then, what Fodor has to say about the nature of the "semantic properties" of mental representations. What Fodor provides by way of an "account of semantic properties for mental representations" is what he calls a "causal theory of content." The motivation for this project Fodor explains as follows: "We would have largely solved the naturalization problem for propositional-attitude psychology if we were able to say, in nonintentional and nonsemantic idiom, what it is for a primitive symbol of Mentalese to have a certain interpretation in a certain context" (Fodor 1987: 98). This theory of "what it is for a primitive symbol of Mentalese to have a certain interpretation" has become progressively less vague in Fodor's work from 1981 to 1990, and Fodor describes the 1990 theory as providing an account of content having "the form of a physicalist, atomistic, and putatively sufficient condition for a predicate to express a property" (Fodor 1990: 52). The 1990 version of this account reads as follows:

I claim that "X" means X if:

1. 'Xs cause "X"s' is a law.

2. Some "X"s are actually caused by Xs.

3. For all Y not = X, if Ys qua Ys actually cause "X"s, then Ys *causing* "X"s is asymmetrically dependent on Xs *causing* "X"s. (ibid., 121)

It is clear from the context that this account is supposed to apply only to mental representations—that is, to be restricted to the cases where "X" indicates a mental representation—so we would seem to be on the right track in looking for an explication of 'means' as it is used of mental representations.

Let us, then, assume that this account of MR-semantic properties can serve as a stipulative definition of the semantic vocabulary as applied to mental representations. We may now substitute this account of MR-semantic properties into CTM's basic schema for explaining the intentionality of the mental, obtaining a Causal Covariation Theory of Intentionality (CCTI):

Causal Covariation Theory of Intentionality (CCTI)

Mental state M mental-means P because

(1) M involves a relationship to a mental representation MR of type R,

(2a) Ps cause Rs is a law,

(2b) some Rs are actually caused by Ps, and

(2c) for all $Q \neq P$, if Qs qua Qs actually cause Rs, then Qs causing Rs is asymmetrically dependent upon Ps causing Rs.

We shall now examine the prospects of CCTI. CCTI is primarily intended as an examination of the consequences of using causal covariation as a stipulative definition of the semantic vocabulary. But of course CCTI could serve as a statement of Fodor's account of mental semantics generally, whether clauses (2a) through (2c) are supplied by definition of semantic terms or merely provide necessary and sufficient conditions. The assessment that follows, therefore, is of interest as a critique of causal covariation accounts, whether they involve stipulative definition or not.

In what follows, I shall argue that this approach to saving CTM has several serious problems. First, even if CCTI provides a consistent theory that avoids the problems of interpretational semantics, it does not inherit much of the persuasive force originally marshaled for CTM, because much of that persuasive force turned upon the intuition that the *same* "semantic properties" could be attributed univocally to mental states, discursive symbols, and mental representations. With this assumption already undercut, it is incumbent upon CTM's advocates to make clear the connection between MR-semantics and mental-semantics in such a fashion that the former can account for the latter— that is, to show how causal covariation is even a *potential* explainer of mental-semantics. This leads to a more fundamental problem about the causal covariation account. What this account seems to attempt to provide is a demarcation account for meaning assignments, not an explanation of meaningfulness. That is, it seems to correlate particular mental-meanings (i.e., meaning-X as opposed to meaning-Y) with certain naturalistic conditions, on the assumption that there is *some* meaning there in the first place. What it does not do is explain why mental states are meaningful (rather than meaning*less*) in the first place, or how causal covariation is supposed to underwrite this fact. In this regard CCTI compares unfavorably to some other naturalistic accounts, but there is also reason to doubt that any naturalistic account could provide an adequate account of the meaningfulness of mental states. Finally, at best, CCTI would provide an account of the semantic primitives of mentalese, leaving the semantic values of complex expressions to be generated through compositional rules. But as we have seen in the last chapter, the only way we know of to provide syntactically based compositionality is through *conventions*. So even if CCTI succeeds in escaping the problems of conventionality at the level of semantic primitives, those problems will still reassert themselves as soon as one is concerned with expressions whose semantic properties are due to compositionality.

8.4.1 WHAT IS GAINED AND LOST IN CAUSAL DEFINITION

Before making a direct frontal assault upon the Causal Covariation Theory of Intentionality, it will be useful first to become clear about what is gained and what is lost in adopting the strategy of defining semantic terminology for mental representations in causal terms. There seem to be three immediate benefits. First, we have clarified the semantic terminology to a point where we seem in little danger of running afoul of the ambiguities in the semantic vocabulary. Second, we are no longer in the embarrassing position of not being able to say what kinds of properties it is that are supposed to explain the intentionality of mental states. Third, we have done so in a fashion that manages to avoid all of the awful problems about conventions and intentions that plagued the semiotic-semantic account. If causal covariation is not free from the taint of the conventional, it is hard to imagine what would be.

On the other hand, it is important to see that a truly vast amount of the *persuasive* strength of the case for CTM is lost in the transition. The case for CTM, after all, traded in large measure upon the intuition that thoughts and symbols have some important things in common: namely, both are meaningful, both represent, both have semantic properties. This is a point to which Fodor repeatedly returns. To take a few sample quotes:

> Propositional attitudes inherit their semantic properties from those of the mental representations that function as their objects. (Fodor 1981: 26)

> Mental states like believing and desiring aren't... the only things that represent. The other obvious candidates are *symbols*. (Fodor 1987: xi)

> Symbols and mental states both have representational *content*. And nothing else does that belongs to the causal order: not rocks, or worms or trees or spiral nebulae. (Fodor 1987: xi)

The reasoning that is supposed to follow from such claims seems quite clear: computational explanation in cognitive psychology makes it seem necessary to suppose that there are mental symbols over which the computations are performed. Perhaps these have semantic properties as well, and it is the semantic properties of the symbols that account for the semantic properties of the intentional states in which they are involved. That is, one is inclined to argue as follows:

(1) Mental states have semantic properties.

(2) Symbols have semantic properties.

(3) There is a class of properties—semantic properties—shared by symbols and mental states.

so,

(4) It seems reasonable to try to reduce the meaningfulness of mental states to that of the representations they involve.

Of course, in light of the distinctions made in chapters 4 and 5, the argument from (1) and (2) to (3) is exposed as a paralogism, since 'have semantic properties' must mean something different in the two contexts (mental- and semiotic-semantic properties, respectively). And without (3), there is much less reason to be inclined towards (4). It is one thing to claim

(A) Mental state M has property P because M involves MR, and MR has P.

It is quite another to claim

(B) Mental state M has property P because M involves MR, and MR has X, and $X \neq P$.

(B) requires a kind of argumentation beyond what is required for (A), because (A) proceeds on the assumption that property P is in the picture to begin with, and just has to explain how M gets it. (B), on the other hand, has to do something more: it has to explain how P (in this case, mental-intentionality) comes into the picture *at all*.

As for the quotes cited above, their interpretation becomes quite problematic once they are read in light of the distinctions between different kinds of "semantic properties." If words like 'semantic', 'represent', and 'content' are *defined* in causal terms for mental representations, claims such as these are irrelevant at best. At worst they are logical howlers. To say, for example, that "mental states and (discursive) symbols both represent" is perilously misleading. As we have seen in chapter 4, there is no one property called "representing" that is shared by mental states and discursive symbols. Instead, 'represent', like other semantic terms, means different things when applied to symbols and to mental states. So the sentence, "mental states and symbols both represent" involves faulty parallelism that disguises a more basic conceptual error.

The same kind of problem occurs if we just define 'refers to' or 'means' in causal terms for mental representations. Suppose "mental representation *MP* refers to *P"* just *means* "mental representation *MP* was caused by *P* in fashion *F*." What, then, would we make of such assertions as "propositional attitudes inherit their semantic properties from those of the representations that serve as their objects"? This assertion, like the claim that mental states and symbols "both represent," is perilously misleading. For the claim implies that there is some set of properties called "semantic properties" that are ascribed both to mental states and to mental representations. If the "semantic properties" ascribed to mental representations are defined in causal terms, however, the semantic properties ascribed to mental states must be defined in causal terms as well, if they are to be the same properties. But surely this is not so. When we say that Jones is thinking about Lincoln, what we mean is surely not precisely that he stands in a particular causal relation to Lincoln. We *certainly* mean nothing of this kind when we say that Jones is thinking about *unicorns* or *numbers*. So if we define semantic terms applied to mental representations in causal terms, it is misleading to speak of the "inheritance" of semantic properties: such properties as might be conferred upon mental states by representations are not the same properties that are possessed by the mental representations themselves. And such arguments for CTM as depend upon a genuine inheritance of the *same* "semantic properties" turn out to be fallacious.

A similar problem can be made for CTM's attempt to vindicate intentional psychology. The strategy for the vindication was to show, on the basis of the computer paradigm, that the postulation of mental representations could provide a way of coordinating the semantic properties of mental states with the causal roles they play in thought processes. Such an argument might be formulated as follows:

Argument V1

(1) Mental states are relations to mental representations.

(2) Mental representations have syntactic and semantic properties.

(3) The syntactic properties of mental representations determine their causal powers.

(4) All semantic distinctions between representations are preserved syntactically.

∴(5) The *semantic* properties of *representations* are coordinated with causal powers (3,4).

(6) The semantic properties of mental states are inherited from the representations they involve.

∴ (7) The semantic properties of mental states are coordinated with causal powers (5,6).

Now consider just steps (5) through (7). If we were to interpret the expression 'semantic properties' univocally, we could recast (5) through (7) as follows:

Argument V2

(5́) There is a strict correspondence between a *representation's* semantic properties and its causal powers.

(6́) A *mental state M* has semantic property *P* if and only if it involves a *representation MR* that has semantic property *P*.

∴ (7́) There is a strict correspondence between a *mental state's* semantic properties and its causal powers.

On this construal we appear to have a reasonable and valid argument. But consider this second construal, which is forced upon us by the recognition of the homonymy of semantic terms:

Argument V3

(5*) There is a strict correspondence between a representation's MR-semantic properties and its causal powers.

(6*) A mental state *M* has mental-semantic property *P* if and only if it involves a representation *MR* that has MR-semantic property *X*.

∴ (7*) There is a strict correspondence between a mental state's mental-semantic properties and its causal powers.

The plausibility of the deduction to (7*) depends in large measure upon the plausibility of (6*). The plausibility of (6*), in turn, will depend upon what MR-semantic properties turn out to be. But whatever they may turn out to be, (6*) lacks some of the immediate prima facie appeal of (6) and (6́), since it depends upon a (contingent) correlation of different kinds of properties, whereas (6) and (6́) involve ascriptions of the *same* properties to two different objects. This kind of contingent correlation is itself in need of explanation.

The upshot of these observations is this: if the "semantic properties" of mental representations are defined in causal terms, the proponent of CTM owes us something that he did not owe us on the assumption that the semantic properties of mental states were the very properties possessed by mental representations: namely, he owes us a plausible account of why having a representation *MR* with certain MR-semantic properties (say, certain causal connections with objects in the environment) should be a sufficient condition for having a mental state with certain mental-semantic properties (say, a belief about dogs). This is significant because the arguments given in favor of CTM seem to assume that the same kinds of "semantic properties" can be ascribed indifferently to symbols, mental representations, and mental states. But if one *defines* the semantic terminology that is applied to representations in causal terms, most of what Fodor says to commend CTM to the reader is patently fallacious.

In summary, then, we may say that defining MR-semantic properties in terms of causal covariations allows us to avoid the major pitfalls presented for earlier readings of CTM, but the case for CTM now seems much weaker than it once did. The reason for this is that originally the road from representations to mental states was a road from semantics to semantics, and the road from semantics to semantics seemed relatively short and straight. If the "semantic properties" of mental states and representations were *the same properties,* there would be no question but that the latter are the sort of things that *could* account for the presence of the former, but only a question about whether such "inheritance" indeed takes place. On the current interpretation, however, the road from representations to mental states is a road from causal covariation to mental-semantics. That road is surely much longer, and there is no small question about whether the roads shall meet at all. It may be that they are like Down East roads: "Ya can't get there from here!"

8.4.2 COVARIATION AND MENTAL-SEMANTICS

The vital question, then, is whether causal covariation is the right sort of notion to provide an explanation of the semantic properties of mental states. I believe that it is not. But in order to see why it is not, it may prove useful to see what it *is* suited to doing and how that falls short of explaining mental-semantics. In order to do this, it will be helpful to make two sorts of distinctions. First, we may distinguish between two sorts of accounts: those that provide *explanations of what it is to be an X,* and those that merely provide *criteria for the demarcation of X's* from non-*X*'s.

Second, we may distinguish between accounts of meaning *assign-ments* (i.e., distribution of meanings) from accounts of meaning*fulness*. The former differentiate things that mean *A* from those that mean *B*, on the assumption that the items in question mean *something;* the latter explain why items mean something rather than nothing. I shall argue that CCTI is suited at best to providing a demarcation criterion for meaning assignments, whereas an account of mental-semantics requires something stronger: an explanation of meaningfulness.

8.4.2.1 EXPLANATION AND DEMARCATION

To begin with, let us distinguish between accounts that give an *explana-tion* of why something is an *X* from accounts that merely provide a *crite-rion for the demarcation* of *X*'s from non-*X*'s. Aristotle's characteriza-tion of humans as featherless bipeds is an attempt at a demarcation crite-rion. It happens to be a poor attempt, since apes, tyrannosaurs, and plucked chickens are also featherless bipeds. But even if humans were, in point of fact, the *only* featherless bipeds, the featherless-biped criterion would at most give us a litmus for distinguishing humans from other spe-cies. If what we wanted was an explanation of *what makes* Plato a human being, the fact that he is a featherless biped is clearly a non-starter. The problem is not that demarcation criteria can be wildly contingent, for in fact they need not be—some demarcation criteria can be metaphysically necessary. Even demarcation criteria that are metaphysically necessary, however, can fail to be explanatory. For example, if you want to know what makes a figure a triangle, the answer had better be something like "the fact that it has three sides." But there are descriptions that distin-guish triangles from everything else that do not provide this information: for example, "simplest orthogonal two-dimensional polygon," "shape of the faces on a regular octahedron," and (worst of all) "Horst's favorite geometric example." (This last, of course, is not metaphysically neces-sary.) If you want to know what makes a figure a triangle, the fact that it has the same shape as one of the faces of an octahedron just will not do as an explanation, though it is necessary and sufficient.

There are relationships between demarcation criteria and explanations. Significantly, things that can serve as explanations are a proper subset of things that can serve as demarcation criteria. On the one hand, an account that explains what it is to be an *X* must also be able, at least in principle, to serve as a demarcation criterion for distinguishing *X*'s from non-*X*'s. On the other hand, the opposite is not true: we have already seen exam-ples of demarcation criteria that lack explanatory power. A corollary of this is that one way of showing that something is not an explanation of what it is to be an X is to show that it does not even distinguish X's from non-X's.

8.4.2.2 MEANING ASSIGNMENT AND MEANINGFULNESS

Let us further distinguish between two aspects of accounting for a token T's meaning-X. On the one hand, one might want to account for why T means X as opposed to meaning something else, treating it as a background assumption that T can mean *something*. When we explain the role of particular morphemes in determining the meanings of polymorphemic words, for example, we take it as a given that words can mean something and confine ourselves to asking, say, how various sorts of affixes interact with the meanings of root morphemes. This provides an account of why words have the particular meanings they have without touching upon the question of how language gets to be meaningful in the first place. But one might ask this second question as well, and it is here that, say, Ruth Millikan's account of truth and meaning for languages is at odds with accounts based on convention or speaker meaning. Such accounts are accounts of meaning*fulness* rather than of meaning *assignment*.[1] Presumably one may offer an account of meaning assignments without thereby offering an account of meaningfulness, and vice-versa.

8.4.2.3 WHY WE NEED AN EXPLANATION OF MEANINGFULNESS

Now what kind of "account of meaning" is required for mental-semantic properties of mental states? Well, if one wants to know how it is that things in the mind get to be about things in the world, one presumably wants to know *both* how thoughts get to be about *particular* things and how they get to be about *anything at all* —that is, one wants accounts of meaning assignment *and* of meaningfulness. Now suppose further that we are interested (as CTM's most notable advocates clearly are interested) in a *naturalistic* account—one that explains mental-semantic properties on the basis of some naturalistic properties ("N-properties"). Here the problem of meaning assignment becomes one of associating particular mental-semantic properties (e.g., meaning "horse") with particular N-properties (e.g., causal covariations with horses). And if all we are interested in is a naturalistic *demarcation criterion* for particular mental-meanings, all the "association" need amount to is strict correlation—some set of N-properties that all and only horse-thoughts (as opposed to cow-thoughts, unicorn-thoughts, etc.) possess. But if we are interested not merely in a demarcation criterion, but in an explanation of what it is to mental-mean "horse," our naturalistic account of meaning assignments needs to be augmented with a naturalistic account of meaningfulness as well. Unless N-properties are sufficient to explain mental-meaningfulness, particular N-properties cannot explain particular mental-meanings either.

If CCTI is to provide an adequate account of intentionality and mental-semantics, then, it must provide an explanation of mental-meaningfulness. I shall now argue, however, that CCTI cannot plausibly be supposed to do this. All it can plausibly be supposed to do is provide a demarcation criterion for meaning-assignments. I shall first argue that CCTI attempts to provide a demarcation criterion for meaning assignments, and then argue that it fails to do more than this.

8.4.3 CCTI AS A DEMARCATION CRITERION FOR MEANING ASSIGNMENTS

There are three main reasons to see CCTI as a demarcation criterion for meaning assignments. First, there is a strong tendency in the literature to see the task of "fixing meanings of representations" as a matter of imposing a suitable interpretation scheme—namely, one that assigns the right meanings. Second, CCTI seems naturally suited to providing a demarcation criterion of the desired sort. Third, the bulk of the discussion of the causal covariation version of CTM has been centered around CCTI's *success* or *failure* at providing a successful such demarcation criterion.

8.4.3.1 DEMARCATION, INTERPRETATION, AND MEANING FIXATION

A reader of the cognitive science literature will have noticed that there is a strong tendency to view the problem of accounting for content of representations as one of imposing a coherent representational scheme. Pylyshyn writes, for example, that the computational approach to the mind involves the assumption that

> there is a natural and reasonably well-defined domain of questions that can be answered solely by examining (1) a *canonical description* of an algorithm (or a program in some suitable language—where the latter remains to be specified), and (2) a system of formal symbols (data structures, expressions), *together with what Haugeland (1978) calls a "regular scheme of interpretation" for interpreting these symbols as expressing the representational content of mental states* (i.e., as expressing what the beliefs, goals, thoughts, and the like are about, or what they represent).... Notice... that we have not said anything about the scheme for interpreting the symbols—*for example,*

whether there is any indeterminacy in the choice of such a scheme or whether it can be uniquely constrained by empirical considerations (such as those arising from the necessity of causally relating representations to the environment through transducers). (Pylyshyn 1980: 116, emphasis added)

Notice two things about this quote. First, semantic properties are discussed in terms of a "scheme of interpretation." Second, the question about this scheme that seems foremost in Pylyshyn's mind is whether the meaning assignments of a given scheme can be constrained so as to be unique. Similar issues arise in Haugeland (1981: intro.; 1985: chap. 3). It seems clear that these writers view the issue of finding a semantics for mental representations as one of finding a way to constrain the specification of an interpretation scheme for representations so that it is unique and so that it gets the causal relationships right—that is, their concern is for providing an adequate demarcation criterion for meaning assignments.

8.4.3.2 THE SUITABILITY OF CCTI FOR DEMARCATION

CCTI also seems well suited to providing a demarcation criterion for meaning assignments. (Or, to be more precise, it seems suited to providing a *candidate* for such a criterion, since there is one question about what it sets out to do and another about whether it accomplishes it.) It is quite easy to see that, whatever else CCTI might be used to do, it at very least purports to be a demarcation criterion for meaning assignments. For it is set up to give sufficient conditions, in naturalistic terms, for particular mental-meanings: the mental states that mental-mean P are the ones that have mental representations that are in a relation of causal covariation with the class of objects or states of affairs designated by P. This account may or may not be true, but if it is true, it provides a way of separating mental states that mean P from those that mean Q: the former involve representations characteristically caused by P's and the latter involve representations that are characteristically caused by Q's.

8.4.3.3 THE PROBLEM OF MISREPRESENTATION

Now there has been a substantial amount of discussion of CCTI in the literature, assessing the merits of causal covariation as a way of explaining mental-semantics. What this discussion seems to center on, however, are *the prospects for causal covariation as a way of providing a demarcation criterion for meaning assignments*. This provides some evidence that supports the conclusion that this is the role that the theory is commonly regarded as performing.

The focus of this discussion has been upon CCTI's ability to account for the possibility of misrepresentation. According to CCTI, those thoughts are about P's that involve representations of a type caused by P's. But it is surely possible to have thoughts about P's that are not caused by P's and, worse yet, to have thoughts that are about P's that are caused by something other than P's—Q's, for example. So, for example, someone visiting Australia might see a dingo and say to himself, "Oh, there's a doggie out back in the outback!" (Dingos are not dogs, etymologically speaking.) This person's thought has the content "dog," but is caused by a nondog, a dingo. And it is even possible for this error to be systematic: someone might *always* mistake dingos for dogs, wrens for sparrows, gnus for cattle, and so on. The problem is that, according to CCTI, thoughts are supposed to be about whatever it is that is the characteristic cause of their representations. But if dingos systematically cause a tokening of the same kind of representation that dogs cause, it would seem to follow that what this kind of representation MR-means is the disjunctive class dog-or-dingo. This has several unwelcome results. First, my dog-thoughts turn out to mean not "dog," but "dog or dingo." (And this quite unbeknownst to me and contrary to what I have assumed all along.) Second, it would seem to be impossible to misrepresent a Q as a P, since the fact that Q's cause the same representations as P's under certain conditions will occasion a change in the "meaning" to be assigned to such representations. (And it just seems wrong to say, for example, that someone who mistakes holograms of unicorns for real unicorns has thoughts that mean "hologram" and not "unicorn.") There are related problems arising from the fact that thoughts about dogs can be caused by things other than distal stimuli entirely—for example, I can think about dogs in dreams or in free fancy. It is hard to see just how a strict causal theory should treat these cases.

This problem, which Fodor likes to call the "disjunction problem," was apparently a significant incentive in his development of the causal covariation account of intentionality from the form in which he articulated it in 1987 to the form it took in 1990. What is new in the more recent account is the addition of a notion of "asymmetric dependence," which is introduced to handle the disjunction problem. Recall the form of the account in Fodor (1990), which we have used here to develop CCTI:

I claim that "X" means X if:

1. 'Xs cause "X"s' is a law.

2. Some "X"s are actually caused by Xs.

3. For all Y not = X, if Ys qua Ys actually cause "X"s, then Ys *causing* "X"s is asymmetrically dependent on Xs *causing* "X"s. (Fodor 1990: 121)

The first and second clauses are already implicit in the older formulation. The notion of asymmetric dependence appears in clause (3). The idea is as follows: a thought involving a given representation R can mean "dog" and not "dingo" or "dog-or-dingo," even if it is regularly caused by both dogs and dingos, if it is the case that the causal connection between dingos and R-tokenings is asymmetrically dependent upon the causal connection between clogs and R-tokenings. And the nature of this "dependence" is cashed out in purely *modal* terms: what it means is that if dogs did not cause R-tokenings, dingos would not either, but not the reverse. (In other words, dingos might fail to cause R-tokenings without dogs failing to do so as well.)

Now I have no interest in contributing here to the already good-sized literature debating the success or failure of this move. What I wish to do is merely to point to what it is a debate *about*. And what it is a debate about is whether CCTI provides meaning *assignments* in the ways we should wish a semantic theory for the mind to do so. It is about such questions as whether such a theory would assign counterintuitive meaning assignments (such as "dog-or-dingo") and whether it can accommodate such patent facts as misidentification, in which one has a thought the content of which does not match the thing one is trying to identify. It may be that the fancy footwork provided by the notion of asymmetric dependence can finesse a way through these problems, but it is these problems that it seems intended to finesse.

8.4.4 WHAT CCTI DOES NOT DO

What CCTI notably does not seem to do is provide *more* than an demarcation account of meaning-assignments. It is not clear that it is *even an attempt* to provide an account of meaningfulness for mental states; and if it is so intended, the account it provides is woefully inadequate. I shall attempt to argue this in two different ways. First, I shall argue that CCTI does not provide so much as a demarcation criterion for meaningfulness (as opposed to meaning *assignments*), and hence cannot provide an *explanation* of meaningfulness, since an account that explains will also provide a demarcation criterion. Second, I shall argue that CCTI lacks the right sort of explanatory character to explain the intentionality of the mental.

8.4.4.1 FAILURE TO DEMARCATE THE MEANINGFUL

While causal covariation may or may not provide a demarcation crite-rion for meaning *assignments*, it does not provide a demarcation crite-rion for meaning*fulness* —that is, for separating things that mean *some-thing* from those that mean *nothing*. For the notion of causal covaria-tion is cashed out in terms of regular causation, and regular causation is a feature not just of mental states and processes, but of objects and events generally. The overall project here is to explain the mental-semantic properties of mental states in terms of some set N of natural-istic properties, and the proposal at hand is that N-properties are causal covariation relations. But *this* set of properties has a domain far broader than that of mental representations: any number of objects and events not implicated in thoughts have characteristic causes, and hence have N-properties. Cow-thoughts are not the only things reliably caused by cows: so are mooing noises, stampedes, and cowpies, to name a few. The CCTI cannot be a viable demarcation criterion of meaningfulness, because it does not distinguish thoughts about cows from stampedes and cowpies. And this is surely a demarcation we should expect a theo-ry that accounted for meaningfulness to entail. So either we must im-pute mental-semantic properties to all kinds of objects and events, en-dowing much of nature with content, or we must allow that something more than N-properties are required to explain mental-semantics.

The obvious strategy for sidestepping this objection is to point out that, while representations may share N-properties with many other sorts of objects, it is only mental representations that take part in the relations char-acteristic of intentional states. There may appear to be a threat of endowing the world with content—namely, with MR-semantic properties. But re-member that the word 'semantic' in "MR-semantic" is not doing much work, since we have *defined* the expression 'MR-semantic properties' in terms of causal covariation. Thus in allowing most of nature to have MR-semantic properties, we have not endowed them with anything counterin-tuitive, even though the word 'semantic' might suggest as much. Moreo-ver, CCTI, as we have formulated it, involves more than causal covaria-tion: it involves explicit reference to the effect that the items that have MR-semantic properties *are also part of an intentional state*. It is this additional fact that differentiates them from objects in nature generally. To use some terminology that has not yet been used here, we might say that *indication* or natural meaning plays a role in the production of mental-meaning *only when the indicator is present in an organism in one of the functional rela-tions characteristic of intentional attitudes*. Or, to put it slightly differently, the domain over which the CCTI is quantified is not all objects, but all objects that are representations involved in intentional states.

There is something appealing about this strategy, but it is important to note that it violates one of the fundamental canons of CTM: namely, that the semantic properties of mental states be "inherited" from the "semantic properties" of representations. According to the formulation in the previous paragraph, however, this is not the case: mental-semantic properties are not explicable solely in terms of MR-semantic properties of representations, but in terms of MR-semantic properties of representations *plus something else*. Worse yet, this "something else" seems to consist precisely in the fact that the representations are elements of an intentional state! But if we must allude to the fact that representations are part of an intentional state to make CCTI proof against the semantification of nature, we have failed to provide a naturalistic explanation of mental-meaning, since part of our account still presumes the intentional rather than explaining it. It is, of course, possible to begin by assuming intentionality, and then asking the question of what kinds of natural properties are involved in the realization of intentional states; and if we do this, we need not worry about the fact that part of what differentiates mental representations from other things that participate in causal covariation is that they also play a role in intentional states. But if we do this, we are no longer seeking an account that provides supervenience or explanatory insight. And this, it would seem, is less than CTM's advocates generally desire by way of an "account of intentionality" (even if it is, in my view, a far more sensible strategy).

The upshot of this is that CCTI does not succeed in providing a criterion for the demarcation of the meaningful from the meaningless. It is not really clear that it was *intended* to provide such a criterion, but it fails to do so regardless. It follows from this a forteriori that it does not provide an *explanation* of meaningfulness, since an explanation would also provide a demarcation criterion.

8.4.4.2 WHY CCTI DOES NOT EXPLAIN MEANINGFULNESS

It is also possible to tackle the issue of the explanation of meaningfulness by way of a frontal assault. And it seems prudent to do this, since someone might be inclined to try to rescue CCTI as a potential demarcation criterion for meaningfulness by way of some clever patchwork, much as Fodor has tried to rescue it as a criterion for meaning assignment by way of the notion of asymmetric dependence. To do so, however, would be to miss a much more serious point. The deep problem with CCTI is not that

I have some clever counterexamples that it has failed to catch in its net, and that might be brought into line with the insertion of an additional clause or two. The deep problem, rather, is that causal covariation is just not suited to explaining why some X is capable of meaning something rather than nothing. Causality is just too bland a notion for that task, and fancy patchwork would only serve to reveal this problem rather than to remedy it.

Now the way I should *like* to be able to proceed here would be to provide a really tight and compelling analysis of explanation and then give a knock-down argument to the effect that CCTI does not fit that analysis if the *explanandum* is meaningfulness. Explanation, however, is a notion that is notoriously difficult to analyze, and I shall have to content myself with a slightly more roundabout course for getting to the same conclusion: I shall attempt to establish one of the crucial "marks" of successful explanations, and then attempt to argue that the account of intentionality offered by CTM lacks this mark.

One characteristic of successful explanation is the kind of reaction it produces: the "Aha!" reaction that comes with new insight. Suppose I have some familiarity with some phenomenon P, with a set S of notable features. Now suppose that I try to explain P by means of an explanation E, cast in terms of some set of entities and relations X. Now E succeeds as an explanation to the extent that understanding X gives me insight into S — that is, to the extent that upon understanding X I become inclined to say, "Ah, now I see why things in S are as they are." Indeed, in the ideal case, understanding of X should be sufficient for me to *infer S*, even if I have no prior knowledge of S. Someone with an adequate knowledge of the behavior of physical particles, for example, would be able to derive the notion of "valence" and the laws of thermodynamics, and hence particle theories provide first-rate explanations for these other phenomena. Of course, in practice the process of explanation progresses in the other direction, but an ideal grasp of the explaining phenomena could be sufficient to allow for the *derivation* of the explained phenomena. This idea that an ideal explanation should allow the derivation of one phenomenon from another (e.g., a more complex one from a simpler one) is part and parcel of the Galilean method of resolution and composition that has informed much of modern science and modern philosophy of science, and is found notably in recent philosophy of science in both reductionist and supervenience accounts.

8.4.4.3 INSTANTIATION AND REALIZATION

I think that the weakest sort of explanation meeting this strong requirement is what Robert Cummins (1983) calls an "instantiation analysis."

(There are stronger sorts of explanation meeting it as well, of course, such as reductions.) Cummins proposes the notion of an "instantiation analysis" as a way of understanding theories that identify instantiations of a property P in a system S by specifying organizations of components of S that would count as instantiations. An instantiation analysis of a property P in a system S has the following form:

(6i) Anything having components $C_1...C_n$ organized in manner O–i.e, having analysis $[C_1...C_n, O]$–has property P;

(6ii) S has analysis $[C_1...C_n, O]$;

(6iii) S has property P. (Cummins, 1983: page 17, numbering preserved from original text)

Instantiation analyses are distinguished from *reductions* (ibid., 22-26) by the fact that a single property can have *multiple instantiations* in different systems, whereas the reduction of a property requires a *unique* specification of conditions under which it is present. But the instantiating property *is* intended to *explain* the presence of the instantiated property. Indeed, Cummins writes that one should be able to *derive* a proposition of the form (6i) from a description of the properties of the components of the system, and that when we can do this we can "*understand how* P is *instantiated* in S" (ibid., 18, emphasis added). That is, from a specification of the properties of the components of the system in the form

(6a) The properties of $C_1...C_n$ are <whatever>, respectively,

we should be able to *derive*

(6i) Anything having components $C_1...C_n$ organized in manner O—i.e., having analysis $[C_1...C_n, O]$—has property P:

Thus, with an instantiation analysis, supplying a description of the interrelations of the components of a system *S* should be enough to show that a property *P* is instantiated in *S*, because one can derive the conclusion that *S* has *P* from a statement such as (6i), and one can, in turn, derive (6i) just from a description components of *S* —that is, from a statement such as (6a). And since one can derive the conclusion that *P* is instantiated in *S* in this way, providing such an analysis should be sufficient to allay doubts that *P* can be instantiated in *S:* given a proper description of the components of *S*, one can, quite simply, *infer* the instantiation of *P* in *S*.

We may also distinguish the notion of an instantiation analysis from that of a weaker sort of account, which I shall call a *realization account*. A realization account provides a specification of how a property P is realized in a system S through the satisfactions of some set of conditions $C_1...C_n$—but *without* any implication that the satisfaction of $C_1...C_n$ provides a metaphysically sufficient condition for the presence of P. I shall give several examples:

(1) There are individual objects that have a particular status, such as the Victoria Crown kept in the Tower of London or the Mona Lisa. One could, in principle, give a complete physical description of the matter through which the Mona Lisa is realized. But meeting that description does not provide a sufficient condition for being the Mona Lisa. Additional objects meeting that description would not be additional Mona Lisas, but perfect forgeries. Likewise, there are object-kinds such as "dollar bill" that must be realized through objects with a particular physical description. But once again, meeting that description alone does not make something a genuine dollar bill. If you or I make one, it is a forgery. Dollar bills are realized through particular material configurations, but no instantiation analysis of dollar bills is possible.

(2) Some kinds of human attributes are realized through a person's behavior without the behavior itself providing a sufficient criterion for the presence of the attribute. For example, Jones and Smith may both give a substantial portion of their resources to persons in need, yet in a very different spirit. It may be that Jones does so because he is generous, while Smith does so only because he believes that it is the sole way of saving himself from the flames of hell. Jones's behavior is a realization of generosity, while Smith's is not, even if the behaviors themselves are indistinguishable.

(3) We have seen that there are certain senses in which a computer may be said to perform such operations as adding two numbers. Such operations may be said to be realized through the processes that take place in the computer's components. But a specification of the processes that take place in the computer's components does not provide a sufficient condition for the computer's overall behavior counting as addition, because it only counts as addition by virtue of meaning-bestowing intentions or conventions of designers, programmers, or users, and these are not mentioned in specifications of the interactions of the components through which the adding process is realized in the machine.

Now there is an important methodological and theoretical difference between instantiation analyses and realization accounts. Realization accounts proceed on the *assumption* that one may sensibly talk about the property P being realized in some system S. They do nothing, and can do nothing, to show that the organization of components of S *would* result in the presence of P. Indeed, it *need not* result in it—a particular set of behaviors might be a realization of jealousy or a realization of a fear of perdition, and a certain configuration of matter only counts as the Victoria Crown or a dollar bill in the context of particular institutional facts and historical acts. Realization accounts do not require even supervenience.

As a consequence, a realization account could not do anything to allay doubts about P 's being susceptible to realization in S: it proceeds on the assumption that P can be realized in S, and hence cannot justify that assumption. In the case of instantiation analysis, by contrast, one can infer the conditions for the ascription of P from a description of the components of S. As a result, providing an instantiation analysis of P in S also serves to *vindicate* the claim that P *can* be instantiated in a system like S. It vindicates it because it shows that it can be so. A realization account, on the other hand, does not in any comparable sense show that a property P can be instantiated in a system S. If someone is inclined to doubt that Jones is capable of generosity, for example, pointing to Jones's sizable donations to various charities will not prove the doubt to be mistaken. The donations might, of course, be realizations of generosity in Jones, but it might alternatively be the case that Jones really *is* incapable of generosity, and is merely giving of his wealth because he is trying to buy his way into heaven. Showing how a property is realized in a system gives us insight into the property and the system in which it is realized, but the resulting description cannot be used to demonstrate that the property *is* realized in the system or even that it *can be*.

8.4.4.4 INSTANTIATION AND THE EXPLANATION OF MEANINGFULNESS

Now I think it should be clear that in order to explain meaningfulness in naturalistic terms, it would be necessary to provide something on the order of an instantiation analysis for meaningfulness—that is, to provide an account such that an adequate understanding of the explaining properties would be sufficient to ground inferential knowledge of the properties explained as well. It also seems clear that, as an explaining property, causal covariation does not come within a country mile of meeting this condition. Causal covariation might very well provide what is needed for seeing why some thoughts are about one thing and other thoughts are about something else. (Then again it might not—I have no interest in taking sides here.) *What it does not do is provide understanding of why causal regularities might contribute to meanings in the case of mental states while failing to do so in all of the other cases of causal covariation occurring in nature.* And it is precisely here that the problem of meaningfulness lies.

Nor will any minor patchwork help in the slightest. Asymmetric dependence, for example, is of no assistance here. That can, at best, explain why my thought does not mean "dingo" or "dog-or-dingo." About why it means "dog"—or, more to the point, why *it* means *something* and *other things caused by dogs do not* (let the reader's imagination run wild)—is in no wise clarified by the notion of causal covariation.

Robert Cummins has suggested to me an alternative way of making this point: theoretical identifications, such as the identification of heat with a kind of motion, are of interest only insofar as they help us to understand something about the phenomena that are being explained. Descartes (*Le Monde,* chap. 2), for example, rejects the Scholastic view that "fire" or "heat" names a kind of substance in favor of the view that fire involves a kind of change of state in the matter of the combustible material, and that heat consists in the increased level of agitation of the matter. Other theorists were impressed by such factors as the ability to convert mechanical force into heat (as when a nail gets hot when it is driven by a hammer) and back again (as in the case of a steam engine). Viewing heat in terms of the motion of matter (and ultimately in terms of kinetic energy) allows us to understand why iron glows when heated and why nails get hot when pounded with a hammer. Now if CCTI is to be of interest as an *explanation* of intentionality, one would at very least expect there to be something about intentional states that we are able to understand better once we view them through the lens provided by the theory. But in fact there seems to be nothing of the sort. There was perhaps once hope of such a result when causal theorists were more inclined to identify content with information, and hence to view the causal chains involved in their accounts as being chains of information transmission. But the incompatibility of strict information accounts with misrepresentation has caused causal theories such as CCTI to abandon this identification. Information at least looked like an intuitively plausible candidate for explaining "aboutness" in a way that causation does not. If there is anything about intentional states that is explained by CCTI, its nature needs to be more clearly shown. In short, it does not

	Meaning-Assignment	*Meaningfulness*
Demarcation	**Attempts** to do this (success or failure unclear at present)	**NO** - Does not differentiate from other causal regularities that are meaningless. ("cow" thoughts vs. cow pies)
Explanation	**NO** — Gives no explanatory insight, epistemically opaque, no instantiation analysis.	

Figure 10

seem that CCTI explains the nature of intentionality; and indeed, it is not clear that there is anything of interest about intentionality that it does explain.

In summary then, CCTI seems at best to supply a demarcation criterion for meaning assignments, and neither an explanation of the same nor any sort of account of meaningfulness (see fig. 10).

8.4.5 SOME TELLING COMPARISONS

The issue might be put into further perspective by contrasting the explanatory power of CCTI with that of some other "accounts of intentionality." There are a number of writers who address the issue of intentionality, either in general or in specific contexts such as visual perception, whose accounts seem to me at least to provide a certain degree of explanatory insight that CCTI fails to provide. The accounts that come most quickly to mind for me in this regard are Ruth Millikan's (1984) explanations of features of mind and language in terms of reproductively established categories with a selectional history, Kenneth Sayre's (1986) and Fred Dretske's (1981, 1988) information-theoretic accounts of intentionality in perception, and David Marr's (1982) account of vision. Each of these accounts is in some sense an attempt to reduce some kind of intentionality to some set of states and processes and relationships that can be specified naturalistically. (Or, if *information* is not a natural but a formal category, each tries to give a nonintentional specification of intentionality.)[2] And in each of their accounts causality plays some explanatory role (in contrast, for example, with Searle's [1983] account, which is largely an ontologically neutral analysis of intentionality). But in each of these accounts, causality fits into the picture *only within the framework of a much richer story* about the mechanisms through which perception and cognition are accomplished.

Now each of these accounts is extremely complex and strongly resists presentation by way of a thumbnail sketch. I shall thus assume that the reader may refer back to the original sources for any details beyond the following brief sketches. Sayre (1986) tells a story of how information (in the technical sense of Shannon and Weaver [1949]) is conveyed, in a well-defined series of stages, from an object perceived to a stage of cognitive processing that might be rich enough to merit the name "intentionality." The account is an attempt to build "information," in the semantically pregnant sense of the term, out of "information" in the technical sense of "reduced uncertainty" or "negentropy," and assumptions about the functions of perceptual systems as describable as processors of information in the technical sense. Dretske employs a somewhat looser sense of "information" to similar ends. Both have stories about what it is for a thought to be about an object, stories that involve answers to questions about, for example, fidelity of perception and about what it is that connects object to intentional state and is common to both.[3] Millikan's account of belief also makes use of causal connections between the intentional state and its object, but these are embedded in a larger story about the function of belief and how it has been selected for within our species. To understand intentional states, on Millikan's view, is to understand a relationship between an organism and its environment that is the product of a history of adaptation and selection within the species. Marr presents an elaborate and detailed account of how the mind transforms sensory input into a three-dimensional visual representation through the application of a series of computational algorithms involving several distinct levels of representation of visual information.

Now these accounts do several things, in varying measures, that could contribute something towards legitimate *insight* into the phenomena they set themselves to discussing. (Of course it only merits the description of insight insofar as it turns out to be correct in the long run, but at least these accounts, *if correct,* yield new insights.) First, they subsume the phenomena to be explained (e.g., intentionality) under more general categories, and thereby provide a characterization, in nonintentional terms, of what *kind* of phenomenon it is. Millikan uses the notions of a reproductively established kind and selection history to do this for intentionality generally. Sayre treats perception and perceptual intentionality as a very rapid kind of adaptation to environmental features (much as learning and evolution are much slower sorts of adaptation), further characterized by a state of high mutual information. Second, these accounts give some insight into what kinds of mechanisms are necessary to the realization of particular kinds of mental states, whether the formal properties of these mechanisms be characterized in terms of algorithms from computer science (Marr) or in terms of the Mathematical Theory of Communication (Sayre). There is, to be sure, a purely empirical component in this latter enterprise, but there is also a component that one might describe as "transcendental." Talk of things such as intentionality of perception is primarily motivated by our own case, and it therefore makes sense to ask what *must* be true of creatures who perceive *as we do,* much as it made sense for Kant to ask what must be true of beings whose only contact with an external world is through sensuous intuitions. Insofar as we take the phenomena going on in our own mental lives as given and try to provide an account of them, we gain substantial insight from accounts that succeed in telling us what sorts of processes must go on for such phenomena to take place.[4]

Now I do not think that any of these accounts goes so far as to provide an instantiation analysis for intentionality or any particular variety thereof. I shall present my reasons for this conclusion in the next chapter. There are, however, ways of providing more or less insight—and hence of coming closer to providing an adequate explanation—short of an instantiation analysis. My intent here has been to indicate that, in comparison with these other accounts, CCTI fares comparatively poorly in explanatory merits. For while the accounts offered by Millikan, Sayre, or Marr may not provide an instantiation analysis for intentionality, they do (if successful) provide at least the two kinds of insight already mentioned. If, for example, the things Millikan says are essentially correct, and I take the time to master her theory, I will have gained substantial insight into the nature of intentionality. As far as I can see, the same cannot be said for causal covariation accounts. It may well be that an adequate account of intentionality would have to involve a causal component, but when I entertain this proposition, I do not have a sense that any fundamental secrets about intentionality have thereby been revealed, or that I have achieved a grasp of even one principal aspect of the nature of intentionality. My own sense is that, if it *is* a fact about intentional states that they (characteristically) involve representations standing in a relationship of causal covariation with the intentional objects of those states, this fact stands with respect to intentionality in a relationship analogous to that in which being the shape of a face of an octahedron stands to triangularity, or perhaps the relation that being a featherless biped stands to being human (that is, if we are talking about intentional states *generally,* and not about specific kinds of intentional states, such as perceptual judgments, in which causal connections do seem to be essential). Causal covariation *might* provide some kind of demarcation criterion, but it seems to me that it provides no insight into meaningfulness, and indeed can be invoked only with the prior assumption of meaningfulness. It does not provide an *explanation* of mental-meaning or intentionality. (I have grave doubts about causal covariation even as a demarcation criterion for meaning assignments. These will be a special case of the arguments against "strong naturalization" in the next chapter.)[5]

8.4.6 THE TENSION BETWEEN GENERALITY AND EXPLANATORY FORCE

Now the consideration of accounts such as those offered by Millikan, Sayre, Dretske, and Marr brings up an additional issue that is worthy of consideration. On the basis of the sample presented by these accounts, it would seem that accounts of intentionality become more plausible as explanations of what it is to be about something or to mean something as they become more detailed in their descriptions of how a system is related to its environment. But as they become more detailed, they become correspondingly *more specific and less general.* But this has the consequence that as they become more explanatory, they stray further from being general accounts of intentionality, and look more like accounts of, say, *the realization of intentionality in the visual perceptual apparatus of human beings.* What would seem to be required for a general account of intentionality or mental-semantics, however, would be a characterization that applied equally well to different kinds of cognizers (human, Martian, angelic, silicon-based) and that was indifferent to the intentional modality (perception, judgment, will, etc.). This kind of generality, moreover, is *absolutely essential* if we want to view cognition as computation over meaningful representations of the sort that Fodor postulates, because the MR-semantic properties of the representations must be *independent of* what kind of propositional attitude they are involved in. (Indeed, even if one   is   *not*   committed   to   computationalism,   this   would seem to be implicit in the familiar attitude-content analysis of intentional states.)

To take an illustrative example, consider the account of the intentionality of visual perception in Sayre (1986). Sayre's account is compelling insofar as it makes a case for how some features of perceptual intentionality could be accounted for by viewing certain environmental conditions and features of the perceptual apparatus in information-theoretic terms. While Sayre's account does not supply logically sufficient conditions for getting semantics out of "information in the technical sense," it is a compelling attempt to show how the realization of perceptual intentionality is accomplished. But the details that make Sayre's account compelling also render it too *local* to be a general account of intentionality. For example, Sayre's account is concerned with mechanisms involved in perception, and hence is oriented towards successful cases of perception and towards *transparent* construals of ascriptions of intentionality. Familiar philosophical problem cases such as brains in vats and Cartesian demons lie far afield of Sayre's paradigm cases, and it is not clear how his model could address the problems they present for giving an account of intentionality that accommodates intuitions about opaque construals of intentional verbs. Second, Sayre's account of perceptual intentionality treats the intentionality involved in perception as directed towards an object rather than a proposition or proposition-like psychological state. It is quite possible that perception *differs* from other intentional modalities in this regard, however, and so the extension of Sayre's account to higher cognitive functions may well require a significantly different sort of account from his account of perceptual intentionality. Third, while Sayre's account is sufficiently abstract to avoid being specific to a species, it does seem to be based upon a construal of the abstract nature of the processes that beings such as ourselves undergo in perception. It is conceivable that other beings might reach a similar goal (perceptual intentionality) by a different path, one not describable by Sayre's story.

Millikan's story about intentionality has features that make it arguably even more local: to explain intentionality you have to tell a story about adaptive role and selection history. And selection history is dependent upon lineage. Indeed, according to Millikan, if a being were suddenly to emerge into existence that was identical with one of us in structure, in input-output conditions, and in subjective experiential states, this being would nonetheless have no beliefs or desires, because, according to Millikan, what it is to be a belief or a desire involves being the product of a certain kind of selection history. This would seem to have the

consequence that we would have to tell separate stories about intentionality in species where the relevant functions did not develop in a
common evolutionary history. (Perhaps even if the histories were completely parallel to one another.) This might not mean that we would
have to tell separate stories for humans and chimps (since the relevant
selection process may have taken place before the species diverged),
but we would have to tell separate stories for humans and Martians, or
even humans and Twin-Earthers. (How we would tell such a story
about beings without an evolutionary history—such as God, angels,
and intelligent artifacts—is quite beyond me.)

Now it is not fully clear what moral one ought to draw from this. One
distinct possibility is that what we have here is evidence that, contrary to
commonsense assumptions, there is no one phenomenon called "intentionality," but several different phenomena which require rather different
sorts of accounts. A slightly more modest moral would be that we have
evidence here that the direction of inquiry ought to be to begin with more
local phenomena that sometimes receive the label "intentionality"—for
example, "intentionality" as it appears in visual perception—and proceed
to an attempt at a general theory only when we have a good understanding of specific kinds of intentionality already in hand.[6]

There is, however, a very different possibility, which will be developed more fully in the next chapter: namely, that the problem may lie not
with the notion of intentionality, but with attempts to provide a "naturalization" of it. In particular, it may be that all a naturalistic theory can *hope*
to do with respect to the mental is to spell out how mentalistic properties
are *realized* in particular kinds of physical systems, in which case it
comes as little surprise (*a*) that what is common to different cases is not
captured by the naturalistic theory, or (*b*) that different kinds of accounts
may be required for different kinds of beings having the same intentional
properties, since the same mentalistic properties might need to be realized through different means in different kinds of beings.

8.4.7 COMPOSITIONALITY REVISITED

Even if CCTI were to succeed as an account of the semantics of the
primitive elements in the hypothesized language of thought, CTM would
not thereby be immune to criticism. For in addition to telling a story
about the semantic properties of the primitives, CTM attempts to tell a
compositional story about the semantics of the complex representations.
Unfortunately, the only way we know of telling a story about composi-

tionality is to tell a story about symbols whose semantic properties, in conjunction with syntactically based rules, generate meanings for symbolic expressions. Now on the one hand it is not clear that there is any real force left to speaking of representations as *symbols* if one is no longer endowing them with *symbolic meaning* (i.e., semiotic-meaning). On the other hand, we still have no nonconventional way of generating meanings for complex expressions (i.e., complex machine-counters) out of concatenations of simple expressions, even if we take the meanings of the simple expressions for granted. At best, the account leaves the fact that there are such compositional functions an unexplained brute fact. What we need, in addition, is some rule that makes it the case that, for example, things of the form x-&-y will mean "X and Y." In overt languages, this is accomplished through convention. It is not clear that it *could* be accomplished in any other way. For it is not clear that there is any other pathway that will yield the kind of specificity of interpretation that we are able to get by dint of arbitrary conventions in a natural language. At the very least, even if advocates of CCTI could make their analysis of semantic primitives stick, they would further need to provide a naturalistic account of compositionality before their account could be regarded as viable. The notion of syntax that yields compositionality is conventional to the core, as argued in chapter 6, and no theory of compositionality has been developed for machine-counters.

8.5 A SECOND STRATEGY: THEORETICAL DEFINITION

If this stipulative definition of the semantic vocabulary will not save CTM's account of intentionality, it behooves us to examine a second possible reinterpretation as well: namely, that the semantic vocabulary employed in CTM is to be understood as a theoretical vocabulary whose interpretation is fixed by the work it does in the theories in which it is employed. The very brief answer, I shall argue, is *no:* if the semantic vocabulary of CTM is defined theoretically, then we do not have an *explanation* of intentionality (and hence no vindication of intentional psychology) until the underlying nature of these properties that are initially specified theoretically is spelled out. Until then, the so-called "explanation" of intentionality by appeal to "semantic properties of representations" really amounts to an appeal to dormative virtues.

Now what do we mean by "theoretical definition"? Sometimes terms employed in scientific theories mean precisely what they meant all along in ordinary language. In other cases, however, scientific theories appropriate ordinary-language terms and use them in new ways. Terms like 'matter' and 'particle' probably at one time had as part of their meaning all of the notions bound up in the Cartesian notion of "extension," such as size, shape, and definite location. Modern physics, however, countenances the use of these terms even for objects that lack one or more of these properties. Whatever the ordinary connotations of 'work', it has a very specific technical definition in physics. And naturally the property of "charm" attributed to quarks has nothing to do with good breeding and etiquette. Of course, science also countenances the introduction of new terms as a part of theories as well. And sometimes these also have their semantic values fixed by the theories in which they play a part. The word 'gene' in biology, for example, was at one time defined only by the theory in which it played a role: a gene was, by definition, the kind of thing, whatever it would turn out to be, that accounted for phenotypes of living things. When Watson and Crick discovered that the locus of this genetic encoding was the DNA molecule, the term perhaps underwent a change in meaning; but before that time it was a *purely theoretical term* —that is, a term whose meaning was fixed solely by the role it played in a theory.

The suggestion I wish to explore is that when CTM speaks of "semantic properties of representations," the words 'semantic properties' express properties that are theoretically defined in much the same fashion. These properties, which we have called "MR-semantic properties," might thus be defined as follows:

> *MR-semantic properties = df* Those properties of mental representations, whatever they turn out to be, that explain the mental-semantic properties of mental states.

The actual nature of these properties is thus left unspecified at the outset, though presumably it may be determined in the course of further research. This reconstruction of the semantic vocabulary employed in CTM provides a new way of interpreting that theory that avoids the problems involving conventions and intentions.

8.5.1 DOES THEORETICAL DEFINITION EXPLAIN INTENTIONALITY?

Let us then look at the claim that the kind of theoretical definition of semantic terms employed in BCTM provides us with an account of the intentionality of mental states. Earlier, we proposed a schematic version of CTM's proposed account of intentionality:

Schematic Account

Mental state M has mental-semantic property P because

(1) M involves a relationship to a mental representation MR, and

(2) MR has MR-semantic property X.

Having specified that MR-semantic properties are defined in theoretical terms, we can substitute our theoretical definition into our schematic account. But there are two different ways of substituting into our definition, which we may think of as the *de dicto* and *de re* substitutions. The *de dicto* substitution simply replaces the expression 'MR-semantic property X' with its theoretical definition as follows:

De Dicto Interpretation

Mental state M has mental-semantic property P because

(1) M involves a relationship to a mental representation MR, and

(2) MR has that property of MR, whatever it is, that accounts for mental-semantic property P.

The *de dicto* interpretation yields a pseudo-explanation of a well-known type. On this reading, MR-semantic properties fail to explain for precisely the same reason that we cannot explain the soporific powers of a medicine by appeal to its "dormative virtues." If saying "mental states inherit their semantic properties from mental representations" amounts to nothing more than saying "mental states get their semantic properties from something that has the property of giving them semantic properties," we do not have a legitimate explanation of semantics or intentionality.

However, it is also possible to substitute our theoretical definition into the schematic account in another way that does not share this problem: namely, by substituting a *de re* reading of the theoretical definition as follows:

De Re Interpretation

Mental state M has mental-semantic property P because

(1) M involves a relationship to a mental representation MR,

(2) MR has some property X,

(3) the fact that *MR* has *X* explains the fact that *M* has *P*, and

(4) *X* is called an "MR-semantic property" because

    (*a*) it is a property of a mental representation, and

    (*b*) it is the property that explains the fact that *M* has *P*.

On this interpretation, there are no dormative virtues lurking in the wings. Unfortunately, as the account stands, *there is no explanation of intentionality either* until we know (1) what the all-important property *X* might be, and (2) how we can derive the intentionality of mental states from the fact that cognitive counters have this wonderful property (the way we can, say, derive thermodynamic laws from statistical mechanics). BCTM does not supply us with this information; therefore BCTM does not supply an account of intentionality. BCTM no more explains intentionality than nineteenth-century genetics explained phenotype. With regard to intentionality, on a best-case scenario (that is, on the assumption that BCTM is on the right track with respect to the functional shape of the mind and the ultimate possibility of explaining intentionality by appeal to the properties of localized states), BCTM is in the position genetics was in before Watson and Crick: it is a functional-descriptive theory in search of an underlying explanation. (Of course, in the worst-case scenario, mental representations and their MR-semantic properties go the way of heavenly spheres and Piltdown man.)

    In short, it seems to me that BCTM makes no progress at all on the semantic front. It does not so much provide an explanation of intentionality as it makes evident the absence of such an explanation. This fact has generally been obscured by confusions that result from assuming that the semantic vocabulary can be applied univocally to mental states, symbols, and representations. If we say, "Mental states inherit their meanings from mental representations," it *looks* as though there is progress on the semantic front, because we have reduced the problem of mental meaning to a problem about the meanings of symbols in the brain. Meaning, at any rate, looks like the right sort of thing to be a *potential* explainer of meaning, because we do not have to explain how meaning came upon the scene in the first place in order to explain mental-semantics. However, if it turns out that the semantic vocabulary applied to representations is a truly *theoretical* vocabulary, the appearance of progress begins to look like smoke and mirrors. As we noted earlier   in   the   chapter,   it   is   one   thing   to   claim

(1) Mental state *M* has property *P* because *M* involves *MR*, and *MR* has *P*.

But it is quite another to claim

(2) Mental state *M* has property *P* because *M* involves *MR*, and *MR* has *X*, and $X \neq P$.

Claim (1) proceeds on the assumption that property *P* is in the picture to begin with, and just has to explain how *M* gets it, while claim (2) has to do something more: namely, to explain how *P* (in this case, mental intentionality) comes into the picture *at all*. CTM simply does not do this, and to describe CTM as "explaining intentionality" is simply a gross distortion of what it actually accomplishes.

8.6 MR-SEMANTICS AND THE VINDICATION OF INTENTIONAL PSYCHOLOGY

The reader will recall that the explanation of intentionality was the first of two philosophical treasures that CTM was supposed to have unearthed, the second being a vindication of intentional psychology. Let us now return to the problem of vindication. Recall how the attempted vindication was inspired by the computer model. In a computer, the semiotic-semantic properties of the symbols are coordinated with the causal role symbol tokenings can play in the system. It is a useful contrivance to speak of the relationship between symbols and causality as being mediated by syntax, but speaking of the "syntactic properties" of the symbols—indeed, talking about computer states as symbols—is largely a matter of convenience. The symbolic and syntactic character of the symbols is conventional in origin and etiologically inert. What matters is that the semiotic interpretations of symbols are coordinated with the functional-causal role they can play. Now the hope CTM presented was that the mind was a computer, and hence it might be that the mental-semantic properties of mental states could be coordinated with the causal roles they play in inference, thus showing that (contrary to appearances) intentional explanation is grounded in lawlike causal regularities.

Notice that purging CTM of dependence upon symbols and syntax has thus far done nothing to weaken the case for the vindication of intentional psychology. For in point of fact, the notions of symbol and syntax

Mental Representations and Intentionality

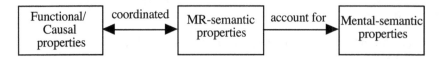

Symbols in a Computer

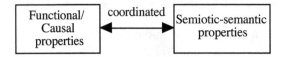

Figure 11

played less of a role in the case of computers than was commonly believed. But notice also that there is an important difference between coordinating the semiotic-semantic properties of symbols in computers with their functional-causal roles, and coordinating the mental-semantic properties of mental states with their functional-causal roles: the former is done directly, the latter is done (according to CTM) by an intermediate step: namely, coordinating the MR-semantic properties of representations with their causal roles. The difference is represented graphically in figure 11.

This illustration reveals several respects in which the computer paradigm itself falls short of providing a vindication of intentional psychology. These are not reasons that one cannot vindicate intentional psychology in the manner suggested, but they do show what more one needs if such a vindication is to proceed as planned.

(1) The computer paradigm shows that semiotic-semantic properties can be coordinated with functional-causal properties. What one needs for CTM, however, is a demonstration that some other kinds of "semantic" properties (immediately, the MR-semantic properties of mental representations) can be coordinated with functional-causal properties. The computer paradigm by no means assures that this can be done. (After all, there might be something *special* about semiotic-semantics.)

(2) The computer paradigm only shows how two sets of properties of one sort of object can be coordinated. CTM needs something more: it needs to show that, by coordinating the MR-semantic properties of *representations* with their causal roles, it can thereby coordinate the mental-

semantic properties of *mental states* with *their* causal roles as well. This would seem to place some additional constraints upon the "vindication" beyond what is involved in saying the mind is a computer.

In what follows, I should like to build a case that each of these problems is potentially very serious. First, there is good reason to hesitate in concluding that other types of "semantic" properties can be coordinated with causal role in the fashion that semiotic-semantic properties are so coordinated in computers. Second, in order for BCTM to license a vindication of intentional psychology, it would have to be able to show that the coordination of MR-semantic properties with causal role would thereby secure the coordination of mental-semantic properties of mental states with causal role as well; *and in order to do this, it would have to supply an instantiation analysis* of mental-semantics in terms of MR-semantics—a realization account is not enough for vindication.

8.6.1 THE SPECIAL CASE OF SEMIOTIC-SEMANTIC PROPERTIES

The computer paradigm shows that a symbol's semiotic-semantic properties can be correlated with the causal role the symbol can play, so long as all semiotic-semantic distinctions between symbols are reflected in syntactic distinctions. What links the semiotic-semantic properties to the marker type, however, are the conventions and intentions of symbol users. So if an adding circuit has the binary pattern 0001 tokened in one register and 0011 in a second and produces a tokening of 0100 in a third as a result, the tokening of the third is accounted for by the functional architecture of the machine and the specific patterns present in the registers, but the overall process is said to be an instance of adding one and three and obtaining a sum of four only because of the interpretive conventions that are being applied.

Now what, in this paradigm, accounts for the "coordination" of syntax with semantics? On the one hand, the functional properties of the system provide necessary conditions for the reflection of semantic distinctions in the syntax. On the other hand, it is the conventions of symbol users that actually establish (*a*) the marker types employed, (*b*) the syntactic types by virtue of which markers can be counters, and (*c*) the semantic interpretation schemes by virtue of which the markers may be said to have semantic properties. The "coordination" of syntax and semantics depends upon the relationship between semantic and syntactic conventions, and so is highly convention-dependent.

I should like to suggest that this convention-dependence is precisely what gives the "coordination" of syntax with semiotic-semantics in computers one of its more useful features, and that we should not *expect* syntax—or, more exactly, functional role and syntactic interpretability-in-principle—to be "coordinated" with non-semiotic-semantic properties in the same sort of way. For one thing that interpretive conventions (or intentions) can do is pick out a *unique* interpretation for each marker that is to serve as a counter. This is significant because (notoriously) any symbol system is subject to more than one consistent interpretation. (Notably, there will always be an interpretation entirely within the domain of number theory.) It is the conventions and intentions of symbol users that account for the fact that a token in a given symbol game means (for example) *dog* and not *the set of prime numbers*. And it is these conventions and intentions that determine *which* semantic properties are coordinated with *which* syntactic properties.

Now there is really something at once unique and mundane about the coordination between semiotic-semantic and syntactic properties of symbols. If someone asks why a given counter type is associated with (i.e., is interpretable as bearing) a particular interpretation, the answer is not at all mysterious: it is associated with that interpretation because there is a convention to that effect among a particular group of symbol users. And if someone asks why it is *not* associated with (i.e., is interpretable as bearing) another interpretation, the answer is that there is no convention linking it to *that* interpretation. It may indeed be surprising that symbol games as *large* as geometry and significant portions of arithmetic can be formalized, and it may be surprising that formalizable systems can be automated in the form of a digital computer, but the basis of the connection between counter types and semiotic-semantic interpretation is not at all arcane.

What would seem to be *unique* about this kind of association between semantic values and marker types is that the relationship between semantic value and marker type is determined by *stipulation* —and it is this that allows for the association of marker types with unique interpretations. Now it might be the case that there are other factors that could determine how syntactic features of *mental representations* are to be connected to *particular* (nonsemiotic) semantic properties and not to others. But it is not at all clear that we ought to *expect* it to be the case. For one might well think that it is *only* the *stipulative* character of semiotic conventions and meaning-bestowing acts that can provide the kind of *unique* correlation of semantic value with counter type that one finds in symbolic representations in a computer. I know of no convincing argument that would absolutely rule out the possibility that some other factor could provide such a unique correlation, but I must say that it seems a bit mysterious just what other kind of factors could provide a unique association between the syntactic properties of any mental representations there might be and their MR-semantic properties. It *must not* be a matter of stipulation, because that would lead to the kind of semantic regress discussed in the previous chapter. But without stipulation, it is unclear how one could get uniqueness of interpretation. The prospects of applying the computer paradigm analogously are thus rendered doubtful, though not precluded entirely.

8.6.2 INSTANTIATION, REALIZATION, VINDICATION

Now even if it is possible to coordinate MR-semantic properties with causal role, this is not enough for the vindication of intentional psychology. For that one also needs it to be the case that coordinating the MR-semantic properties of representations with their causal roles secures the further coordination of the mental-semantic properties of mental states with their causal roles. Presented in the way the case was originally presented, when we assumed that the "semantic" properties of mental states were the very same properties as those of their representations, securing this further coordination seemed almost trivial. The argument for it is expressed by this argument presented earlier in the chapter:

Argument V2

(1) Mental states are relations to mental representations.

(2) Mental representations have syntactic and semantic properties.

(3) The syntactic properties of mental representations determine their causal powers.

(4) All semantic distinctions between representations are preserved syntactically.

(5) There is a strict correspondence between a *representation's* semantic properties and its causal powers.

(6) A *mental state M* has semantic property *P* if and only if it involves a *representation MR* that has semantic property *P*.

∴(7) There is a strict correspondence between a *mental state's* semantic properties and its causal powers.

But of course once one has distinguished different kinds of semantic properties, the argument has to be adapted as follows:

Argument V3

(1) Mental states are relations to mental representations.

(2) Mental representations have syntactic and MR-semantic properties.

(3) The syntactic properties of mental representations determine their causal powers.

(4) All MR-semantic distinctions between representations are preserved syntactically.

(5*) There is a strict correspondence between a representation's MR-semantic properties and its causal powers.

(6*) A mental state M has mental-semantic property P if and only if it involves a representation MR that has MR-semantic property X.

∴ (7*) There is a strict correspondence between a mental state's mental-semantic properties and its causal powers.

The issue here turns upon (6*), the claim that mental-semantic properties of mental states can be coordinated with MR-semantic properties of representations, and the inference to (7*), the claim that mental-semantic properties of mental states would thereby be coordinated with causal powers. In order for (6*) to be true, the mental-semantic properties of mental states would have to be at least correlated with the MR-semantic properties of representations. But in order for this argument to provide a vindication of intentional psychology, something more is required: one must be able to *show* that the MR-semantic properties of representations determine the mental-semantic properties of mental states. For in order to vindicate something, one must *show* that it could be the case. To *vindicate* intentional psychology, one would have to *show* that the mental-semantic properties of mental states can be coordinated with causal roles, and not merely show what benefits *would* be derived if they *were* so coordinated. Given that we can show that MR-semantic properties of representations can be coordinated with causal roles, we would still have to show that, as a consequence, mental-semantic properties of mental states would be coordinated with causal role as well.

Now what sort of account of mental-semantic properties would be needed to achieve this end? What is required is an instantiation analysis of mental-semantics in terms of MR-semantics—a realization account is not enough. For recall a key difference between instantiation and realization: since an instantiation account provides conditions from which one can infer the instantiated property, it provides a vindication of existence claims for that property, given that the instantiating properties are satisfied. But with a realization account, no such benefit accrues: since the realizing properties are not a sufficient condition for the realized property, they do not provide proof for someone who doubts that such a property can be realized. Now we are seeking an account that vindicates the claim that the mental-semantic properties of mental states can be coordinated with their causal powers. An account of how mental-semantic properties are *instantiated through* the MR-semantic properties of representations could provide such a proof, because one would be able to *infer* the mental-semantic properties of the mental states *from* the MR-semantic properties of the representations. A realization account, on the other hand, merely *presupposes* that there is some special relationship between the properties picked out in the intentional idiom and those picked out by the functional-causal account, without either specifying the nature of the relationship or showing why it obtains. Such a presupposition may have great advantages if you are doing empirical psychology, because you can do your research without waiting for definitive results of debates about dualism, reduction, supervenience, or psychophysical causation. But for this version of the vindication of intentional psychology to work, we *must not assume* such a special connection, because *the possibility of such a connection is precisely what has been called into doubt*. If someone doubts that the semantic and intentional properties of mental states can be coordinated with naturalistic properties, and one gives a realization account for the intentional and semantic properties of mental states that just *assumes* that they are specially connected to some naturalistic properties, one has not assuaged the doubt so much as begged the question.

8.7 SUMMARY

The general conclusion of these past two chapters is that CTM does not, in fact, provide an account of intentionality. It provides the *illusion* of such an account by saying that the semantic properties of mental states are inherited from those of mental representations. But on closer inspection, we have not found any properties of "mental representations" (i.e., our hypothetical cognitive counters) that could serve to explain mental-semantic properties of mental states. Semiotic-semantic properties, as we saw in the last chapter, fail on a number of grounds, including the fact that they render the explanation circular and regressive.

One focus of this chapter was upon the possibility that the kind of causal covariation account of semantics championed by Fodor might actually be able to serve as a stipulative definition of semantic terms as applied to representations. I have serious doubts that this was Fodor's intention. But if one were to make such a move, it would seriously undercut the persuasive force of Fodor's apologia for CTM, since that involved explicit and implicit arguments that turn out to be blatantly fallacious if notions such as meaning and intentionality are defined in causal terms for mental representations. Moreover, causal covariation stories do not go very far towards providing an account of what it is for a mental state to be mental-meaningful or mental-intentional—they don't provide an *explanation*. First, the causal covariation story just seems like the wrong kind of "account": it appears to give a demarcation criterion that does not explain, and it seems to distinguish states that have different meanings instead of distinguishing the meaningful from the meaningless. That is, it seems to *assume* that it is dealing with meaningful entities, and then asks, "How can we distinguish the ones that mean X from the ones that mean Y?" In addition, I have tried to make a case that, if the notion of causal covariation is too bland a notion to provide an explanation of intentionality or meaningfulness, this blandness seems the price one must pay for generality: naturalistic accounts become more explanatory as they become more detailed, but in the process they lose the generality one would want from an "account of intentionality." Finally, I have argued that even if CCTI were to succeed as an account of semantics for the primitive representations, it would need to be supplemented by a naturalistic account of compositionality as well, and it is hard even to imagine how such an account might proceed. The upshot of this is that causal covariation does not provide us with a notion of representational meaning that can explain mental-meaning or vindicate intentional psychology.

The theoretical definition of the semantic vocabulary for representations fares no better. On one construal (the *de dicto* construal), it provides a fallacious pseudo-explanation that appeals to dormative virtues. On another (the *de re* construal) it provides no explanation at all. This, I think, is as far as CTM can be made to stretch: it is a theory of the form of mental processes that stands glaringly in need of an account of semantics to supplement it. We saw as well that we cannot "vindicate" intentional psychology in the way envisioned by CTM's advocates unless we have such an account—and indeed a naturalistic account—of semantics and intentionality in hand. In the next chapter, we shall explore the prospects for such a "naturalistic theory of content." In the final section of the book, we shall explore an alternative way of looking at the computer paradigm in psychology that renders unnecessary both the naturalization of the mental and its vindication.

CHAPTER NINE

Prospects for a Naturalistic
Theory of Content

In the previous chapters it has been argued that CTM does not itself provide the explanation of intentionality that is claimed for it, and as a result it cannot produce the kind of "vindication" of intentional psychology it set out to perform. At best, a bowdlerized version of CTM might provide a way of describing the form of mental processes, and this in turn might form a part of a larger theory that would supply an independent theory of content. In our project of assessing CTM, it would not be completely unjust to leave the matter where it now stands. It is strictly speaking false that CTM explains intentionality, and this belies much that is commonly said about it. With the imposture unmasked, we could go straight to the credits and the final curtain without being truly unjust. However, one does not have to look very hard to see that, while BCTM does not itself supply an account of intentionality, it could be a part of a larger theory that does so if it were only to be supplemented by what is commonly called "a theory of content for mental representations." Indeed, in at least some places (e.g., the introduction to *RePresentations*) Fodor himself seems to view the situation in this way. And however you slice the pie, the *overall* explanatory agenda for the computational-representational project is pretty much the same. Perhaps the more common (if mistaken) interpretation has been that the semantics of mental states have been explained by appeal to meaningful symbols, but now the meaningfulness of the symbols needs explanation, and that is what calls for a naturalistic theory of content (see fig. 12). If you take the semantic vocabulary for mental representations to be theoretical in char-

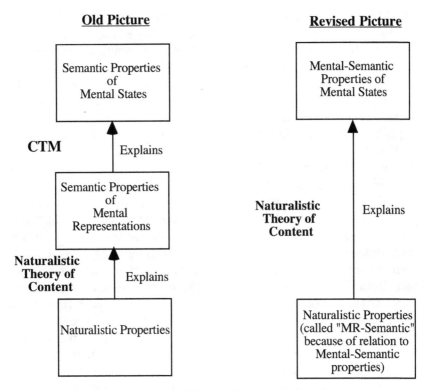

Figure 12

acter, the middle level simply falls out, and you need an account that directly ties the meaningfulness of mental states to some unknown properties of functionally delimited proper parts of the mind or brain that are sufficient to explain mental-semantics. Either way, you ultimately need a pathway from nonsemantic and nonintentional properties of your "representations" or cognitive counters to mental-semantic properties of intentional states. All that is lost in moving from Fodor's narrative to BCTM is the (paralogistic) illusion of having made some progress on the semantic front along the way. So the idea that BCTM is really a theory of the form of mental states and processes that is still in search of an explanation of semantics and intentionality might not be all that repugnant to many in the computationalist camp. The burgeoning industry of naturalizing content, after all, is keeping plenty of philosophers employed, and holds out the hope of someone playing Watson and Crick to Fodor's, Putnam's, or Pylyshyn's Mendel.

It thus behooves us to give at least a brief examination of the prospects of completing this project by explaining the mental-semantic properties of mental states in nonintentional terms in a fashion compatible with

BCTM. This is a big undertaking, and it is very different in character from the rest of this book. The preceding sections have been concerned with assessing the limitations of a particular theory. A complete assessment of the prospects for a naturalistic account of intentionality, by contrast, would require us to examine not only all those theories that have actually been proposed (variations on which seem to multiply by the hour) but also all possible theories that have not been thought of as well. Quite a daunting task, really, and definitely beyond the intentions of this book.

What I propose to do in this chapter is much more modest. I shall endeavor to do four things: First, I shall distinguish weaker and stronger ways of "giving an account," which I will refer to respectively as "weak naturalization" and "strong naturalization" of the mental. Second, I shall point to some different classes of mental states to which the word 'intentionality' is applied and make a case that what needs to be explained in these different classes may indeed be very different (e.g., broad versus narrow content, phenomenology versus functional relations and behavior). Third, I shall try to make a case that at least some kinds of "intentional states" (the ones with a phenomenology) have properties that it seems unlikely that we shall be able to naturalize. And finally, I shall make a case that, with the remainder of "intentional states," it seems dubious that the explanation of meaningfulness (as opposed to the demarcation of meaning assignments) will focus on localized cognitive counters, as required by BCTM, but rather will require an examination, at the very least, of an entire thinker or organism, and very likely its situation in its social and ecological environment as well.

9.1 STRONG AND WEAK NATURALIZATION

We are thus brought to the question of evaluating the prospects for a naturalistic theory of content that could be grafted onto BCTM. In recent years it seems to have become almost a kind of religious commitment in some corners of the philosophy of mind that one believe that there can be a naturalization of content. Upon closer inspection, however, it becomes clear that naturalism is not only loosely argued for, but loosely defined as well. For even among people espousing a commitment to "naturalism" or "naturalization" you will find enormous disagreement about what would count as a naturalization of the mind, including differences as to what is constitutive of the "natural" (is it the domain of physical objects? of causal interactions? of lawful causation? the non-normative and nonteleological?)   and   differences   as   to   what   kind   of "account" or "theory" is at issue. Is it enough to count as naturalization if you *specify* brain states (or abstract states realized in brains) with which content varies without specifying any relationship stronger than logically contingent covariation? Or does a naturalization of psychology require something more: say, a *metaphysical* relationship such as reduction or supervenience, or an *explanatory* relationship such as conceptual adequacy? Just getting a grip on the different possible moves here is a daunting task, and would probably require a book entirely devoted to that topic. What I wish to do here is to make a kind of first Dedekind cut that will separate two very different kinds of projects.

First, consider an ambitious form of naturalism: a naturalism that seeks to bring the mind wholly within the realm of nature by showing how it is possible to subsume our special discourses about thought within the framework of the natural sciences. As a model for the kind of strong explanatory relationship such a project seeks, we might take such strong intertheoretic relationships as the famous proofs that thermodynamics can be derived from the mechanics of particle collisions, or the ability of the atomic theory to explain features of the periodic table and combinatorial laws of nineteenth-century chemistry. Statistical mechanics provides a kind of explanation of thermodynamics that has important properties both metaphysically and as explanation. Metaphysically, the mechanical laws are logically sufficient for the thermodynamic laws: that is, basic mechanical laws, in combination with necessary truths of logic and mathematics, are enough to entail the thermodynamic equations. Moreover, this entailment is epistemically transparent: a person with an adequate understanding of mathematics and mechanics could derive the thermodynamic equations even if she lacked a prior acquaintance with thermodynamics as a branch of physics. I call this kind of explanation "conceptually adequate explanation." *A* is a conceptually adequate explanation of *B* just in case the conceptual content of *A* is enough to derive the conceptual content of *B* without the addition of contingent bridge laws.[1]

I shall refer to the project of explaining the mind in a fashion that is in similar fashion metaphysically sufficient and conceptually adequate as *strong naturalization*. A strong naturalization of an intentional property *I* would explain *I* by appeal to some "naturalistic" properties *N*, where the term 'naturalistic' implies at least (*a*) that the properties that comprise *N* are themselves nonintentional, and (*b*) that they do not presuppose intentional properties. (For example, conventions are not themselves intentional, but arguably presuppose intentional states.) Obviously, important candidates for the properties in N are properties found in the discourses of sciences such as neurology and biology, but I have deliberately left the description of the "natural" open to possibilities that properties of natural objects that are not relevant to the other sciences might prove important for psychology.[2]

In contrast with strong naturalization, consider a much weaker kind of project: that of specifying, so far as possible, the mechanisms in the nervous system through which mental states are "realized"—where "realization" implies some special connection whose metaphysical nature may be left vague. (Such a project need not confine itself to relations between minds and single organisms—it could also, of course, specify any crucial relationships between the thinker and her social or ecological environment with similar metaphysical neutrality.) Such a project need not produce intertheoretic relationships that are necessary or sufficient, and the naturalistic properties specified need not *explain* the mental properties to which they are linked. This kind of account suffers no lack of precedent. The relationships between variables within a given theory are generally of this sort (though they are sometimes explained by an additional theory that provides a microexplanation), as are bridge laws and statements such as that of the wave-particle duality of matter. The psychophysical regularities in such a theory would serve as a kind of contingent bridge law between an intentional psychology and a nonintentional neuroscience.

We might call this kind of project in psychology *weak naturalization* in contrast to the "strong naturalization" described above. However, it is with some misgivings that I apply the name "naturalization" to it at all, as (*a*) most people calling themselves "naturalizers" seem to have strong naturalization in mind, and (*b*) many people who would normally be considered something other than naturalists could subscribe to this "weak naturalization" project as well. Indeed, it is a project in which Descartes was an important pioneer, to which Spinoza explicitly subscribed, and which even Berkeley might have been able to endorse in connection with empirical research. As a result, I am sometimes more inclined to refer to it as the "Neutral Project."

BCTM can be located, with minor variations, within either kind of project: strongly or weakly naturalistic. However, a *strong* naturalization of the mental is required if CTM is to accomplish either of the two philosophical goals that it has set out for itself. To account for the intentionality of mental states, it is not enough to specify some contingent correlations between mental-state type and some physical or abstract property. For this would not explain why meaningfulness appears on the scene at all; and that, after all, is the primary puzzle for the naturalist. Contingent correlations are simply not explanatory. And to *vindicate* intentional psychology, it is necessary to *show* that mental states can be understood in a way that meets the desired criteria. And this, in turn, requires explanation that is epistemically transparent.

Machine computation shows, for example, that for formalizable domains, the semiotic-semantic properties of the symbols can be linked to the physical-causal properties of the machine. The physical-causal properties of the machine, indeed, *entail* its description (or describability) in terms of a machine table (though not uniquely). Yet the physical-causal properties of the machine do not explain the semiotic-semantic properties, because these depend upon conventions as well. I think that this much is likely to prove to be much the same in the case of mental states. Where the two situations diverge (and this is what affects vindication) is the fact that, in the case of symbols in computers, we can make it transparent that the objects of the semiotic description are the *very same objects* as the objects of physical-causal description (the series of bistable circuits and whatnot), whereas identity ascriptions between mental and physical states are at best mere guesswork. The reason you can see this in the case of symbols in computers and not in the case of mental states turns upon the fact that there is something about the notion of a *symbol* that *entails* that a symbol have criteria involving a physical pattern. A token signifier is necessarily a token marker, and a token marker is necessarily a token physical object. But there is no similar connection with material objecthood built into the notion of a mental state. The connection between symbolhood and physical objecthood is conceptually necessary. That between mental states and physical objecthood is contingent at best. And to *show* the compatibility of mentalism with materialism, you need more than guesswork; you need to make the identity transparent. Otherwise there is no *proof* of compatibility, hence no vindication. This only makes a difference to those who are really sold on the premise that intentional psychology is in need of vindication, but it should matter quite a lot to them.

9.2 WHAT IS "THE MENTAL"?

If assessing the possibility of "naturalizing the mental" requires some discussion of the notion of naturalization, it is equally in need of some discussion of its intended domain, "the mental" and even "the intentional."

Thus far, with the exception of a few hedges in chapter 1, we have proceeded as though there were a clear and shared understanding of the population of the intentional bestiary and of the "ordinary" or "pretheoretic" notion of intentionality. However, I have become convinced in recent years that this is not so. There are really several different kinds of things that are called "mental" and even "intentional states." Most important, I think, is the distinction between conscious episodes like perceptual experiences, conscious judgments, and episodes of recollection on the one hand, and dispositional states like beliefs and desires on the other. Their salient properties are very different from one another, and hence require very different accounts. Moreover, different groups of philosophers take different classes of states as their paradigm examples and, as a result, operate under very different assumptions about what a "theory of mind" or an "account of intentionality" would have to explain.

9.2.1 FOUR KINDS OF "MENTAL STATE"

I have argued elsewhere (Horst 1995) that we may usefully distinguish four kinds of entities that go under the name of "mental."

(1) *Conscious Occurrent Episodes (judgments, perceptions)*. Until fairly recently, people interested in the mental in general and intentionality in particular tended to concentrate on episodes of conscious thought in which some object or state of affairs was, as it were, "before the mind's eye." It seems quite clear that this is the sort of thing that the pioneers of modern work in intentionality like Brentano and Husserl had in mind, and it is surely true as well of work on the mind by most of the Early Modern philosophers such as Descartes, the British empiricists and Kant, as well as living philosophers such as Geach (1957), Nagel (1986), Goldman (1992, 1993a, 1993b), and Searle (1983, 1992). Such states would include things like perceptual gestalts, in which an object or scene is presented visually, occurrent judgments ("By gum! That's a dingo!"), conscious wishes ("Oh, that Rhett would come back to Tara!"), recollections, imagination, free fancy, and so on. Such things are events, they are conscious or at least consciously accessible, they have a phenomenology, and there is a quite palpable sense in which it makes sense to say they are "directed" towards something and have an "intentional object" that need not be a real object. In these cases the mind in some sense not only *intends* the object, but *attends* to it as well. Such episodes are, to a certain extent (and not infallibly), susceptible to introspection, and are certainly

not "purely theoretical" in the sense that protons are theoretical or that Pluto was theoretical before its existence was confirmed by telescopy. (That is, we have direct, quasi-observational evidence for their existence as well as retroductive evidence.)

(2) *Dispositional States (beliefs, desires)*. Most recent writers in cognitive science have concentrated, by contrast, on things like beliefs and desires, generally construed (with varying degrees of strictness) in dispositional terms. Dispositions are by definition unobservable. And where 'belief' means something other than "conscious judgment" (which it is sometimes used to mean), it does seem to indicate something that is truly theoretical and indeed cannot be confirmed through direct observation. Perhaps some dispositions have a phenomenology—say, believing that there is a loving God fosters a sense of inner peace and believing that the Mob has put out a contract on you produces a sickening anxiety—but the connection between the dispositional belief or desire and its phenomenology is far less direct (and arguably less essential) than that between occurrent states and their phenomenology. The "aboutness" of a perceptual gestalt is very closely related to the fact that I am appeared to in a fashion that involves an image of a dog, presented from a particular perspective (say, from behind), and under a particular interpretation (i.e., "That's *Marco's* dog, and she's chewing on my shoe!"). And all of this has a phenomenology. For the most part, beliefs only acquire a distinctive phenomenology when they eventuate in conscious episodes.

(3) *The Freudian Unconscious*. Freud speaks of "unconscious" mental states. These seem to be built on the model of conscious states, and are taken to be of the same kind, with the sole proviso that they are repressed. They can (it is said) be brought to conscious awareness in therapy. I do not intend to pursue Freudian theory here, but merely to point out that such events start out as theoretical entities, and particular ones may cease to be purely theoretical when made conscious. They may have a vague and extrinsic phenomenology that manifests itself in some of the complaints that bring the patient to the therapist's couch, but these are not particular to the content of the state in the way that, say, the phenomenology of perception is connected to how I am thinking of the object of perception (for a similar view, see Searle 1992).

(4) *Infraconscious States*. Finally, cognitive scientists often speak of things lying below the level of the consciously accessible in mentalistic terms. We hear talk of cognitive subsystems, for example, cashed out in terms of "beliefs" and "desires" of the subsystems. Such states are surely nonconscious, have a phenomenology only incidentally, and indeed may

bear no more than analogical relations to other things called "beliefs" and "desires," as argued by Searle (1992). Such states are also clearly theoretical in the strong sense that protons are theoretical. (In other words, our only warrant for believing in them is that doing so gives us a certain amount of explanatory payoff.)

Now clearly, when you are asking for an account of "mental states" it will make a great deal of difference what kinds of "mental states" you have in mind. In fact, there are plenty of people who are committed to one or more of these categories while remaining skeptical about others. Many people think Freudian psychology is bunk, for example, but believe in conscious states or beliefs; and outside of cognitive science it is common to find people who agree with Searle and myself that many of the attributions of "beliefs" to infraconscious states and processes are true only if interpreted metaphorically. Indeed, some of us think that nothing could be more clearly real than conscious states but harbor deep-seated misgivings about dispositional beliefs and desires. Conversely, some people seem not to understand talk of phenomenology and subjectivity at all (perhaps in the way some people do not experience imagery), and others think that the conscious experience of mental states is merely a gaudy epiphenomenon that is irrelevant to the "real" (i.e., causal) nature of beliefs and desires.

What you choose as your paradigm examples will have a significant impact on what you consider "essential" to the "mental" and hence what stands in need of explanation. Perception, imagination, recollection, judgment, conscious yearnings, and the like all involve a kind of directedness of the sort reported by Brentano, which in turn involves at least the possibility of consciousness, a phenomenological "what-it's-like," a perspectival character of the object-as-presented (we see and think about objects under only some of their aspects), and a kind of subjectivity (this experience is essentially *my* experience). All of this seems to be bound up in what writers like Brentano, Husserl, Searle, and Nagel *mean* when they talk about intentionality in particular and the mental in general. If this is how you are using those words, on the one hand, it is only natural to assume that an "account" of "the mental" or of "intentionality" should account for all these features. If your paradigm example of the mental is a dispositional belief, on the other hand, you are unlikely to include such features in your list of things needing explanation, and rightly so.

I happen to think that these distinctions explain a lot of the contemporary impasses in the philosophy of mind. People who think mental states are "theoretical" tend to be thinking of dispositional beliefs, the unconscious, or the infraconscious. People who are thinking of perception and judgment regard characterizations of the mental as "theoretical" as outrageous. People in the occurrent-state camp also tend to regard phenomenology, subjectivity, and consciousness as crucial to the mental in general and to intentionality in particular, while those concerned with beliefs and desires often do not. It seems to me (see Horst 1995) that there is room for a dissolution of these impasses that saves face for all: namely, that things like judgments, imagination, and perception are not theoretical entities, and do essentially involve phenomenology, subjectivity, and consciousness, while dispositional states and infraconscious states are theoretical in character and do not involve these features, except incidentally.[3]

9.2.2 INTENTIONALITY AND DIRECTEDNESS

I think that there is likewise some variety in the literature in how the words 'intentionality' and 'intentional state' are used. When the word was reintroduced into philosophical parlance by Brentano (1874), it seems clear that he meant 'intentionality' to denote a feature of certain kinds of whole mental states (and not their proper parts). Indeed, Brentano speaks of intentionality as being the distinctive feature of his "mental" as opposed to "physical" phenomena, but it is clear on closer inspection that his "physical" phenomena are not physical objects but qualia! This may seem mysterious at first glance, but the mystery is resolved when one recognizes that Brentano is starting from the empiri*cist* starting point of examining the contents of the mind from the first-person perspective (see McAlister 1974, 1976). His "phenomena" are literally "things that appear"—some of which (those he unfortunately calls "physical") involve only sensation, others of which (those he calls "mental") involve the presentation of some object as an object. Brentano's empiricist foundations, as well as his examples, make it clear that he is dealing with mental episodes in which one is conscious of some "intentional object" as it is presented, as it were, "before the mind's eye." The reason for speaking of the "directedness" of such states is quite palpable: when I have perceptual experience of a dog, or imagine a dog, or have a recollection of the family dog, my mental gaze is, as it were, directed towards the object of my thought. And famously, of course, this kind of "directedness" does not require the existence of an extramental object corresponding to our ideas. From the empiricist standpoint, or under Husserl's phenomenological "bracketing," "directedness" is a feature of experience itself—

the fact that it is an experience that presents us with a putative object and not just a sensation—and not a relation to extramental reality.

So in Brentano, the "mental states" that are characterized by his notion of "intentionality" are conscious episodes and not dispositional beliefs or desires. Indeed, it is not clear that the kind of "directedness" one finds in Brentano's examples can be applied to unconscious dispositions. Brentano also uses the term 'intentionality' to apply to whole mental states, and not to their proper subparts. This leaves the exact application of the term open to some interpretation. Writers like Husserl and Searle have taken the notion of intentionality to include the whole phenomenologically rich network of mental states that is involved in the directedness of conscious thoughts. When my thoughts are directed towards an object, there is a conscious experience in which

—I am present as the subject of the thought,

—an object is presented under certain aspects and not others, and

—the experience has a phenomenology.

Someone starting from this vantage point will naturally expect an "account of intentionality" to explain all of the salient aspects of such states, including their phenomenological feel and subjectivity.

Through the middle part of the century, however, discussions of intentionality interbred with discussions of the semantics of linguistic entities, with the result that many people now seem to use the word 'intentionality' or the 'directedness' of mental states to be more or less equivalent to the linguistic notions of meaning and reference. And those influenced by the view of formal semantics argued against in chapter 6 may be inclined to view both simply in terms of *whatever* establishes a mapping from words or thoughts to world. This notion of intentionality, unlike its predecessor, seems applicable to beliefs and desires as well as to conscious mental episodes. And it seems natural, if you use the word 'intentionality' in this way, not to view things like consciousness and subjectivity as being essential to intentionality.

9.2.3 BROAD CONTENT, NARROW CONTENT, PHENOMENOLOGICAL CONTENT

Significantly, the problem of accounting for "content" shapes up differently depending on which tradition you are starting from. In recent years, analytic philosophy has given a great deal of discussion to "broad" versus "narrow" content. But the natural construal of "content" from the phenomenological standpoint does not exactly map onto either of these. There the natural distinction is between what we might call the "intentional character" of mental states (the features that are invariant over all possible assumptions about extramental reality) and "veridicality" (hooking up to the world in a felicitous way). The notion of "content" that is a part of intentional character is neither wide nor narrow content exactly.

The basic idea behind the distinction between broad and narrow content is that at least some words and concepts depend for their semantics upon things outside of the mind. Writers like Kripke (1971) and Putnam (1975) have argued, for example, that it is part of the semantics of our notion of "water" (and likewise the word 'water') that it refer to H_2O, and that it did so even prior to the discovery that water was H_2O. Indeed, on this view, "water" would have referred to H_2O even if we all believed that water was of some other molecular type. If there were beings on Twin Earth who were phenomenologically, functionally, and physically identical to us but were exposed to some other compound XYZ in the same contexts we are exposed to water, their concept "water" would mean not H_2O but XYZ. (Of course, to make this work, you have to bracket the problems that arise from using a substance that comprises most of our body weight for the example. I suggest substituting another kind of substance if this distracts you.) A second kind of argument is raised by Burge (1979, 1986), who claims that many words, such as 'arthritis', are often used by people who do not know their full sense. According to Burge, we may use such words felicitously even without knowing their sense because we are tied into a social-linguistic network with experts who do know the sense of the words: when I say 'arthritis', I intend to refer to whatever condition it is that the experts refer to when they employ the word. "Broad" content—or perhaps the broad *notion* of content—is thus something that depends on mind-world relations. This kind of "externalist" view comes in two varieties: the "ecological" kind, which ties semantics to the thinker's environment through relations like causation, adaptation, learning, and selection, and the "social" kind, which embeds semantics within a social, and particularly a linguistic framework. "Narrow" content (or the narrow notion of content), by contrast, is often characterized as what is "in the head." It is often said that molecular (or functional) duplicates (quaintly called *doppelgängers*) would necessarily share narrow content, though they might differ with respect to broad content due to being thrust into different social and natural environments.

From the phenomenological starting point, however, the natural distinction to make is not the distinction between broad and narrow content, but between those properties that are contained within the experience itself, regardless of the relation of the experience to extramental reality, and those properties that depend upon extramental reality as well. Thus Husserl invites the reader to perform an *epoché* or "bracketing" of everything that is dependent upon extramental reality in order to study intentional states as they are in their own right. And Chisholm and others resort to turns of phrase like "seeming to see a tree" or "being appeared-to-treewise" to distinguish the sense of verbs like 'see' that merely report the character of the experience from those that imply a kind of success as well. I shall mark this distinction by speaking of the notion of intentionality that implies a correspondence with extramental reality as *veridical intentionality*. The aspect of intentionality that does not vary with assumptions about extramental reality I shall call the *intentional character* of the mental.[4] What I mean by this latter expression are those aspects of an intentional state that do not vary with variations in extramental reality. And there are two kinds of invariants here: invariants in *modality* and invariants in *content*.

Let us consider an example of an intentional state. Suppose, for example, that I experience a perceptual gestalt of a unicorn on my front lawn. That is, I have an intentional state with the intentional modality VISUAL PRESENTATION and the content [unicorn on my front lawn]. Now there are certain things that one can say about such a mental state that do not depend upon issues such as whether there really is a unicorn there (or anywhere) or what causes me to have the experience that I have. Regardless of whether there is a unicorn there (or anywhere), it remains the case (*a*) that my experience has the intentional modality of VISUAL PRESENTATION (*it appears to me as though* there is a unicorn on my lawn), and (*b*) that my experience has the content of presenting a beast of a certain form and with certain associations (it appears to me as though there is *a unicorn* —rather than a cat or a rock—and it appears as though *it is on my lawn*). Each of these aspects of my experience has a certain phenomenology to it. There is a "what-it's-like" to having a perceptual gestalt, and it is different from what it is like to have a recollection, however vivid, or to have a desire accompanied by imagery, and so forth. Perhaps there are pathologies in which such distinctions are lost, and in some cases we may not differentiate adequately between modalities (e.g., between different strengths of conviction of belief or between imagination and perception); but in ordinary cases, we can quite simply *tell* what intentional modality is at work. Imagine how much more compli-cated life would be if we were systematically unable to distinguish ex-periences that were perceptual gestalts from those that were memories!

There is likewise a "what-it's-like" for having an experience with the content [unicorn on my front lawn], and it is very different from what it is like to be presented with an experience having the content [cat on my front lawn]. To determine whether I am having a gestalt of a cat or a unicorn, I do not have to consider my behavioral dispositions or the functional relations of my state of mind to other states of mind, any more than I have to do so to identify the feeling of pain as pain.[5] There is simply a difference in what different kinds of intentional states are like. So occurrent states have an *intentional character* that arguably dispositional beliefs do not have, and the notion of "content" that emerges from this perspective—which we may call *phenomenological content* —is a proper part of intentional character, which also involves an intentional modality as well.

It should be clear that phenomenological content is not equivalent to broad content, since the former partitions the mental in a way that is insensitive to relations to extramental reality while the latter depends essentially upon such relations.[6] The relationship between phenomeno-logical content and narrow content is more difficult. Narrow content is sometimes associated with the notion of "methodological solipsism" (Fodor 1980), which seems to imply slicing the intentional pie according to things that are invariant for the thinker qua thinker. (It seems hard to see how a *third* -person functionalist approach could merit the name of "solipsism"!) This would seem to imply in turn that narrow content is just phenomenological content. But narrow content has also become associated with characterization in terms of what is (necessarily) shared by physical or functional doppelgängers, and that seems to be different from phenomenological content. After all, it seems epistemically possi-ble both that I do have a body and that I do not (the Cartesian demon scenario). Similarly it seems conceivable, hence logically possible, that there be a being that is my phenomenological doppelgänger but not my physical or functional doppelgänger, and vice versa. In the absence of any way of *deriving* a particular phenomenology from a particular physi-cal or functional description (or vice versa), it seems to me we should assume that these notions diverge—perhaps in real cases, but certainly in counterfactual ones. I suspect and hope that talk of narrow content is re-ally a way of getting at phenomenological content, with incorrect assumptions being made about the necessity of relationships between the two. But for purposes of clarity, I shall treat the notion of narrow content here as though it were *defined* in terms of what physical or functional duplicates would necessarily share in common.

9.2.4 THE PLAN OF ATTACK

My plan of attack on naturalistic theories of content, then, is as follows. There are different issues about explaining the phenomenologically pregnant notion of directedness associated with occurrent states, on the one hand, and explaining the broad and narrow content of dispositional states like beliefs and desires, on the other. I shall argue that, if one is concerned with things like perceptions, recollections, and judgments, then explaining the directedness of these does involve one in explaining their subjectivity, perspectival character, and phenomenology, and that writers like Searle and Nagel are right in saying that these features cannot be reduced to a third-person naturalistic discourse. Moreover, no naturalistic discourse can provide necessary or sufficient conditions for the invariants distinctive of intentional character and phenomenological content. But these arguments do not transfer directly to beliefs and desires. There I shall argue not that no naturalistic theory can provide an account of content (though I happen to believe it), but merely that the likely form of any such theory, were it to emerge, would not place the explanation of meaningfulness where BCTM says it ought to be—namely, in the so-called "representations." This is fairly obvious in the case of broad content. I shall argue that it is very likely true of narrow content as well.

9.3 PHENOMENOLOGY AND THE MENTAL

Our first aim, then, is to examine phenomenological content and the phenomenologically rich properties of consciousness, perspective, aspect, and subjective "feel." In what follows, I wish to separate three major sorts of issues concerning phenomenologically typed mental states. First, we shall examine the *legitimacy* of the phenomenological approach: whether the phenomenological features are *real*, whether they are *essential* to intentional states (or particular kinds of intentional states), and whether they make for a viable *classification* of mental states. Second, we shall examine the question of whether phenomenological properties, however *legitimate* or *real* they might be, are likely to play much of a role in the formation of a scientific psychology. Finally, we shall consider whether phenomenological properties are the sorts of things that can be strongly naturalized.

9.3.1 THE LEGITIMACY OF THE PHENOMENOLOGICAL APPROACH

It is one of the strange turns of twentieth-century philosophy that the phenomenological properties that provided the epistemic bedrock of seventeenth- and eighteenth-century philosophy are now thought by many to be in need of legitimation. There are really a number of separate issues here. One important issue is that of the connection between phenomenology and science. That will be considered in a later section. In this section we shall consider the following questions:

(1) Are phenomenological properties
—real as opposed to unreal?
—observational as opposed to theoretical?
—accurately described as opposed to inaccurately described?
—fundamental as opposed to nonfundamental?

(2) Are phenomenological features such as subjectivity, perspective, and "feel" essential to the occurrent conscious states to which they attach themselves, and more particularly, are they essential to the *intentionality* of those states?

(3) Does the phenomenological approach provide the basis for a *classification* of mental states (especially a classification according to "phenomenological content")?

9.3.2 THE REALITY OF PHENOMENOLOGICAL FEATURES

First, let us consider whether phenomenological features are *real* features. But "real" as opposed to what? They are certainly not unreal in the sense that fictions are unreal. I suppose that it is possible that there are people who do not have the kinds of phenomenological properties that I have, or that they do not have any at all, in much the way that it appears likely that some people do not experience any mental imagery while others do so very vividly. But for those of us who do report phenomenological properties, it seems as clear as anything could be that there is a what-it's-like to, say, seeing a dog in the yard, and that it's different from the "feel" of imagining the same scene or seeing something different. Likewise, subjectivity and perspective seem to be indubitably legitimate features to attribute to my experience. For those of us who report a phenomenology, the claim that phenomenological features fail to be real the way fictions fail to be real is clearly a nonstarter.

It is quite another matter, however, if the issue is one of whether particular claims *about* phenomenology, or even particular *descriptions* of it, are as accurate as they might be. People who complain about phenomenology are often really concerned only about claims of special access that imply *incorrigibility*. But this is a red herring. I do not know any major philosopher in the phenomenological camp who has claimed that phenomenology was easy, or that we could not be mistaken about it, especially at the level of abstract characterization. Husserl was continually stressing the difficulty of phenomenological description to the point of describing himself as a "perpetual beginner" at it; and contrary to the common libel, Descartes acknowledged that we could be quite mistaken about our mental states, even in such seemingly straightforward cases as pain (see *Principles* 1.67 [AT VIIIA.32-33]). I am not aware of *anyone* who seriously thought that a thoroughgoing phenomenological account could be naively "read off" from introspection of one's own experience. (Though British empiricists and common sense philosophers sometimes spoke this way.) If the *existence* of phenomenological features is indisputable, it is equally clear that we have no definitive word on the topography of phenomenological space, nor even firm evidence that such a definitive description might be forthcoming.

I think, however, that there is an important sense in which this implies that our talk about phenomenology is "theory-laden," but also an important sense in which phenomenological properties are not "theoretical." There is a weak sense of "theory-ladenness" which implies only that the way we describe a thing (*any* thing) is set against a set of background assumptions about the world and a network of interrelated concepts or words. If this kind of network theory of meaning is true of language generally, it is surely true of our language for describing our own minds as well (unless, perhaps, one embraces the kind of phenomenalist atomism that Russell espoused at one point). But there is also a stronger sense of "theory" that implies *retroduction*, and this has implications about the kind of epistemic access we have to a thing. An entity or property that is "theoretical" in this strong sense is one supposed to exist just because this supposition explains something else. Pluto was "theoretical" in this sense until it was observed with a telescope. Protons are still "theoretical" in this sense. But it seems clear to me that phenomenological properties are almost by definition *not* "theoretical" in this strong sense (unless perhaps to someone who has heard about them but not experienced them, if that is indeed possible). If you *experience* phenomenological properties, it cannot be the case that your only access to them is inferential. You may, of course, hold some theory-laden beliefs about them (especially if you are a philosopher), just as we may still hold many theory-laden beliefs about Pluto (or, for that matter, about rocks and rabbits). But they are not *retroductive* in origin or warrant.

Finally, questions about the "legitimacy" of phenomenological categories are sometimes questions about whether such categories "cut nature at the joints." In particular, one might wonder if they are (*a*) fundamental as opposed to derivative properties, and (*b*) relevant to the systematic description of the world characteristic of science or merely epiphenomenal. Now I think that raising the question of whether phenomenological properties are fundamental is important and appropriate at some point. But it is surely a ridiculous issue to bring up early in the game as an attempt to discredit phenomenology. Cartesian physics taught that light, magnetism, and gravitation were derivative from mechanical collision. Newtonian physics treated gravitation, light, and mechanical force as separate fundamental forces. Many people objected to the Newtonian view on the grounds that it seemed to involve action at a distance. And perhaps they were right and perhaps they were wrong to do so. But *no one* (at least no one whom we remember) suggested that the irreducibility of gravitation to contact interactions would undercut the legitimacy of the *phenomenon* (as opposed to the theory) of gravitation. To do so would have been sheer madness, not to mention bad scientific practice. Science aims at being systematic and universal, but it does so by integrating discourses that are initially local and particular. If we should arrive at a unified field theory in physics, it will be because we first had serious theories of mechanics, gravitation, electromagnetism, and strong and weak force. We reduced chemistry to physics because we first had a serious chemistry. Likewise, if phenomenology is reducible to something else, the only way we will discover this is by taking phenomenological properties seriously in their own right, and this means countenancing the possibility that they might be fundamental in the sense of not being derivable from nonphenomenological properties. A posteriori arguments on this subject are for the endgame, not the outset. I have never heard a vaguely plausible a priori argument to the effect that mental properties must not be fundamental.

9.3.3 IS PHENOMENOLOGY ESSENTIAL TO SOME MENTAL STATES?

Next, let us consider whether phenomenological properties are essential to certain kinds of intentional states. Questions of essentiality are always difficult, but we might approach the issue by considering some examples of conscious mental episodes and then ask whether they could remain the same kind of episode if deprived of their phenomenology. Consider first a simple kind of perceptual experience, such as having a perceptual experience of a square, where the expression 'perceptual experience of a square' is interpreted in that distinctively intentionalistic way that does not imply a relation to an actual square. Of course, one never simply has *perceptual* experiences; they are always perceptual experiences in some particular modality—a *tactile* experience, say, or a *visual* experience. So let us say the experience in question is one of VISUAL PRESENTATION[square]. Normally, such an experience has a particular kind of phenomenology, both in terms of its qualitative elements (not just *any* configuration of qualia can be constituted as a square) and its conceptual ones (squares have a different "feel" from circles or triangles).[7] Normally, such experiences have very complicated relations to environmental and behavioral counterfactuals as well. Our natural-language attributions tend to be based on assumptions about such normal cases. But suppose that a being were to have states that were very similar to ours in its relations to the environment and behavior, but a radically different phenomenology or no phenomenology at all. We might well say that it was in some kind of perceptual state, but would we want to say that that state was VISUAL PRESENTATION? The answer, I think, is not easy.

Consider first that we can ourselves have perceptions of the same things, and behave in similar ways, on the basis of several perceptual modalities. We can feel squares as well as see them, and blind humans can form most of the same concepts and negotiate most of the same environments as sighted humans. It is just that none of their perceptual states is *visual* in nature. The same goes for echolocation in bats: presumably, echolocation plays a very similar role in bat navigation that sight plays in human navigation, but it is a different modality and presumably has a different phenomenology.

But to make the point more clearly, Sur, Garraghty, and Roe (1988) performed experiments with ferrets in which the optic nerve was severed and reconnected to nonvisual tissue in the brain. The ferrets were able to respond to visual stimuli in a striking display of equipotentiality. Suppose that the same thing could be done with human beings: the evil Dr. No rewires your nervous system so that your optical signals do not go to the visual cortex, but somewhere else. Now the human brain is probably significantly more specialized than are ferret brains, which lessens the probability that the special-purpose functions of the human visual cortex could be duplicated by other tissue; but it is at least worth entertaining the possibility (*a*) that visual stimuli would produce, say, auditory qualia, and (*b*) that you could be conditioned to distinguish some kinds of objects on the basis of these stimuli, thus forming a new kind of perceptual gestalt. Your experience might have the content [square], but would be accompanied by acoustical rather than visual qualia. Now ordinary language might well describe such an experience as "hearing shapes" or the like, but a more sober assessment would probably be that the victim of such rewiring was in fact experiencing a new kind of perceptual experience. Even if the process could be done so seamlessly that the patient could respond to the full panoply of visual stimuli that normal humans do with the same range of behaviors, I think most of us should be loath to call his experiences VISUAL PRESENTATION, precisely because of the differences in qualia. Indeed, even if someone's brain were wired like a normal human brain, I should be disinclined to call his states VISUAL PRESENTATIONS if I somehow came to believe that their phenomenology was acoustical.

Likewise with other intentional states. Suppose I have a recollection of my first day at college. This may or may not be accompanied with visual or auditory imagery; but in order to be a RECOLLECTION it must be presented *as* something that happened to *me* in the *past*. This is really a bit tricky, though. It is possible to become so engaged in memories, imagination, and particularly dreams that one mistakes them for current experiences. However, it is important to distinguish two different issues here. Sometimes, calling something a "memory" reports its causal history. Memories are experiences whose contents are dredged up out of previous experiences, whereas, say, perceptions are caused by one's environment. Thus the distinction between memory and perception can be a distinction of the *source* of the experience. But one might also use the same words to mark a distinction in the *kind of experience* involved: that is, a difference in *intentional character*—and more specifically in modality. In the ordinary cases, experiences that are dredged up from memory have the modality of RECOLLECTION and those caused by our environments have the modality of PERCEPTUAL PRESENTATION. In pathological cases and in dreams, however, this need not be so. We may take an image from memory in a dream and have it presented under the modality of PERCEPTUAL PRESENTATION. (That is, we mistakenly believe that we are having veridical perceptions when in fact we are replaying old imagery under the modality of PERCEPTUAL PRESENTATION.) Likewise it is possible for imagination to cause states with the modality of PERCEPTUAL PRESENTATION. And of course it is possible to have states of RECOLLECTION that are *false* memories, or episodes presented as FREE FANCY that are in fact images that are remembered, and so on. So when I say that, say, states of recollection have a distinctive phenomenology, I mean precisely that states that present themselves as recollections do so, and not that states that in fact draw upon memory share a phenomenology.

The same may be said for many other intentional states. Some, for example, have a particular emotional phenomenology. I cannot experience *remorse* about some action of mine, for example, without having certain *experiences,* regardless of how I act. A sociopath might fake remorse even if he cannot feel it. Likewise, I cannot feel remorse over an action unless I represent it as *my* action, and so on. The point here is that if we take away the experiential character of such states, or change it too drastically, we are no longer left with the same kind of state. Let me hasten to caution the reader, however, about several things that are *not* implied by this.

(1) The phenomenological properties of such states need not be *noticed* or *attended to.* One can, for example, *see* features of a scene that one does not actively *notice.* One sign of this is the ability to notice later things about a previous experience that were not noticed at the time. One notices a square and later realizes that it was set against a lighter background.

(2) Not all psychological distinctions need be reflected in phenomenological distinctions. It is not clear, for example, that different kinds of judgment—judgment with certainty, conjecture, scientific hypothesis—are distinguishable by phenomenological features for everyone.

(3) Phenomenological typing need not be the only valid typing of psychological states, and states that differ with respect to phenomenology may be grouped together under a different typing. For example, there are undoubtedly typings that group together psychological mechanisms we share with other species regardless of whether animals are experiencing subjects. There is nothing particularly out of the ordinary for two objects or events to share one typing and diverge with respect to another, nor for two divergent typings each to be useful for a different kind of inquiry.

9.3.4 DOES PHENOMENOLOGY YIELD A
CLASSIFICATION OF THE MENTAL?

There are really a variety of questions here. It certainly seems true that at a certain level of granularity of description, our natural distinctions between conscious mental states (e.g., between judgments and perceptual gestalts and imaginings) are accompanied by corresponding phenomenological differences. Likewise, it seems clear that we are in a significantly different *epistemic* position with respect to states that have a phenomenology and those that do not, such as beliefs and desires. If the latter are truly dispositional in nature, there is arguably a significant ontological difference there as well. It is far less clear that all meaningful psychological distinctions, even between states that have a phenomenology, are reflected in phenomenological differences. For the ordinary language classification of mental states is likely to prove as much a mixed bag of phenomenological, behavioral, and theoretical features as is the ordinary language classification of speech acts, which includes lots of cognitive, social, and emotional features as well as distinctions in illocutionary force. The project of taxonomizing speech act verbs turned out to be a mare's nest because of this (see Austin 1962, McCawley 1973, Vendler 1972, Fraser 1981, Bach and Harnish 1979, and Searle 1969 and 1971), and the same may hold true of the commonsense list of mental states. The difference between, say, speculating and hypothesizing may not consist in something that has a phenomenology, but upon something like our social conventions about kinds of thinking.

The really vital question for our purposes, however, concerns the typing of intentional *attitudes* and *contents* according to experiential invariants. Now, whatever experiential invariants there are, it seems clear that they will yield *some* partition of possible worlds: for example, between those in which I (or my counterpart) have exactly the same phenomenological properties that I actually have and all the rest. The issue is not whether phenomenology yields *some* classification, but whether the classification it yields is a *good* one. But good for what? It is certainly a good one for describing the mental from a first-person viewpoint. (What kind of classification could be better for that?) And if you think that phenomenology is crucial to the mental, this is itself good reason for liking this classification. But there is also another reason for liking it: it seems really crucial to all the other ways we have of classifying the mental.

It seems to me that all of the talk about "functional classification of the mental" is deeply misleading at best. People *speak* of functional classification of intentional modalities and even of contents. But you never see such a characterization produced. I think this is quite ironic, as one of the stock arguments against the behaviorists turns upon exactly the same inability actually to produce a single definition of the sort their theory depends upon. When characterizing intentional modalities, rather, writers like Fodor appeal to the kind of mental state we are in when we think, as it were, "Lo! a horse!" But this is clearly an appeal to something on the model of a conscious occurrent state. We all know what kind of mental state is meant, but only because we associate the description with a kind of state we have experienced. It might be the case that an ideal psychology could produce a psychological Turing table from which one could derive characterizations of each kind of mental state holding the rest as constant. But this is surely not how we actually go about classifying the mental—probably not even in the case of beliefs and desires, and *certainly* not in the case of perceptual gestalts and judgments and imaginings. Rather, phenomenology gives us at least a rough initial classification to start with, and we test this against observations of people's behavior and try to systematize and refine it through rigorous modeling (including computer modeling). It is not as though the "functional classification" of the mind implied by some discussions of narrow content was carried out in isolation from a phenomenologically based starting point. (Indeed, it is not as though such a classification has ever actually been carried out at all—a point that is missed with shocking regularity.) Any functional classification of the mental there might be is a distillation of a classification that started out in phenomenology—and which, I shall argue in the next section, must answer to phenomenology as well. The notion of "narrow content" is really a kind of theory-laden abstraction from phenomenological content. (And, I expect, the functional notion of belief is ultimately a theory-laden abstraction from conscious judgments as well.)

As for broad content, that certainly goes beyond what is present, strictly speaking, in phenomenology (that is, in intentional character). But, first, it contains implications of intentional character: a veridical perception is, among other things, a perceptual gestalt. And, second, it seems to me that writers like Husserl have been correct in saying that intentional states in some sense carry with them their own "conditions of satisfaction." Having a veridical perception of a dog requires us to be in the right causal relationship with a dog. Why? Because that is built into the notion of the intentional modality of PERCEPTUAL PRESENTATION. This is no great empirical discovery. It is simply an explication of what is implicit in the phenomenology of this particular intentional modality. Likewise, if the broad content of "water" is fixed by something in the environment, it is because the intentional character of the state implies that it should be so. So in short it seems to me that it is simply bootless to deride phenomenological classification in favor of some other kind of classification, since the other kinds of classification that have been proposed turn out to depend heavily upon our prior phenomenological understanding.

9.4 PHENOMENOLOGY AND SCIENTIFIC PSYCHOLOGY

As often as not, those who minimize the role of phenomenological properties (or, for that matter, of the mental in general) do so not so much as a rejection of the *reality* of such properties or of their utility in commonsense predictions as they do as a rejection of the idea that such properties will play a role in an explanatory science of psychology. Of course, it is not uncommon on the current scene for a concept's inclusion in the theoretical vocabulary of a science to be held up as a standard of its ontological legitimacy—a view I shall argue against in chapter 11—but really that is a stronger position than one need take here. It is enough for the moment to say that phenomenological properties, albeit real, are not the sorts of properties that enter into causalnomological relations (except perhaps insofar as they are reliably produced as epiphenomena of brain events), and that phenomenological typing will correspond to the typing of a mature psychology accidentally if at all.

I think that there are certain things that are right about this view, but many more that are mistaken. On the one hand, it is surely right that there are large domains of psychology that cannot be explained in terms of conscious mental states at all, much less in terms of their phenomenological features. While perception eventuates in conscious states with a phenomenology, the processes that produce this product are almost entirely infraconscious. Likewise memory and imagination have conscious products, but also involve mechanisms that must be of an entirely different sort. And while there are conscious processes of reasoning, association, and inference, there are also nonconscious processes that go by the same names—and even the conscious ones must have their own nonconscious mechanisms which support them. So if the issue is one of whether conscious states with a phenomenology can provide the bulk of

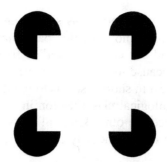

Figure 13. From Gaetano Kanizsa, "Subjective Contours," *Scientific American* 234 (April 1976): 51. Copyright 1976 by Scientific American, Inc. All rights reserved.

the explanatory resources needed by psychology, the answer is surely *no*.

On the other hand, there are clearly *some* kinds of explanation that do call for appeal to states with a phenomenology. Notably, when we ask why a person spoke or acted in the manner that she did, we will often appeal not just to dispositional beliefs and desires, but to conscious judgments and perceptions—and in particular, we will appeal to the phenomenological content of her judgments and perceptions. Why did Jane pick up the flyswatter? Because the thing flying around *looked like a fly to her*. Note that questions of broad content are irrelevant here—the explanation is unaffected if all of Jane's fly-gestalts were caused by midges. Likewise narrow content, if defined in purely functional-causal terms, does us no good here: it won't do to say that Jane picked up the flyswatter because she was in the kind of mental state caused by flies and resulting in flyswatter grabbing behavior.

Perhaps even more clearly, we need to appeal to phenomenological content to explain why people behave the way they do in the case of optical illusions like subjective contour figures, in which the subject "sees" a figure that is "not really there" in the sense that there is no objective reflectance gradient that makes up a figure of the type that is seen. For example, a subject seeing the Kanizsa square (fig. 13) will report seeing a light square against a slightly darker background, and will experience borders making up the edges of the square, even though there is no reflectance gradient to be found in those positions in the stimulus (see Kanizsa 1976 and 1979). Now suppose we ask our subject to respond in one way when she sees a square and another way when she sees a figure that is not a square. When presented with the Kanizsa square, she behaves as though presented with a square. How are we to explain this? What unites the cases of being presented with an actual square with the cases of being presented with the Kanizsa square, and hence unites the behaviors involved? I submit that it is precisely that they share a certain phenomenology—namely, the phenomenology of experiences having the intentional character of VISUAL PRESENTATION[light square against darker background].

In short, it seems to me that, whenever it is necessary to appeal to conscious states like judgments or perceptions to explain behavior, it will very likely be a typing according to phenomenological content that will be relevant. Now whether typing by phenomenological content will produce the kinds of regularities needed for something systematic enough to count as a nomological science is still to be determined (as is the distinct yet related question of whether we could catch such regularities if they were there). But it does seem plausible that at least some such explanations will resort to phenomenological typing.

But there is another connection between phenomenology and scientific psychology that is, to my mind, far more important. If phenomenological properties make up a relatively small portion of the *explanatory apparatus* of psychology, they comprise a significantly larger portion of the *phenomena that a scientific psychology needs to explain*. That is, they make up much of the *data* of psychology. I think that the case can be made most forcefully here in the case of the relationship between psychophysics and theoretical work in perception. Psychophysics, which is viewed by many as the one area of psychology that has already attained some of the benchmarks of scientific maturity, is largely concerned with the measurement of relationships between stimuli and the percepts that they produce. The properties of the stimuli to be studied include things such as the objective intensity of the stimulus and the spatial and temporal patterns of intensity in stimuli. Percepts, however, are experiences—they are phenomenological in character. They involve properties like how intense a stimulus *seems* or whether one *seems* more intense than another, or the way the percept is organized into a perceptual gestalt. I shall discuss two well-known experimental results from the psychophysical literature and show how phenomenology is essential to psychophysics.

First, consider the Weber laws that describe the relationship between stimulus intensity and percept intensity in terms of a logarithmic law (Fechner 1882), or, in alternative versions, a power law (Stevens 1951, 1975). Here the relata are an absolute property of a stimulus (say, its luminance) and the subjective property of a percept (the word 'brightness' is often used in contrast with 'luminance' for this subjective property).

Now, first of all, it seems just inescapable here that the phenomenon we are after involves a phenomenological property. Take away the property of perceived brightness and there is no Weber law left. Second, it seems equally clear that the kind of description of human perception that the Weber law presents is exactly the sort of thing that we should require of our theories of perception. A model of perception that does not obey the Weber law or that does not produce the optical illusions that humans experience is, to that extent at least, a bad (or at least an incomplete) model (see Todorovic 1987). Qualitative phenomenology is essential to psychophysical data such as the Weber law and provides much of the data for theories of vision and other perceptual modalities.

The same can be said for phenomena involving at least simple forms of intentionality. Consider again the Kanizsa square. Here the psychophysical data show that there is a certain class of conditions under which we "see" something that "is not there"—in this case, we perceive a square where there is no square and perceive it as brighter than its background when in fact the "interior" of the "square" and its "background" are actually equal in luminance. This kind of mismatch between the "objective" features of the stimulus and the "subjective" features of the percept tends to be what makes a given stimulus-percept pair an "effect" and renders it of particular psychological interest. (You can't make your reputation in experimental psychology by finding that people see squares when they are presented with squares; if they see squares when presented with circles, you get to have an effect with your name in front of it.) And the ability to reproduce such effects is precisely the sort of thing that can be used to test the adequacy of a particular model of how perception works in human beings. Again, our data involve a relationship (a mismatch) between an objective property of luminance distribution and a phenomenological property of seeing a particular kind of figure. Take away the phenomenology and there is no effect. Take away the effects and there is no psychophysics. Take away the psychophysics and there is nothing for theoretical psychology of perception to explain.

Unlike the Weber law, moreover, subjective contour features involve at least a primitive form of intentionality. The subject does not merely experience more and less intense qualia—she constitutes them as a figure of a particular kind and shape and constitutes the figure as being in a particular relationship to its background. Moreover, this kind of illusion vividly illustrates what Chisholm has cited as the cardinal property of intentionality and intentional objects: the subject can "see" a square when there is no square there to be seen.

I think that some other phenomenological properties likewise provide data that set tasks for psychological theories. It seems clear, for example, that the object-directed character of intentional states is something that needs to be mirrored in any successful theory. (It has surely motivated much work in artificial intelligence.) Likewise the perspectival character of conscious experiences: a theory of thinking must do more than provide for the fact that we think about objects; it must provide for the fact that we think about them under particular aspects and from particular points of view. It must, for example, account for the fact that we can infer the hidden edges of familiar three-dimensional objects, or move between different things we know about an object we are viewing without keeping all of its known properties before the mind's eye at once. The ultimate source of our knowledge that thought has these properties is phenomenological, and so once again phenomenology sets constraints on the form of a scientific psychology.

9.5 WHY PHENOMENOLOGY CANNOT BE NATURALIZED

A number of kinds of arguments have been offered over the years to the effect that some one or more features of the mental cannot be naturalized—features such as subjectivity, the what-it's-like of experience, the first-person perspective, and consciousness. I shall examine some variations on arguments of this sort in this section, as well as adding one of my own at the end.

9.5.1 THE ARGUMENT FROM EPISTEMIC POSSIBILITY (CARTESIAN DEMONS REVISITED)

The kinds of epistemological issues involved in old-style thought experiments involving Cartesian demons stem from the phenomenological perspective on content. The Cartesian demon experiment is, if nothing else, a marvelous tool for driving a wedge between the intentional character of my mental states and all questions of their veridicality. As Descartes points out, I can be mistaken about the causes of my experiences and about whether they correspond to extramental reality, but I cannot be mistaken in the same way about what kind of ideas I am experiencing.[7] I can be sure that I am experiencing a particular kind of perceptual gestalt, but I cannot be similarly sure, for instance, that there is indeed a cat before me.

Thought experiments involving such exotica as brains in vats and Cartesian demons do not enjoy the popularity that once they enjoyed. There are no doubt a number of factors contributing to their decline. One would probably be the shift away from epistemological interests in the philosophy of mind. Another would be a shift in interest from providing accounts involving logically necessary and sufficient conditions to finding accounts that are empirically adequate. Considerations of necessity and sufficiency do seem to be in order with accounts that purport to provide a strong naturalization, though. If mental-semantic properties are to supervene upon naturalistic properties, those naturalistic properties must provide sufficient conditions for them. And if the resulting account is to be an account of the nature or essence of mental-semantics or intentionality, it had best be necessary as well: if an object could have a property A while lacking B, then B cannot be essential to A.

Now I think that some of the traditional thought experiments are well suited to showing that naturalistic properties are neither necessary nor sufficient for intentionality or mental-semantics. Let us begin with necessity. The notions of supervenience and of instantiation analyses themselves claim nothing about the necessity of the conditions they provide. If A supervenes upon B, it does not follow that B is a necessary condition for A; and if A is given an instantiation analysis in terms of B, it similarly does not follow that B is a necessary condition for A. But this is in some ways very misleading. When people say that the supervenience of A upon B does not involve a necessary relation from A to B, what they tend to be concerned with is the lower-order physical properties through which a mental property is realized—with the fact that it does not matter whether the underlying structure is wetware or hardware or whatever. But when people try to give a naturalistic account of intentionality, they tend not to be specifying the instantiating system at that low a level, but in terms of notions such as causal covariation, adaptational role, or information content. These notions form an intermediate level of explanation that is neutral as to underlying structure. And theorists who propose such theories generally *do* take it that the conditions they articulate at *these intermediate levels* are necessary conditions for intentionality and mental-semantics. Millikan, for example, is quite clear about this: a being that does not share our adaptational history not only does not share our particular beliefs, it does not have beliefs at all! Similarly strong views might be imputed to causal covariation theorists. In Fodor's account, it is a necessary condition for a representation of type MR to mean "P" that MR's are sometimes caused by P's. Similarly, with

Dretske's account, a representation cannot mean "*P*" if its type was never caused by a *P* in the learning period. So while the language of supervenience and token physicalism suggests that naturalistic explanations do not provide necessary conditions, this is belied by actual practice of theorists. *Either accounts in terms of causal covariance and adaptational role are not naturalistic accounts, or the best-known contemporary naturalistic accounts of intentionality involve a commitment to providing necessary conditions.* And this seems quite appropriate in a way, since such theorists claim to provide accounts of the *nature* or essence of mental-semantics and intentionality.

This being said, I think that there is good reason to believe that naturalistic accounts of these sorts do not succeed in providing necessary conditions, for reasons that may be developed by way of some familiar sorts of thought experiments. Consider the Cartesian scenario of a being that has experiences just like ours, not because he is in fact coming into contact with elm trees and woodchucks, but because he is being systematically deceived by a malicious demon. Such a scenario is clearly imaginable, since one cannot reach Cartesian certainty that it is not in fact an accurate description of one's own case. (There is, after all, no experiment one can perform to determine whether one's experiences are veridical or systematically misleading.) And there seems little reason to deny that such a scenario is logically possible. Now a being in such a state would be in many of the same sorts of intentional states that we are in-that is, states with the same attitude and the same phenomenological content. (Whether you have perceptual gestalts or recollections, after all, does not depend on whether you turn out to be the victim of a Cartesian demon.) But it would not share most of our naturalistic properties. In particular, the intentional states it has would not be hooked up to the world in the ways called for by a respectable naturalistic psychology. Thoughts about dogs are not caused by dogs, nor are beliefs about elm trees caused by elm trees, and the being may not even have the ancestors requisite for an adaptational history. All of his beliefs are demon-caused (although they are not about demons).

Here we have an example of a being that has meaningful intentional states but does not share the naturalistic descriptions that apply to us. A fortiori, it is possible for a being to be in a state with a mental-semantic property *M* while lacking naturalistic property *N*. Therefore N cannot be a necessary condition for *M*. Therefore naturalistic properties cannot be necessary conditions for mental-semantic properties.

It remains to consider sufficiency. In order for there to be an instantiation analysis of some mental-semantic property M in terms of some naturalistic property N, it must be the case that N is sufficient for M. Indeed, it must be the case that someone who had an adequate understanding of N would be able to infer M from N. So if there can be cases of an entity possessing N but lacking M, N is not a sufficient condition for M, and hence one cannot have an instantiation analysis of M in terms of N.

Let us now bring some modal intuitions into play. It seems to be imaginable, and hence plausibly metaphysically possible, that there might be beings who were completely like us in physical structure and in behavioral manifestations, yet lacked the kind of interiority, or first-person perspective, that we have. When one stubs her toe, she says "Ouch!" and withdraws her foot, but she has no experience of pain. When one is asked to comment upon Shakespeare, she utters things that sound every bit as intelligent as what a randomly selected human being might say, but she never has any mental experiences of pondering a question or hitting upon an insight. If one could come up with a talented telepath, the telepath would deliver the verdict that nothing mental is going on inside this being. These beings, by stipulation, share all of our natural properties, yet they do not enter into any of the paradigm examples of mental states. Hence naturalistic properties do not provide sufficient conditions for intentional states, either.

9.5.2 AN OBJECTION: METAPHYSICAL AND "NOMOLOGICAL" SUFFICIENCY

One concern I can expect this argument to raise would be that people interested in supervenience accounts tend to view the kind of sufficiency involved not as *logical or metaphysical* sufficiency, as I have assumed, but as something called "nomological sufficiency." I must confess to some puzzlement about what is meant by "nomological sufficiency." It must mean something more than *material* sufficiency, since materially sufficient conditions may be completely unrelated to what they are conditions for. If the tallest man who ever lived was in fact married to the first woman to climb Everest, and was her only husband, then being married to the first woman to climb Everest is materially sufficient for being the tallest man who ever lived. But surely nomological sufficiency amounts to more than this. Perhaps nomological sufficiency amounts to something like "material sufficiency in all possible worlds that have the same natural laws as the actual world." But, according to the thought experiment above, the world described is like the actual world in all physical laws. If these assured that the psychophysical relationships must be the same the way fixing your statistical mechanics fixes your thermo-dynamics, we should be able to *derive* this fact the way we can do so in the case of thermodynamics. But this seems plainly to be impossible. It seems, then, that naturalistic conditions would not be *nomologically* suf-ficient for intentionality either. But perhaps nomological sufficiency does not apply to *all* worlds with natural laws like our own, but only ones specified by a certain counterfactual. But which counterfactual? And how do we know that a world like the one described above does not fall within the scope of it? Indeed, how does one know that the *actual* world meets the desired criterion? But perhaps nomological sufficiency is mate-rial sufficiency in all worlds sharing *psycho*physical laws with the actual world. This stipulation, however, would be inadmissible for two reasons. First, this violates the condition of strong naturalism that the relation be metaphysically necessary and epistemically transparent. Second, we do not know that the naturalistic criteria are met in the actual world.

Finally, let us be quite clear about separating the question of logical possibility from the question of warranted belief. No one is claiming that it is reasonable to believe that one is, for example, in the clutches of a Cartesian demon. And while some people do claim that there are nonmaterial thinking beings, their use in this kind of example is not based upon the likelihood of their existence, but upon their possibility. If one has an account of what it is to be in a meaningful mental state, it had better apply to all possible beings that could have such mental states. Regardless of the *likelihood* of Cartesian demons or nonembod-ied spirits, if they are *possible,* then an account of the nature of inten-tionality had best apply to them too.

9.5.3 THE PHENOMENOLOGICAL "WHAT-IT'S-LIKE"

A number of writers have argued that at least some mental states (the conscious ones) have an experiential quality for the subject of the expe-rience that is not captured in any third-person "objective" characteriza-tions. This point now seems widely accepted with respect to qualitative states such as pain: even if we know that C-fiber firings are the physio-logical basis of pain, a complete knowledge of the neurology of C-fiber firings could not yield an understanding of what pain *feels like.* To know what pain feels like, you have to feel it. And likewise for other qualia: a blind person who knows state-of-the-art theory in electromagnetism, optics, and the physiology of vision will not thereby gain a knowledge of how magenta looks, and so on (see Jackson 1982). Thomas Nagel has developed this point famously in an article entitled "What Is It Like to Be a Bat?" (1974), in which he points out that a sensory modality like echolocation would, like vision, have its own phenomenology; and lacking this faculty, we cannot imagine what it would be like to have it.

While many writers in philosophy of mind acknowledge that there is a problem for naturalistic theories in trying to explain qualia, it is less often recognized that there is a similar problem for intentional states, which also have a phenomenology. Take perceptual experiences, like seeing a dog in the yard. There is a what-it's-like to seeing a dog in the yard, and it is different from what it's like to see a pine tree in the yard (change of content) and from what it's like to imagine a dog in the yard (change of intentional modality). And the differences here are not just differences in qualia. Suppose you are at the wax museum. You turn the corner and see a familiar face and say, "My gosh! That's Bill Clinton!" You have an intentional state of the form: VISUAL PRESEN-TATION[Bill Clinton]. But then you remember where you are and cor-rect yourself. "Oh," you say, "that's just a wax replica of Bill Clinton! Boy am I a dope!" Your intentional state changes from VISUAL PRESENTATION[Bill Clinton] to VISUAL PRESENTATION [wax statue of Bill Clinton]. The qualia have not changed; it is just the con-tent of the gestalt that has changed. But part of that gestalt is conceptu-al, and that conceptual part has a phenomenology. The difference be-tween having an experience of seeing Bill Clinton and that of seeing a replica of Bill Clinton is not just a functional difference in how they relate to behavior and other mental states—they are different *as experi-ences* as well. Likewise in perceptual illusions like the Necker cube and the faces-vase illusion: the qualia remain the same while the interpreta-tion changes; but clearly there is a difference in what it is like to see the faces and what it is like to see the vase.

The same point can be made with Nagel's bat. Perceptual modalities are among the sorts of things that have a phenomenology. But this phe-nomenology is not confined to individual qualia. There are ways of con-stituting things as objects in visual perception, in touch, in hearing; and in perception one situates oneself relative to the objects one constitutes as being in one's presence. A person lacking one of the sensory modalities is indeed unable to understand the qualia associated with that modality; but she is also unable to understand what it is like to constitute objects under that modality. For example, there are people blind from birth who have had operations that restore the integrity of the visual pathway and who, as a consequence, suddenly experience visual qualia. Many such people are already competent at identifying objects and persons by sound and touch, but this ability does not translate to the formation of visual gestalts. The person with restored visual pathways suddenly knows what visual qualia are like, but not visual *perceptions*. (In fact, they tend to feel quite disoriented by vivid but uninterpretable visual qualia.) Perception is characterized by a particular kind of *intentional* as opposed to *qualitative* experience that essentially involves constituting something as an object. Object experiences involving a sensory modality involve object-constituting operations that are modality-specific. Presumably the same would hold true with echolocation. We could perhaps build prosthetic devices that would duplicate the function of the bat's vocal cords and ears and surgically connect their output to some portion of the human brain. Perhaps the subject would even experience some new qualia. But this in itself would not add up to echolocation until there were also experiences corresponding to the conceptual representation of objects under particular aspects within this sensory modality. To know what it is like to be a bat, it is not enough to know what it is like to have the bat's qualia; we would also have to have the bat's experiences of constituting objects on the basis of those qualia as well.

Nagel and others urge upon us the idea that the what-it's-like of experiences cannot be accounted for in nonexperiential terms. In some cases, the argument appears to be an epistemic one: Jackson (1982), for example, appears to argue as follows:

(1) A person could know the neurophysiology of a mental state but fail to know what mental state it was.

(2) If you can give an account of P in terms of Q, then an adequate knowledge of Q should let you know you were dealing with P.

∴ (3) You cannot give an account of mental states in terms of their neurophysiology.

Searle and Nagel, however, claim that their point is metaphysical as well: namely, that the phenomenological what-it's-like is a *property* of conscious mental states. Searle points out, for example, that some things have a what-it-feels-like while others do not, and argues further that the uniting feature for those that do is consciousness:

> The discussion of intentionality naturally leads into the subjective feel of
> our conscious states.... Suffice it to say here that the subjectivity necessari-
> ly involves the what-it-feels-like aspect of conscious states. So, for exam-
> ple, I can reasonably wonder what it feels like to be a dolphin and swim
> around all day, frolicking in the ocean, because I assume dolphins have
> conscious experiences. But I cannot in that sense wonder what it feels like
> to be a shingle nailed to a roof year in and year out, because in the sense in
> which we are using the expression, there isn't anything at all that it feels
> like to be a shingle, because shingles are not conscious. (Searle 1992: 131-
> 132)

In a sense, though, the real crux of the matter is neither purely epistemological nor purely metaphysical: the real issue is whether you can
give an *account* of the experiential what-it's-like in third-person naturalistic terms. If the kind of "account" you want is a strong naturalization, you need logical sufficiency and conceptual adequacy. And it
does not look as though you are going to get either of those things. A
person who did not have a commonsense notion of heat could still
derive thermodynamic laws from the mechanics of particle collisions.
But a person who did not know what a visual gestalt was like could
not derive that from a knowledge of optics and the physiology of vision, or indeed from any list of sciences you might give. The sciences
as we know them just do not seem to have the right conceptual resources to generate the necessary concepts. To be sure, the physiology
of vision can explain why our phenomenological color space has
some of the properties it has. (Given contingent relations between
particular qualia and particular bodily states, it can explain why certain forms of color blindness occur, how color perception is affected
by saturated lighting, why particular optical illusions occur and not
others, etc.) Likewise, an account of the visual cascade through the
visual cortex may explain why we can detect certain primitive shapes
and not others and why we are subject to certain illusions. And they
will hopefully tell us what brain processes are involved in the very experiences we describe in phenomenological terms. What they do not
seem to have the resources to do is explain the phenomenological
"feel" of those experiences. It is, of course, risky to make arguments
about what cannot be done. On the other hand, it seems clear at this
point that any assurance that we *can* derive phenomenology from neuroscience the way we can derive thermodynamics from statistical mechanics places a great deal of nonempirically based faith in the idea that

a particular paradigm of explanation can be applied universally. This kind of naturalism seems to be more ideology than well-argued position.

9.5.4 PERSPECTIVE, SUBJECTIVITY, AND THE LOGICAL RESOURCES OF NATURAL SCIENCE

Next, let us consider two other features of intentional states that some writers think render them insusceptible to naturalization. First, Searle points out that intentional states are *perspectival* in character:

> My conscious experiences, unlike the objects of the experiences, are always perspectival. They are always from a point of view. But the objects themselves have no point of view.... Noticing the perspectival character of conscious experience is a good way to remind ourselves that *all intentionality is aspectual*. Seeing an object from a point of view, for example, is seeing it under certain aspects and not others.... Every intentional state has what I call *an aspectual shape*. (Searle 1992: 131)

Second, an experience always involves a first-person perspective. And that first-person perspective is one of the identity conditions for the experience. You can have an experience just like mine, but you cannot have my experience. Even if you were a telepath or empath like the ones depicted in science fiction stories, you would not be experiencing my thoughts and emotions, but reproducing them in your own mind under some intentional modality distinctive to telepaths or empaths. Or, as Searle puts it, "For it to be a pain, it must be *somebody's* pain; and this in a much stronger sense than the sense in which a leg must be somebody's leg, for example. Leg transplants are possible; in that sense, pain transplants are not" (ibid., 94).

Here again it is possible to interpret the case in epistemic or in metaphysical terms. But here again I think the real issue lies in the possibility of *explaining* subjectivity and aspectual shape in third-person, "objective," naturalistic terms. And there is a weaker and a stronger variation of the case against naturalization here. First the weaker one. The project of explaining intentionality in naturalistic terms is one of uniting two bodies of discourse—the languages of two sciences, if you will. (Or, if you do not think discourse about experience is scientific, a science and a non-science.) Let us call the language of our naturalistic discourse N and that of our phenomenological psychology P. The question is, does N have the right kind of conceptual resources for us to derive P from N

in the way, say, that we derive thermodynamics from statistical mechanics, or perhaps even the way we "derive" arithmetic from set-theoretic constructions? And there are features of aspectual shape and subjectivity that give us reason to suppose that the answer may well be *no*.

The reason subjectivity and aspectual shape pose problems for the would-be naturalizer is that a discourse that encompasses subjectivity and aspectual shape would seem to require logical features that do not seem to be present in the languages used for the natural sciences. This, I think, is what Searle is after when he says that "the world itself has no point of view, but my access to the world through my conscious states is always perspectival, always from my point of view" (Searle 1992: 94-95) and "my conscious experiences, unlike the objects of the experiences, are always perspectival. They are always from a point of view. But the objects themselves have no point of view" (ibid., 131). But if Searle is right about the basic issue here, he is wrong about the specific form it takes with respect to aspectual shape. It is true of course that objects themselves are nonperspectival; but it is also true that all of the sciences do represent objects under particular aspects: say, as bodies having a mass or as living beings. The problem is not in getting a perspective into our discourse, but with the fact that discourse about mental states requires that we build a *second layer* of perspective into that discourse: to attribute an intentional state to someone is not merely for *us* to represent an object under an aspect, but to represent a person as representing an object under an aspect. And it is not at all clear that the resources for this are present in the kind of discourse found in the natural sciences.

Likewise with subjectivity. The special problem here is that, in order to talk about my experience as experience, I have to talk about it as essentially *mine,* as experienced from a first-person perspective. And this seems to require a language that has resources for expressing first-person as well as third-person statements. But the languages of the natural sciences arguably lack such resources. As Nagel argues, a complete description of the world in third-person terms, including the person I happen to be, seems to leave out one crucial kind of fact: the fact that that person is *me.* I interpret Nagel to mean by this that third-person discourse, even third-person *psychological* discourse, lacks a way of linking itself into the first-person discourse that is vital to our description of our mental lives.

This seems to me to be a powerful objection to the project of strong naturalization. If the kinds of discourse employed in the natural sciences lack the logical and conceptual resources to generate the kind of discourse needed to talk about subjectivity and aspectual shape, then these features of our mental lives cannot be strongly naturalized. And if these features are part and parcel of the phenomenon we call "intentionality," then intentionality cannot be strongly naturalized either.

9.5.5 THE OBJECTIVE SELF AND THE TRANSCENDENTAL EGO

An even more radical variation on the same sort of claim is, I think, to be found in the writings of Kant, Husserl, and Wittgenstein. These writers seem to note that every intentional thought requires an *analysis* that involves at least three features: (1) a thinker (the "transcendental ego"), (2) a content (meaning, or *Sinn*), and (3) an object aimed at (the "intentional object"). However, it is important to note—as Kant, Wittgenstein, and Husserl do and many other writers do not—that these "features" in the analysis of intentional states do not function in experience as three *things,* but as *aspects or features of a seamless unity.* Wittgenstein expresses this as follows in the *Tractatus:*

> 5.631 There is no such thing as the subject that thinks or entertains ideas.
>
> If I wrote a book called *The World as I found it*, I should have to include a report on my body, and should have to say which parts were subordinate to my will, and which were not, etc., this being a method of <u>isolating the subject, or rather of showing that in an important sense there is no subject; for it alone could *not* be mentioned in the book.</u>—
>
> 5.632 The subject does not belong to the world: rather it is a limit of the world.
>
> 5.633 Where *in* the world is a metaphysical subject to be found?
> You will say that this is exactly like the case of the eye and the visual field. But really you do *not* see the eye. And nothing *in the visual field* allows you to infer that it is seen by an eye.
>
> 5.64 Here it can be seen that solipsism, when its implications are followed out strictly, coincides with pure realism. The self of solipsism shrinks to a point without extension, and there remains the reality co-ordinated with it.
>
> 5.641 Thus there really is a sense in which philosophy can talk about the self in a non-psychological way.
> What brings the self into philosophy is the fact that 'the world is my world.' The philosophical self is not the human being, not the human body, or the human soul, with which psychology deals, but rather the metaphysical subject, the limit of the world—not a part of it. (Wittgenstein, 1961, italics his, underlining mine)

Husserl similarly speaks of intentional experience as a unity encompassing subject, meaning, and object. He writes that

> the experiencing Ego is still nothing that might be taken *for itself* and made into an object of inquiry on its *own* account. Apart from its "ways of being related" or "ways of behaving," it is completely empty of essential components, it has no content that could be unravelled, it is in and for itself indescribable: pure Ego and nothing further. (*Ideas* §80)

Kant likewise speaks of the transcendental ego only in the context of the transcendental unity of apperception—that is, the possibility of the "I think" accompanying every possible thought (*Critique of Pure Reason,* Sec. 2, §16, B131).

The reason this distinction seems important is that, if writers like Wittgenstein and Husserl are right, the great divide lies not so much between *mental and physical objects* as between discourse about the (logical) *structure of experience* and discourse about *objects* generally (including thoughts treated as objects). On this view, when one comes to a proper understanding of thinking, what one finds there are not several interrelated *things* (the self, the intentional state, the content, and the object-as-intended), but a single act of thinking that has a certain logical structure that involves it being (*a*) the thinking of some subject (*b*) aiming at some object (*c*) by way of a certain content being intended under a certain modality. It is possible, of course, to perform an act of analysis whereby one directs one's attention separately to self, content, modality, and intentional object. And when one does that, each of these things comes to occupy the "object" slot of another intentional act. Indeed, from the perspective of the analysis of experience, *what it is to be an object is to be a possible occupant of the object-slot of an intentional act.*[9] But if this is so, then the logical structure of intentional states is in some sense logically prior to the notion of object, and the tags 'experiencing self', 'content', and 'object', as they are applied to moments or aspects of experiencing, are not names of interrelated objects. Indeed, they are not *objects* and hence are not *related* (since relations can only relate objects).[10]

Now if this is right, the task of *relating* objectival and experiential discourse becomes all the harder: relations are things that obtain between objects. If the "I" and the content that appear in experiential analysis do not appear there as objects, there can be no question of relating them to things appearing in discourse about objects. There can be no question of objectival-experiential *relations,* because in the experiential analysis, the experiencing "I" and the content do not appear as objects at all. Nor is it possible to "cash out" the logical structure of intentional experience in terms of relations between objects, for reasons already described. (Or, as Husserl suggests, at least doing so necessarily involves a distortion of one's subject matter.) The only other way to bridge the Cartesian divide between mind and nature, it would seem, would be to find a way to subsume objectival discourse within experiential discourse, as Husserl tries to do in his transcendental phenomenology. I shall not pursue this possibility here, but shall point out that it seems right in at least one regard: namely, that intentional character is in a certain way conceptually anterior to the notion of an object in the world. For it is the content of an intentional state that lays down the satisfaction conditions determining what kind of object or state of affairs would have to exist in order for the state to be fulfilled. It is the content "unicorn" that specifies what criteria something would have to fulfill to be a *real* unicorn, and not vice versa. (It is, of course, possible simply to live with the dissatisfying result that there is an unbridgeable gap between two disparate realms of discourse. To those uneasy with such a gulf, I heartily recommend a careful consideration of the kind of combination of transcendental idealism and transcendental realism advocated by Husserl.)

9.5.6 THE ARGUMENT FROM THE
   CHARACTER—VERIDICALITY DISTINCTION

Finally, it seems to me that there is a fairly straightforward argument to the effect that intentional character cannot be accounted for in naturalistic terms. Intentional character was defined in terms of the aspects of intentional states that are invariant under alternative assumptions about extramental reality. Hence, it should be clear that *any* analysis we might give of intentional character must not depend upon anything *outside* the domain of experience. Notably, it must not depend upon any presumptions about (*a*) correspondence to extramental objects, (*b*) the causes of the intentional states, or (*c*) ontological assumptions about the mind. For having an experience with the character of, say, VISUAL PRESENTA-TION[unicorn on my front lawn] is compatible (*a*′) with there being or not being a unicorn there, (*b*′) with the experience being caused by a unicorn under normal lighting conditions, a dog under abnormal conditions, LSD, or a Cartesian demon, and (*c*′) with materialism, dualism, transcendental idealism, Aristotelianism, and Middle Platonism, to name a few possibilities. And it seems to follow straightforwardly from this that any account of intentionality that is not similarly neutral cannot serve as an account of intentional character because such an account would have to be valid for all possible instances of the phenomenon it explains. In particular, an account framed in terms of assumptions about the actual nature of physical world, including human physiology, cannot be broad enough to cover all possible cases that would share a particular intentional content. Hence one cannot have a naturalistic theory of content—at least if by a "theory of content" one means something like "an account of the essential features of intentional character" as opposed to, say, "a specification of the natural systems through which intentional character is realized."[11]

9.5.7 SUMMARY OF PROBLEMS FOR NATURALIZING PHENOMENOLOGY

In short, then, the prospects for strongly naturalizing the phenomenological properties of mental states appear to be rather dim. Thought experiments about brains in vats and Cartesian demons cast significant doubt on whether there could be metaphysically necessary relations between phenomenologically typed states and naturalistic states. And properties like subjectivity, perspectival character, and the "what-it's-like" alluded to by Nagel do not seem to be susceptible to conceptually adequate explanation in naturalistic terms. Moreover, typing by intentional character necessarily classifies mental states in a way that is insensitive to extramental realities, so that it is impossible for a naturalistic theory to capture the same invariants. And finally, there is the tantalizing suggestion that discourse about "the experiencing self," "the thought," and "the intentional object" is not really discourse that relates *objects* at all, in which case it is hard to see how naturalistic discourse could have the right sorts of logical-grammatical resources to subsume it. If the kind of "content" we wish to naturalize is the kind that is delimited along phenomenological lines, weak naturalization (i.e., mathematical description and localization) is the best we are entitled to hope for.

9.6 NATURALIZING BROAD CONTENT

A very different set of issues confronts us when we turn to the broad notion of content. There has been a great deal of discussion in recent philosophical publications about the implications of broad content for a representational theory of the mind. It is not my intention to canvass these or to discuss this already large conversation in any detail. Instead, I wish to focus on a very specific point. Unlike the arguments in the previous section, I shall not try to argue that broad content cannot be naturalized. (Though I suspect this is so.) Instead, I shall argue that the kind of theory that would be needed to naturalize broad content would not be able to focus its explanation on the properties of localized representations, as required by CTM, but would have to appeal to broader relations involving the entire organism and its environment as well.

Suppose, for example, that we want a naturalistic account of why a particular kind of thought means "arthritis," and that we accept Burge's contention that this story will have to be dependent upon the way the individual language user's words and concepts are tied in with those of expert users. We ask, "Why does this mental state M have the mental-semantic property (call it P) of meaning (broad sense) 'arthritis'?" And here we might be asking one of two things: (1) why does M mean "arthritis" as opposed to meaning something else (the problem of meaning assignments), or (2) why does M mean "arthritis" as opposed to not meaning anything at all (the problem of meaningfulness)?

9.6.1 MEANING ASSIGNMENTS

First, let us consider how CTM's explanation of semantics could be applied to the assignment of broad content. The schematic form of CTM's explanation of mental-semantics went as follows:

Schematic Account

Mental state M has mental-semantic property P because

(1) M involves a relationship to a mental representation MR, and

(2) MR has MR-semantic property X.

Now under the assumptions we have adopted, this schematic might be unobjectionable if the word 'because' were replaced by 'iff'. But in fact the schematic above is not intended merely as a biconditional but as an explanation. And as an explanation of broad content assignment, it seems to be barking up the wrong tree. The key issue in broad content assignment is what makes it the case that cognitive counters hook up with particular objects and properties and not with others. Whatever answer we give must explain the web of relations between organism and environment that accounts for the relations that are thus established. It may be true that my brain has a state that has the broad content "arthritis" and my counterpart has a structurally identical state that means "osteoporosis" due to the fact that people on Twin Earth say "arthritis" when they want to refer to osteoporosis ('water' for XYZ, etc.). We might even say that my brain state has a property of meaning-arthritis (meaning-H_2O, etc.) while his has a property of meaning-osteoporosis (meaning-XYZ, etc.). But the fact that there is such a property would not *explain* broad content assignment, but be a by-product of such an explanation. The explanation of broad content assignment would have to focus not on localized properties of cognitive counters, but on the network of relations between organism, cognitive counter, and environment that endows those cognitive counters (and the mental states in which they play a part) with broad content. Properties of representations, in and of themselves, are just not the right sorts of things to explain broad content assignment. Moreover, once we have to appeal to properties of whole organisms and their environments to explain (simultaneously) the broad content of the cognitive counter and that of the mental state, it is no longer clear why we should try to localize the mental-semantics of thoughts to some properties of their proper parts in the first place.[12]

9.6.2 MEANINGFULNESS

CTM fares no better with the explanation of broad-meaningfulness. Here the issue, above and beyond the issues involved in narrow-meaningfulness, is that of how thoughts manage to attach themselves to particular real-world objects and properties in a way that is underdetermined by their sense and by other factors internal to the organism. Now one might think that because cognitive counters are themselves internal to the organism, this should disqualify them from being explainers of broad contentfulness. But this is not exactly right: even though the cognitive counters themselves are internal to the organism, their properties may nonetheless be relational properties whose relata include ecological and social factors.

The problem, rather, lies once again with the focus of the explanation. If we ask why a particular thought is about *water,* as opposed to just consisting of a bunch of descriptions, it would seem that what we need is a story that shows us how to get broad content out of an amalgam of (1) the organism, (2) its narrow-contentfulness, (3) the cognitive counters, and (4) the environment. The explanation cannot focus solely on properties of the cognitive counters, as CTM would seem to have it, because the properties that can be explained by looking *just* at these entities remain constant over the cases of me, my counterpart on Twin Earth, and indeed over counterparts who may fail to have broad-contentful states at all.[13]

Let me make it clear what points I am and am not trying to make here. I am not saying that broad content cannot be strongly naturalized. (Though I happen to believe that it cannot.) Nor am I saying that some form of CTM—say, BCTM—cannot be making true assertions about the form of mental processes. Rather, I am saying that even if we *grant* both of these assumptions, and grant that semantic properties of mental states covary with properties of local states of cognitive counters, we cannot *explain* broad content merely by looking at the properties of these localized units, but must look back at the larger system embracing the whole thinker and her environment. And once we have done this, we might do well to reassess what is bought by trying to "reduce" the mental-semantic properties of mental states to MR-semantic properties of their proper parts.

9.7 NATURALIZING NARROW CONTENT

Finally, let us consider the question of whether narrow content can be strongly naturalized. The first problem we face here is in determining just what narrow content is supposed to be. An intuitive way of looking at narrow content is that it is the kind of content, or the portion of content, that is not dependent upon extramental factors such as ecological and social relations. This, however, sounds a great deal like phenomenological content. So one hypothesis about narrow content would be that it is the same thing as phenomenological content—that is, two mental states have the same narrow content just in case they are indistinguishable to the experiencing subject, and nothing lying outside of experience can be constitutive of a difference in narrow content. This also seems consistent with some discussions of "methodological solipsism" in the philosophy of mind (see Fodor 1980). However, most discussions of narrow content have concentrated not upon invariants of experience but invariants of structure and function. Narrow content is characterized as the property that molecular or functional duplicates would necessarily share. Now this would be consistent with the thesis that narrow content is phenomenological content if it could be shown that molecular or functional duplicates were necessarily phenomenological duplicates as well, but that thesis is contentious at best. So rather than make assumptions about the nature of narrow content, I shall explore four possibilities: (1) narrow content is phenomenological content, (2) narrow content is defined in terms of the properties molecular duplicates would share, (3) narrow content is defined in terms of properties functional duplicates would share, (4) narrow content is defined in terms of some other property of cognitive counters.

First, if narrow content is just phenomenological content, then all of the problems of accounting for phenomenological content accrue to it as well. To account for phenomenological content, your theory has to explain subjective feel, consciousness, subjectivity, and the perspectival character of intentional states. We have argued in the previous section that there are significant obstacles to this kind of explanation, and they would simply carry over as problems for strong naturalizations of narrow content as well if narrow content is phenomenological content.

If narrow content is defined in terms of structural properties, one is faced with a number of messy problems. First, it is not transparent why something defined in structural terms ought to be called "content" at all. The mentalistic overtones require explanation here. Second, if narrow content is *defined* in structural terms, it is trivial that it can be strongly naturalized, as presumably structural descriptions are already cast in a discourse that is patently naturalistic. But, third, if this is the case, it is not clear that any explanation of the mental has taken place. Defining narrow content in structural terms would simply shuffle the mystery around—the mystery would then be how you get from structurally defined narrow content to the *mental*. To arrive at a strong naturalization of the mental, you need to have a road from your naturalistic description (in this case, a structural description) to your mentalistic description that enjoys metaphysical necessity and explanatory transparency. As anyone who has tried to supply such explanations knows, this is no trivial task. Defining narrow content in structural terms does not solve the problem here, it merely presents us with a special variation on the problem.

Similar problems arise if we define narrow content in functional terms. First, we must be careful to specify what use of the word 'functional' is operative here. At one level of abstraction (the "rich" level), we can look at the mental, qua mental, as a math-functionally describable system. At a second level (the "sparse" level), we can look at the functional description in purely formal terms, abstracting from the fact it was originally a description of mental states. A richly construed functional model of the mind does not really *define* mental states in functional terms, though it may provide a unique characterization of each particular kind of mental state in terms of its functional relations to the others and to stimuli and behavior. A rich model assumes some knowledge of what content is, and does not explain its nature in functional terms, any more than, say, Newton's laws explain what gravity is. But as we have seen, a sparse functional model cannot serve as a definition or explanation of the mental, on the grounds that (1) the same functional description can apply to many things (e.g., abstract number-theoretic entities) that are not mental, and (2) nothing about the functional description has the conceptual riches to generate the distinctively mental character of intentional states.

So it looks as though narrow content will have to be defined in some other way if it is to be a viable notion at all. I wish that I had a candidate for such a definition, but I don't. The intuitive notion of content that I work with is the notion of phenomenological content. Perhaps other people operate with a different notion, but if they do they have not made it very clear, beyond the constraints (1) that it is not to be broad, (2) that it is not to be phenomenologically based, (3) that it is somehow to map things in the head onto things in the world, (4) that it is necessarily to be shared by molecular or functional duplicates, and (5) that it is to be unproblematically mental in nature. I am not sure that there is anything that does fit this bill. (Similar skepticism about the category of narrow content is voiced by Baker [1987] and Garfield [1988].) But let us consider the possibility in any case, however vaguely specified.

First, note that this notion of content bears a peculiar dialectical relation to the project of strong naturalization. If one of the defining features of narrow content is that molecular duplicates must share narrow content, then the metaphysical side of strong naturalization is assured. Either there is such a thing as narrow content and narrow content assignments are implied by physical description, or else there is no such thing as narrow content. The other side of strong naturalization, however—the explanatory side—is probably less easy to come by. Whatever narrow content is supposed to be, a strong naturalization must not merely bind it to naturalistic conditions by contingent bridge laws, but demonstrate its presence from some lower-level theory. The viability of this will of course turn heavily upon what one means by "narrow content," but it is a tall order to fill in any case.

Second, it seems to me that there is a problem for explaining narrow content in terms compatible with CTM in just the way there was a problem for explaining broad content with these constraints. If we assume for purposes of argument that there are some naturalistic features of cognitive counters that covary with narrow content, we still do not have a strong naturalization of narrow content unless we can show why the fact that a proper subpart of an intentional system has particular naturalistic properties makes it the case that the system has a thought about objects or states of affairs in the world. I am not at all convinced that such an explanation is to be had at all, but if it is to be had, it seems clear that the place to look is not in properties of the cognitive counters themselves, but in relations between the overall system and its environment. Again, the point here is not that BCTM must be wrong as a theory of the form of mental states. Rather, the point is that, even if BCTM is *right* about its functional description of the mind, and cognitive counters are the things that covary with meaning assignment, we need a theory with a very different focus to account for meaningfulness. If the question is "Why does this thought mean X rather than Y?" it may be appropriate to look at cognitive counters. If the question is why the properties of cognitive counters are so intimately associated with the mentalistic property of meaningfulness, it seems that we shall have to look elsewhere, at a theory that embraces a larger system.

9.8 CONCLUSION

This chapter has been a very quick examination of the prospects for what is commonly called a "naturalistic theory of content." We have seen that there are many issues lurking in the wings here—issues about what counts as "naturalization," what counts as a "theory," and what kind of "content," "intentionality," and "mental states" are at issue. What I have tried to argue here may briefly be summarized as follows: (1) If we are talking about conscious thoughts (the paradigms embraced by writers like Brentano, Husserl, and Searle, among others), these states do indeed seem indefeasibly to have properties like phenomenological feel, subjectivity, and the like, and there do seem to be serious obstacles to strongly naturalizing such properties. (2) If we are talking about dispositional states like beliefs and desires, and about their broad and narrow content, it may or may not prove possible to strongly naturalize these properties. But if it is possible to do so, it looks as though the kind of explanation we need will not focus on local properties of cognitive counters, as does CTM, but will appeal to relational properties of cognitive counters, the entire organism that is doing the thinking, and its environment.

I wish to draw both a weak moral and a strong one from this. The weak moral concerns where the burden of proof should lie. Much of the current discussion in the philosophy of mind assumes the possibility of a fairly strong sort of naturalization. But once we have distinguished the projects of strong and weak naturalization, we can see that strong naturalization calls for some fairly exacting (and rare) kinds of connections between discourses: namely, metaphysical sufficiency and conceptual adequacy. And once we look closely at the prospects of "explaining" the mental in these very stringent senses of "explanation," there seem to be some very large and glaring problems, especially if we add the further constraint of locating the nexus of meaning in properties of localized representations. Thus it seems to me that the burden of proof ought to be on the would-be strong naturalizer: we have reason to believe in strong naturalism when we see it accomplished and not before.

There is also a stronger moral one might draw, one that I want to state in a more assertive voice. All in all, it looks dubious that we shall ever have a strong naturalization of the mental. It looks even less likely that we should have one of the sort called for by CTM. CTM certainly does not provide such a theory of intentionality, and without it, those who doubt the propriety of intentional psychology have not been refuted. It is hard to see how to view computational psychology as a success if its success is to be judged by the standard of how well it naturalizes intentionality and vindicates intentional psychology. But perhaps these are not the right standards by which to judge computational psychology in the first place. Perhaps it is possible to separate issues about computational psychology as an empirical research programme from its relationship to more purely philosophical problems about the mind. In the final section of this book I shall attempt to present the beginnings of an alternative philosophical approach to computational psychology that frees it from the constraints of strong naturalization and vindication.

PART IV

An Alternative Vision

CHAPTER TEN

An Alternative Approach to Computational Psychology

The preceding chapters have presented an extended argument to the effect that CTM does not make good on either of its principal philosophical claims—that is, it provides neither an explanation of intentionality nor a vindication of intentional psychology. At best, we are left with a weakened, "bowdlerized" version of CTM (really a strong version of machine functionalism) that purports to describe the form of mental processes while treating the nature and legitimacy of intentional categories as a background assumption. Deprived of its impressive philosophical claims, CTM will seem to some to have lost its appeal entirely. And perhaps even more importantly, some writers seem drawn to infer that the failure of CTM as a *philosophical* project entails the bankruptcy of *computational psychology* as an *empirical research strategy* as well. Computationalism's critics and defenders alike often seem to assume that a successful scientific psychology ought to solve philosophical problems (like the mind-body problem) as well. On this view, the strong naturalization of the mental becomes a criterion for a successful scientific psychology, and even for the legitimacy of intentional categories. As a result, the *failure* of CTM to meet the philosopher's difficult metaphysical and explanatory criteria is thought to impugn the *scientific* enterprise of computational psychology out of which CTM arose as well.

I think this view is distinctly wrongheaded. (And, incidentally, I have yet to find a practicing scientist who shares it.) I make no claims about whether computational psychology *will* turn out to provide the foundations for a mature science of cognition, but it seems clear that what is needed for a successful scientific research programme is substantially weaker than what is needed for a solution to the mind-body problem or a strong naturalization of the mental. I said in an earlier chapter that in the end we would need to distinguish between the claims of computational psychology as an empirical research programme and CTM's distinctly philosophical claims. I intend to make good on that claim in this final section of the book by presenting the outline of an alternative interpretation of the importance of computational psychology as an empirical research programme. On this view, what the computer metaphor tries to provide for psychology are two features that have widely been considered cardinal virtues of mature sciences: namely, (1) mathematically exact descriptions and explanations, and (2) strategies for connecting our discourse about mental events with other kinds of discourse—particularly "lower-level" descriptions such as those provided by neuroscience. It is presently unclear whether computational psychology will *succeed* in either of these goals, or whether it will perform better than other competitor theories arising from neuroscience or neural network approaches. But were it to succeed in these goals, even without strongly naturalizing the mental, this would count as substantive scientific progress, and in ways that have important precedents in the history of science. And all that is needed for this kind of progress is the kind of *weak* naturalization called for by BCTM. *Strong* naturalization, I shall argue, is not a requirement internal to the practice of science, but rather an externally imposed criterion deriving from a particular philosophical ideology. From the standpoint of the practicing scientist, it is not necessary to produce an instantiation analysis of mental states—a realization account is enough. And likewise for the scientist it is not necessary to *vindicate* things like perceptual gestalts—such things are assumed as data to be explained, and their status is never called into question.

10.1 A STORY ABOUT THE MATURATION OF SCIENCES

I should like to begin by pointing to two features that seem to be common to the sciences that have traditionally been regarded as "mature" and "hard." The first characteristic of such sciences is that they have achieved a certain degree of rigor in their explanations. In particular, they have discovered mathematical expressions that capture, with greater or lesser degrees of exactitude, the relationships among the objects forming the domain of a given science insofar as they take part in the phenomena that that science seeks to explain. The second characteristic of such sciences is that they involve what are sometimes called "structural explanations" or "microexplanations"—or, more generally, connections between domains of discourse. These explanations relate phenomena that occur at one level of description $L1$ to the objects, relationships, and processes at a more basic level of description $L2$ that are ultimately responsible for phenomena at $L1$. Microexplanation can occur wholly within the bounds of a single science, and it can also occur across the boundaries of sciences, as in the case of the explanation of the combinatorial properties of the elements (chemistry) by reference to the behaviors of charged particles (physics). The issue of whether features such as these are *necessary* for scientific maturity is an important locus of contention in philosophy of science. Notably, there have been heated disputes about the status of biology in this regard. I wish to sidestep such issues here: I embrace Newton-Smith's (1981) idea that scientific theories can enjoy a plurality of "good-making" qualities. Mathematization and connectivity are two such qualities, and happen to be ones that have been emphasized in the "modern" view of science. My claim here is not that they are essential to a discipline's scientific status, nor that they are the *only* virtues relevant to scientific maturation, but merely that computational psychology may fruitfully be seen as an attempt to endow psychology with these virtues; and that if it succeeded in doing so, this would be a significant achievement.

A third feature that is often closely connected to the maturation of a science is the occurrence of a conceptual revolution that involves seeing the phenomena a science sets itself to describing and explaining in a fundamentally new way. The use of *metaphor* often plays a crucial role in such conceptual revolutions, though as often as not the metaphor is abandoned once rigorous mathematical description of the domain in its own right has been achieved. I am inclined to regard conceptual revolution and the use of guiding metaphor more as a feature of crucial stages in the *process* of maturation rather than a feature of mature sciences as such.[1] (After all, conceptual change and the use of metaphor are just as much a part of attempts at science that never get off the ground as they are of successful science.) Thus the conclusion of previous chapters that computing machines provide only metaphorical inspiration for computational psychology is in keeping with the role of metaphor in other sciences as well.

A few examples of mathematization and microexplanation may be of use in setting the stage for a discussion of psychology and computer science.

10.1.1 COPERNICUS, GALILEO, NEWTON

The emergence of the "new science" of the sixteenth and seventeenth centuries is sometimes referred to under the heading of the "Copernican revolution" in physics. And it is true that Copernicus played a crucial role in starting a conceptual revolution in astronomy that paved the way for the development of what was to become Newtonian mechanics. It is important to see, however, that Copernican astronomy in its own right is only the first step towards a mature physics.[2] Copernicus's own concerns were still largely those of an *astronomer*. He was concerned with finding a description of planetary motion. His own model of that motion, however, was highly influenced by his Ptolemaic predecessors: a system of circular orbits and epicycles around a point close to the sun.[3] (The sun was not at the center of Copernicus's system; it, like the planets, orbited another point in space. It is also worth noting that Copernicus's model contained more epicycles than Ptolemy's. The virtue of this model lay neither in its elegance nor in its predictive accuracy [see Kuhn 1957: 169-171].) Kepler, by contrast, was engaged in a project of finding a kind of mathematical description of planetary motion that would at once be elegant and exact. One important breakthrough—the one we probably imprinted upon when we learned about the progress of modern physics—was Kepler's discovery of the fact that planetary orbits are elliptical and that orbital speed can be determined on the basis of the area of the ellipse subtended by a portion of the orbit.[4] But when we learned this, we probably overlooked what was really important about this discovery. The fact that orbits take the form of an ellipse rather than some other conic form is really irrelevant to the progress of physics. What is crucially important is that the motions of the planets can be described exactly by mathematical expressions, regardless of which ones, and that they can all be described by the same kinds of expressions.[5] Physics would have done just as well if planetary orbits had been of a different, yet precisely describable shape. Celestial mechanics would have gotten nowhere so long as the only descriptions of planetary motion were in terms of a motley batch of epicycles having no discernible overarching pattern.

The further progress of modern physics was facilitated by the emergence of two other mathematical innovations: the development of algebraic geometry allowed for the possibility of performing algebraic calculations upon the motions of the planets through time, and the calculus provided for the possibility of making calculations about acceleration. The culmination of these advances was Newtonian mechanics, which summarized the interactions of gravitational bodies in a set of extremely elegant mathematical equations that came to be known as "Newton's laws." Newtonian mechanics unified the fields of astronomy, celestial mechanics, and sublunary physics under one set of mathematical descriptions, and stood as the standard for scientific theories until displaced by relativity theory (which was, itself, dependent upon the development of differential geometry for its descriptions of space and time).

It is worth emphasizing that Newton's achievement lies in his having left us a rigorous and general *description* of the effects of gravitational bodies upon one another. His ambivalent attempts to address the *why* of gravity (his much-discussed flirtation with "forces") add nothing to the picture, and his failure to solve the "why" of gravity detracts not in the least from the power and the utility of Newtonian mechanics.[6] One might well suspect that gravity amounts to something more than the empirical regularities of how bodies move in relation to one another, and it is appealing to seek some insight into this "something more," but such insight is not needed in order to make Newtonian mechanics "good science."

In brief, Newtonian mechanics provides for the mathematical maturity of a large portion of physics without providing any microexplanation for gravitational attraction in terms of some subgravitational level of explanation. Gravitation is treated as *fundamental*. And the lack of such an explanatory connection is not generally viewed as a fatal objection to Newtonian mechanics as good science, even though it is in some ways dissatisfying. Moreover, it displaced a Cartesian physics which *did* offer a microexplanation of gravitation in terms of mechanical interactions of particles.[7]

10.1.2 CHEMISTRY

A second example can be supplied by chemistry, which experienced distinctly separate stages of progress towards mathematical and connective maturity. There was a time when chemistry was largely independent of physics. In fact, chemistry attained a remarkable degree of mathematical maturity with very little help from physics, and it is possible to learn large portions of chemistry with little or no knowledge of physics. (Indeed, I believe it is still the practice in teaching chemistry in the schools first to present a chemistry that involves little or no physics before moving on to those parts where physics becomes crucial.) For there was substantial progress in understanding basic laws governing combinations of the elements before there was any underlying physical theory about what sorts of microstructure might account for these laws. The periodic table (Mendeleev around 1869), the notion of valence (Frankland in 1852), and a remarkable set of laws governing combinations of elements (as early as Lavoisier's work published in 1787) were developed long before these notions were further grounded in a theory of subatomic particles in the twentieth century. With the development of the periodic table, the notion of valence, and laws of combination, chemistry achieved a significant degree of mathematical maturity. In this respect, the age of Lavoisier made a significant step beyond the procedures and wisdom of previous chemists and alchemists, however great their technical skill, because there was, for the first time, a rigorous and systematic description of how the elements reacted in combination.

The major progresses in theoretical chemistry since the time of Mendeleev have been in the connections, the border marches, between chemistry and other disciplines such as physics and biology. In order to understand reactions between large molecules, for example, it was necessary to understand something of their physical structure. The propensities of molecules to combine in certain ways (some of which seemed anomalous) called for an explanation in terms of underlying structure, an explanation supplied by such notions as electron orbitals, ionic and covalent bonding, and the postulation of charged and uncharged subatomic particles. In the process, chemistry became increasingly connected to physics. At the same time, it became evident that many biological phenomena could be accounted for by chemical explanations: the bonding between hemoglobin and oxygen accounts for the transport of oxygen through the circulatory system to cells throughout the body (and hemoglobin's preference for bonding with carbon monoxide explains the ease of carbon monoxide suffocation); important parts of processes such as the Krebs cycle are chemical in nature; and of course the basic element underlying genetics, the DNA molecule, is typified by a particular molecular structure. With the advent of discoveries such as these, chemistry—which already had a high degree of mathematical maturity—acquired a large amount of what we might call *connective maturity* as well. It is worth noting, however, that most of these connections were not made until the latter half of the twentieth century, more than a century after chemistry had gained significant mathematical maturity. Some things take time.

It seems safe to say that sciences tend to become both more powerful and more firmly established as they become more intimately connected with one another. Chemistry raises questions that physics has to answer, and provides answers to questions raised by biology. Astronomy provides a lab for physics to study things that cannot be reproduced here and now, and physics provides a lab for testing hypotheses about things that are too far away to investigate firsthand.

10.2 THE APPEAL OF A MATURE PSYCHOLOGY

It should be abundantly clear that any developments that could bring all or part of psychology towards either or both of these kinds of maturity would be of major importance. Indeed, in the case of psychology, the absence of these kinds of maturity has been a large factor contributing to the widespread sentiment that psychology is not and perhaps cannot be a mature science. Consider first the matter of mathematization. Psychological explanation has traditionally been among the least systematic bodies of explanation among those disciplines that aspire to the name of science. Even higher-level disciplines such as economics have a stock of mathematical laws that describe their subject matter, even if only under "ideal" conditions. But while psychology has made inroads in terms of measurement of abilities (particularly in perceptual psychophysics), and seems susceptible to statistical generalizations over populations, the kind of explanation that takes place about and in terms of cognitive states has been notoriously resistant even to rough generalization, much less mathematization.[8]

The situation is little better with connectivity. While it is the case that some higher-level disciplines such as economics proceed on assumptions about cognition (e.g., rational decision making), the connections between psychology and *lower*-level disciplines such as neurology and biology (not to mention physics) have been at once contentious and unedifying. On the one hand, there has long been almost universal agreement that there are systematic and "special" connections between mind and brain. Even Descartes, notorious to many as the arch-dualist, attributed a wide array of psychological processes to the brain and nervous system, reserving only language, reasoning, and the will for the immaterial soul.[9] Descartes also viewed the connection between soul and body as extremely intimate—far more so than that between a pilot and the ship he steers—and probably sui generis (*Meditation* VI [AT VII. 81]). On the other hand, the nature of such a "special connection" has been elusive both philosophically and empirically. The philosophers cannot seem to agree on what the precise nature of the "special connection" might (or must) be, and the empirical scientists have been hard pressed to discern what the elements on the neurological (physiological, physical) side of the relationship might be. If one of the marks of a mature psychology would be having discoveries of the form "Mental phenomenon M bears special relation R to neurological phenomenon N," there seem to be two problems: the scientists cannot discover what N is, and the philosophers cannot decide what R has to be. To put it very mildly, it would be great progress if one could find a way beyond this old and frustrating impasse.

Now modern psychology has, in fact, made some progress on some fronts. There has been some significant quantification of perceptual psychophysics, and quantification of at least some of the observations in cognitive psychology. At the same time, neuroscience has emerged as a distinct offshoot of physiology that can draw upon other formal and empirical disciplines. Problems that were known to Helmholtz but unsolvable in his day are now solvable due to advances in mathematics (see Grossberg 1980). And the localization of mental functions in the brain has been greatly aided by more exacting and less intrusive observational techniques, such as those supplied by magnetic resonance imaging. But until recently the domain of cognition seemed largely untouched by these advances.

10.3 COMPUTATION, MATHEMATIZATION, AND CONNECTIVITY

It is here that the computer paradigm may prove to be of considerable worth. What computer science provides is a rigorous set of terms and methods for talking about certain kinds of systems: systems whose distinctive characteristic is their functional organization. What can be characterized in functional terms can be described rigorously by computer science. Now in order for this to be of use to psychology, several things must be the case. First, psychological phenomena must be functionally describable. And here the sense of "function" is the technical mathematical sense. To put it differently, psychological phenomena must be such as to he describable by an algorithm or effective procedure expressible in the form of a machine table. Here computer science supplies two things: first, a language (or set of languages) for the rigorous specification of algorithms; and second, an assurance that a very large class of algorithms (the finite ones) have a structure that can be instantiated by a physical system. So the computer paradigm might do two things here for psychology: it might provide a rigorous language for characterizing the system of causal interrelations between psychological phenomena, and at the same time provide assurance that this characterization can be realized in a physical mechanism that does not simply flout every law of nature. In short, the computer paradigm might provide the right tools for the mathematization of at least some part of psychology.

If computer science might *directly* provide the right tools for psychology to progress towards mathematical maturity, it might thereby *indirectly* provide an important contribution towards connective maturity as well. Of course, computation is not the right sort of notion to provide *everything* needed for connective maturity. Computation is an abstract or formal notion, and is therefore neutral, in important ways, about what sorts of things it describes. This is not to say that it does not itself specify functionally delimited kinds, but rather that in so doing it remains absolutely agnostic about (*a*) what the nature of these kinds may be, apart from their formal interrelations, and (*b*) how these functions are realized. A single computational description could apply equally well to a set of silicon chips, a network of cells, a structure of gears and levers, a set of galaxies, or the changes in affections of a Cartesian immaterial substance. Hence, even the best imaginable computational description of cognition would, in and of itself, do nothing about connecting psychology with other disciplines. For all that computational description buys us, it might still turn out that the things so described are not brain processes after all, but processes in an immaterial soul without even any analogous processes taking place in a brain. To be sure, the fact that computational structures *can* be physically instantiated is "bracing stuff" to someone who feels committed both to cognitivism and to materialism. It shows that the evidence for intentional realism may not be evidence against materialism, and vice versa. But the claim that cognitive processes are functionally describable has no consequences for the debate over whether materialism is correct in its ontological claims.

Nor does any computational description of cognitive processes have any direct consequences for how they are realized in the nervous system. This is, of course, a famous benefit of the computational approach: it allows for the possibility of the realization of equivalent functions in vastly different architectures—human, angelic, Martian, or Macintosh. The strongest constraint the computational description might place upon the realizing system is that it share the functional structure characteristic of the realized cognitive process. That is, if cognitive phenomenon C is realized through a realizing system R, and C is characterized by functional structure F, it must be the case that R is also characterized by F.

But while this does not directly connect psychology with, say, neuroscience, it may provide just the sort of link that is needed to forge a connection between the two. The brain, after all, is a complex and bewildering set of interrelated units, and those who wander in its tractless wastes are constantly groping to discern what are the significant units and relations. The availability of careful characterizations of cognitive processes is the sort of thing that might serve, if not as a Rosetta stone for the brain, at least as a hastily scribbled map. Indeed, the grand appeal of the functionalist strategy in empirical psychology lies largely in the fact that starting "top-down" and unlocking black boxes one stage at a time has often seemed to be the *only* way one can proceed if one is interested in phenomena lying at a higher level than, say, on-center offsurround structures. As a somewhat idealized characterization, sometimes the only way to proceed is to get as clear as possible on the form of the process you wish to describe and then look for some candidate realizing system that has the right "shape" to match it.

It thus appears that progress towards mathematical maturity is one of the more likely roads towards connective maturity as well. The link between the two is not hard and fast: one might get a good descriptive functional psychology without making much progress in seeing how the functional structures are realized in the brain, much as we have no microexplanations for gravity or magnetism. But then again, progress in mathematization might bring connective progress in its wake, as combinatorial chemistry was eventually supplemented by a structurally oriented chemistry that is strongly linked to physics. One simply does not know in advance how the cards will fall.

10.4 THE IMPLICIT FORM OF COGNITIVE PSYCHOLOGY

These considerations suggest an outline of how cognitive science proceeds and how it is related to intentional psychology. It is perhaps worth a brief digression to emphasize a few basic points. First, cognitive science is not so much in the business of *justifying* psychological phenomena as trying to bring them to some clarity. In particular, it is concerned with developing models of mental states and processes that get their formal properties right, and in a fashion that is precisely statable. Second, to employ this sort of strategy, it is necessary to proceed from some precomputational understanding of mental states and processes. To be sure, the process of modeling often alters our precritical understanding of our subject matter (be it psychology or fluid dynamics); but we have to start from something like commonsense belief-desire psychology (or one of the precomputational attempts to make it more rigorous). The application of the computer paradigm is an attempt to clarify a mode of description and explanation we already use, and intentional states are involved in what one would normally take to be both the explanatory posits of psychology and the data to be explained.

Third, research tends to proceed top-down, from behavior and consciously entertained intentional states and processes, to hypotheses about underlying intentional states that could explain them, to mechanisms that could support such states and processes. Fourth, the initial specification of underlying mechanisms emphasizes their formal properties rather than their physical nature. Eventually one would wish to reach a stage where the formal properties necessary to explain some higher-order process are precisely those of simple physical mechanisms like neurons or even complex mechanisms like fields of interconnected neurons. But it is not clear how many intermediate formal "information-processing" levels are needed to mediate between intentional description and neurological description.

Finally, there are clearly distinct projects of mathematization and connection, and it might be possible to make progress in one without progress in the other. Notably, it might be possible to achieve considerable insight into the formal structure of cognition through computer modeling without thereby achieving much progress towards knowing how that structure is realized through brain tissue. It is therefore conceivable that computational research in psychology could produce mathematical maturity without connective maturity. On the other hand, it is similarly possible that neuroscience and connectionist research would produce models that would, to the great surprise of many, exhibit emergent formal properties that are much like those independently desirable for description of intentional states and processes, in which case progress in mathematization and connection might come together.

10.5 INTENTIONALITY

Let us now make our discussion somewhat more concrete by looking at how one might apply the resources of computational psychology to the description—and, so far as possible, the explanation—of intentionality. In so doing, we shall take careful note of what kinds of connections are forged between domains and what kinds of explanation are actually likely to arise. It will be helpful to distinguish several different kinds of "accounts" that might be given, or perhaps several different kinds of description that might enter into a general account of intentionality. I wish to suggest that we may distinguish three separate components that an account of intentionality might have: (1) a "pure logical analysis" of intentionality, which describes the necessary structures of intentional states, (2) an abstract description of the formal properties of what is given in the logical analysis, and (3) an account of how the properties described in the vocabulary of the logical analysis are related to the realm of nature. These initial descriptions are necessarily a bit unclear, but will be expanded upon in the following pages.

10.5.1 THE PURE LOGICAL ANALYSIS OF INTENTIONALITY

The topic of intentionality has received a great deal of attention in the century or so since Brentano (1874) reintroduced it into the European philosophical milieu. Much of this attention (e.g., by writers such as Brentano, Husserl, and most of the continental tradition, and writers like Chisholm and Searle in the English-speaking world) has been devoted to the examination of what one might call the "logical structure" of intentionality—that is, of properties that intentional states have just by virtue of being intentional states, or by virtue of being intentional states of a particular sort (e.g., judgments, conjectures, perceptual gestalts). A number of such properties stand immediately to the fore.

- All intentional states involve an attitude-content structure.

- Every intentional state is "directed towards" something—its "intentional object"—whether anything actually exists corresponding to that object or not.

- Every intentional state is the intentional state *of* some intending subject.

- Intentional states can have other intentional states as their intentional objects.

- Every intentional state presents its object in some fashion or under some description (as being *thus*) and under some intentional modality (judging, hoping, desiring, etc.).

- Every intentional state has properties that determine its "conditions of satisfaction"—that is, that determine what would have to be true of the world in order for that state to be felicitous. And so on.

Features such as these are features of intentionality per se, and not features of intentionality that accrue to it specifically as it occurs in some particular kind of being. So whereas, for example, the claim that *all desires are realized in brains* is at best a contingent truth (it seems logically possible that things with different bodies—or perhaps even no bodies—could have desires), it is a necessary (and indeed analytic) truth about desire that every desire is a desire *for* something. The process of clarifying such features seems to be more a kind of *analysis* than empirical inquiry, and seems to be in large measure concerned with what might be called the "logical form" of intentional states—that is, the fact that they have an attitude-content structure, the fact that they posit an object or state of affairs under some description, and so on.

These "logical" properties of intentionality were given some attention by Brentano, and have been more carefully developed by writers like Roderick Chisholm (1957, 1968, 1984b), John Searle (1983), and particularly Edmund Husserl (1900, 1913), who devotes several volumes to the explication of intentionality.[10] It might be appropriate to call this kind of description of intentionality "pure logical analysis" of intentionality. Husserl's expression "pure phenomenology" is also appropriate, though it may prove misleading to readers who associate the word 'phenomenology' with things having to do with qualitative feels and not with intentional states. What is properly suggested by the term involves the claims that (1) (occurrent) intentional states are things we experience, (2) they can also become the objects of our inquiry and analysis, (3) such intentional states "have a phenomenology" in the sense that features such as the attitude-content structure of intentional states are part of the "what-it's-like" (see Nagel 1974) of intentional states, and (4) these features can be discovered by phenomenological reflection. The "what-it's-like," of course, is not a *qualitative* "what-it's-like" (a "what-it-*feels*-like") but a *logical* "what-it's-like" (a "what-form-it-has"). Chisholm's linguistically based approach to intentionality is an attempt to attain greater clarity about mental *states* by attending to the logical forms of *sentences* used to report them. (Popular myths to the contrary notwithstanding, the focus of Chisholm's interest is the intentionality of mental states, and he approaches them through an analysis of the sentences used to report them because of the difficulty of addressing the topic of mental states directly. Chisholm, like Husserl, thinks that phenomenology is difficult and elusive.)

In addition to the analysis of features common to *all* intentional states, this kind of pure analysis could reveal features peculiar to *particular kinds* of intentional states.

- It is part of the very nature of PERCEPTUAL experiences that they set conditions of fulfillment involving a state of affairs in which something corresponding to the intentional object actually caused the state.

- It is part of the essence of states having the modality of RECOLLECTION (things that present themselves as memories) that they be founded upon previously experienced immediate experiences of PERCEPTUAL PRESENTATION or some other intentional modality, and so on.

This kind of analysis of intentionality stands in some ways *prior* to the kind of investigation of the mind undertaken by computational psychology and BCTM. Computer modeling and artificial intelligence might, of course, provide very useful tools in pursuing such an analysis, as computer science provides ways of talking about inference and data structures that can greatly enrich one's ability to talk about logical form and conceptual relationships. It may also be that certain ideas that have emerged out of computer science (procedural representation, to name one notable example) may provide tools for the logical analysis of intentionality that would not otherwise have been available. But by and large, it is our intuitions about our mental states that constrain our computational descriptions, and not vice versa. If computational description is useful, it is useful in furthering a project to which we are already committed when we undertake the analysis of intentionality.

10.5.2 THE FORMAL DESCRIPTION OF INTENTIONALITY

There is, however, a second level of description at which computational description might really add something new to an account of intentionality. For while traditional logical analyses yield numerous essential insights into intentionality, they tend to do very little to give an overarching model of how these insights fit together, and in particular they do not give the kind of model that would seem to be of much use in the project of building an empirical theory. Here the resources of computer science may be of some use precisely in their ability to supply descriptions of the formal properties of certain kinds of systems. And the insights gained through logical and phenomenological analysis might be interpretable as formal constraints placed on a mathematical description of the "form" of intentional states and processes. This line of thought has been pursued by writers like Dreyfus and Hall (1984) and Haugeland (1978, 1981, 1985), who have seen a certain continuity between the Husserlian approach to intentionality and computer modeling. I shall not go into detail about where I agree and disagree with the analysis presented by these writers but shall supply a few examples of how I think this sort of intuition might be fleshed out.

(1) One insight to be gained from the logical analysis of intentionality is that intentional states can be about other intentional states. I can, for example, wish I could believe that my neighbor was trustworthy (WISH [BELIEF [my neighbor is trustworthy]]), or remember once having believed in the lost continent of Atlantis (RECOLLECTION [BELIEF [Atlantis exists]]). And such an insight is all very well and good, not to mention true. This same insight, however, can also be cashed out as a more interesting claim about the possible structures of intentional states: namely, that the structure permits of recursion. Or, to put it differently, if we were to give a formal description of the form of intentional states, it would have to involve a rule that allowed for recursion by embedding reference to one intentional state within the content of another. And since we have formal ways of talking about recursion, we have now taken a small step towards being able to say something about the abstract formal properties of intentionality. Such an insight might also provide the basis for other hypotheses—such as that the distinction between competence and performance can be applied to this embedding of intentional states, and that there might be general rules governing what intentional states can take particular other intentional states as arguments.

(2) Some insights gained from logical analysis take the form of either normative or productive rules concerning intentional states. For example, an analysis of the intentional modality of recollection reveals that it presents its object as having been previously experienced in some other intentional mode (e.g., perception). This sets normative constraints on the satisfaction of such a state: you cannot *felicitously* remember seeing Y unless you have at some previous time had a perceptual gestalt of Y. You can, however, experience a state whose intentional modality is RECOLLECTION and whose content is that of oneself having seen Y without actually having had a perceptual gestalt of Y in the past. (There are false

memories, after all.) So we would wish to describe our intentional processes in such a fashion that

(1) it is possible to experience RECOLLECTION[self having seen Y] without having previously experienced PERCEPTUAL PRESENTATION[Y], but

(2) the satisfaction conditions for RECOLLECTION[self having seen Y] cannot be fulfilled unless PERCEPTUAL PRESENTATION[Y] has previously been experienced.[11]

Such rules can, of course, be characterized in terms of purely formal relationships expressed in the form of normative licensing rules (which set constraints on satisfaction conditions) and productive rules which describe what combinations of intentional states actually result in the generation of particular new intentional states.

(3) To take a somewhat different example, the analysis of intentionality may show us how to separate the issue of "being about something" in the sense conveyed by the opaque construal of intentional verbs from the issue of the fulfillment of such states in veridical intentional states. There is, I think, a good case to be made to the effect that, once this is done, we already have a mathematical format for talking about the fidelity of at least some intentional states (e.g., the perceptual ones): namely, the Mathematical Theory of Communication (MTC). Even if one is wary of the claims made by Sayre (1986) that one can build semantic *content* out of the technical notion of information employed in MTC, it nonetheless seems that MTC might be telling a perspicuous story about the difference between veridical perception and perceptual gestalts that result from illusions, hallucinations, and the like.

Now it is important to see how this story differs from some other stories about computers and the mind. The point here is not that intentional states *are* just functional relationships to symbols and hence precisely analogous to computing machines. The point, rather, is that *there is a system of abstract properties* to be found in the system of intentional states and processes, and these might very well be the same abstract properties that are being explored in computer science, in much the same sense that the calculus provided an appropriate set of mathematical forms for problems in classical mechanics. The question is that of finding the right description for the formal features of intentionality, and not that of whether anything sharing those formal features would thereby have intentionality as well. The answer to that latter question is surely *no:* there will always be purely abstract objects having any given formal structure, and these do not have intentionality. And in general we should not expect any two isomorphic systems to be identical in all properties: for example, thermodynamics and Mathematical Theory of Communication share a formalism, but have different subject matters. The "intentionality" of symbols in computers may seem to track the intentionality of mental states, but only because symbols in computers are representations that have semiotic-meanings and hence are designed to express the mental-meanings of mental states.

To put matters somewhat differently, if we start with an analysis of intentionality and add the resources of computer science, we might end up with a useful set of formal constraints upon the shape intentional systems can take. On the other hand, if we merely start with formal properties, we will never develop notions such as mental-meaning out of those, and hence will never get *intentionality* as opposed to getting the formal shape shared by intentionality and perhaps any number of other things. Moreover, we need to start with our intuitions about intentionality to know which formal properties are relevant. There are many possible formal descriptions which might be interesting but are not viable as descriptions of cognition. The only way to get a formal description of *intentionality* is to start top-down from our intuitions about the intentional states we already know about—namely, our own—and study their formal "shape" by a process of abstraction.

10.5.3 INTENTIONALITY AND THE REALM OF NATURE

We have thus far discussed two possible components of an account of intentionality: a pure logical analysis and a more mathematically perspicuous description of the formal relations revealed in the logical analysis. The remaining portion of such an account—and the portion that has seemed to be of greatest interest to writers in the philosophy of cognitive science—is an account of how intentional phenomena relate to the natural world. Of course, what many people really want is a way to see intentionality as itself being a natural phenomenon; but as we do not at this point know whether that is possible, it seems a bit strong a desideratum to set for an account of intentionality. It seems a more sober approach to begin by asking what *can* be done to relate intentionality to natural categories and then assess the relation between our conclusions and our previous metaphysical and methodological commitments.

It is, I think, agreed by almost everyone who believes in mental states

at all that there is *some* sort of special and intimate relationship between mental states and particular kinds of bodily states. What is open
to investigation and dispute are the following three questions: (1) What
is the right inventory of mental states? (To what extent is our commonsense inventory accurate? Are there "states" called "beliefs"? Was
common-sense explanation ever intended to imply that there were, or is
this an error of philosophical analysis, as implied by Wittgenstein and
some in the continental camp?) (2) What bodily states are thus "specially related" to particular mental states? And (3) what, precisely, does
this "special relationship" consist in? I shall say very little about the
first question here. Let the reader simply consider the remaining questions with regard to those mental states she does feel committed to.

Now the question of what bodily states particular mental states are
"specially related" to seems to present a reasonable agenda for empirical psychology without shackling the psychologist to a burden of metaphysical proof. Empirical psychology can show such things as that
there is a special relationship between C-fiber firings and the experience of pain. It cannot derive the qualitative state from a description of
the physiology of C-fibers, nor from a description of how they interact
with the rest of the body. And the result that C-fiber firings are "the
physiological side of pain" is agreeable to philosophers who fall into
very different metaphysical camps. Where they differ is on what to say
about the precise nature of the relationship between C-fiber firings and
the experience of pain: whether they are contingently identical, or that
one supervenes on the other, or one causes the other, or that they are
causally unrelated but perfectly coordinated by some preestablished
harmony, and so on. And it seems quite clear (*a*) that scientists do not,
by and large, care about these further issues, and (*b*) that, qua scientists,
they are right not to care. (Consider what a burden upon science it
would be if scientists waited until all the metaphysical disputes could
be resolved!) And while the opacity of qualia to scientific analysis (i.e.,
the fact that you cannot "derive" qualitative states from neuroscience
by way of something like an instantiation analysis) may seem a dis-
tressing anomaly to some, it is an anomaly that philosophers are, by
and large, deciding that we have to live with.

I think the situation is very much the same with respect to intentionality. To spell matters out more explicitly: (1) Intentional states have a phenomenology, a "what-it's-like," though it is not a *qualitative* "what-it's-
like" but a *logical* and *semantical* "what-it's-like." (2) Psychology might,
in principle, be able to identify bodily states that are "specially

related" to intentional states such as occurrent judgments or perceptual gestalts. (3) If it can do this at all, it can do so in a scientifically respectable way without settling questions about ontology. (4) As argued in the previous chapter, the phenomenological element of intentional states is not subject to the kind of "strong naturalization" involving an instantiation analysis. And, hence, (5) intentionality is not subject to strong naturalization.

There are, of course, important differences between qualia like pain and intentional states. First, intentionality has a rich logical structure that pain lacks. And it is for this reason that a simple quality such as pain can be realized by a physiological mechanism with so few dimensions of freedom as the firings of certain kinds of nerve cells. A phenomenon such as judgment could not, even in principle, be realized through the firings of particular cells, because the physiological phenomenon involved does not have the right logical structure to support the logical structure of judgment. And here we have an important link between the formal analysis of intentionality and any account we might give of its realization: namely, that *the analysis of intentionality places formal and in some cases causal constraints upon the kinds of mechanisms through which intentional states can be realized.* This is the sort of issue that was being explored through the work done in "knowledge representation" by artificial intelligence researchers during the 1970s. It is also of fundamental importance to psychology, for the mathematical description of the mechanisms both specifies the functional properties that constitute it as a mechanism (and not an accidental by-product) and gives an important clue to identifying the tissue in which it is realized and the kinds of activity in that tissue that are of interest. (There can, or course, be other clues, such as evidence of activation through magnetic resonance scanning and the topology of neural connections.) And here too computer modeling (both conventional and connectionist) can be of crucial importance in determining whether a given architecture can support the formal features necessary to a particular kind of state or process.

10.5.4 BCTM AND ACCOUNTING FOR INTENTIONALITY

Given the foregoing analysis, what can BCTM and computational psychology do by way of providing an "account" of intentionality? The first thing they might be able to do is to supply a way of taking a pure logical analysis of intentionality of the sort offered by Brentano, Chisholm, Husserl, or Searle and teasing out a more rigorous description of the formal properties of intentionality. This would be the kind of project that would move (intentional) psychology towards mathematical maturity. This, however, holds a further possibility: the analysis of intentionality places constraints upon the formal and causal features that a physical system must have in order to realize intentional states, and this might be of use in the project of providing a realization account for intentional states, thereby providing a measure of connective maturity for psychology as well.

Of course, whether this connective maturity could *actually accrue* to psychology is an empirical question. For a formal specification of intentionality would open the doors to a number of alternative possibilities. It seems to me that any of the following could turn out to be the case:

(1) Intentionality has formal properties that can be physically realized, and we can find mechanisms in the body that share those properties and whose activation is correlated with the experience of the corresponding intentional states.

(2) Intentionality has formal properties that cannot be realized by any physical system.

(3) Intentionality has formal properties that can be physically realized, but not by a digital machine (hence we need a noncomputational psychology if we are to provide a realization account for intentionality).

(4) Intentionality has formal properties that are not in fact shared by any mechanisms in the body, and hence at least some intentional states are not realized through bodily states.

(5) Intentional states are individually matched with physiological states sharing their formal properties, but this typing of states is not relevant to causal regularities.

I suspect that there are a number of other possibilities as well. But this selection should be sufficient to show that empirical study of intentionality could have *some* ramifications for metaphysics, albeit not definitive ones. If intentional states and physiological states are nicely correlated in a way that preserves causal regularities, a great number of ontological possibilities remain open. If intentionality has formal properties that cannot be realized by any physical system, intentionality and materialism are incompatible, and most dualists are likely to be surprised as well. (Perhaps Platonists or Kantians would find this possibility less jarring; I am not sure.) It might be the case that something like the frame problem could be made to pose a case for something like (3), in which case we need some noncomputational approach to psychology. Possibility (4), again, opposes intentional realism and materialism, though again it might surprise dualists as well. And (5) might well be very welcome to both interactionists (who might want individual thoughts to have physical correlates through which the body is influenced while reserving the causal regularities for the nonmaterial soul) and epiphenomenalists. The kind of analysis I suggest thus does some limited metaphysical work, but not in a way that is question-begging and ideological: it is only by getting the best analysis of intentionality we can get, and seeing how it might match up with natural phenomena, that we really know what is at stake metaphysically in an account of intentionality.

The research programme associated with BCTM thus might do something very significant by way of providing an "account" of intentionality: it might render the logical analysis of intentionality formally perspicuous, and it might provide the key to a realization account of intentionality as well. What it does *not* do, of course, is produce an account of the *nature* of intentionality—of what it is to be an intentional state—in terms of some other kinds of categories (for instance, naturalistic ones). The key notions of "aboutness" and "(mental-)meaning" are left unexplained even if there should turn out to be some particular naturalistic relationships through which they are realized.

10.6 A REORIENTATION IN THE PHILOSOPHY OF COGNITIVE SCIENCE

It is my belief that the thumbnail sketch of computational psychology presented above presents an interesting scientific research programme that might turn out to produce theories with the two important scientific "good-making" qualities that I have described. The preceding sections should, I think, make it clear that it might endow psychology with these virtues even *without* providing a strong naturalization of the mental. Mathematization distills the form of a process by abstracting from the nature of the things that are related, and hence the mathematically reduced description does not provide sufficient conditions for the properties and objects it relates. Moreover, mathematization is methodologically dependent upon the prior assumption of the phenomena to be related by the mathematical descriptions.

It is likewise possible to form linkages between domains of discourse that fall short of metaphysical sufficiency and conceptual adequacy. We can accept the wave-particle duality of matter without being able to derive one from the other, and we can accept that the detection of oriented lines is realized through columns in the V1 region of the visual cortex without being able to derive phenomenological properties from neural states, or vice versa. Such a situation leaves questions open, of course, and they are nagging questions. But (1) it is not clear at the outset whether they are really *scientific* questions or *philosophical* questions that ultimately turn upon the way we have misconstrued the problem, and (2) in the meantime, the existence of these questions does not impugn the progress that has been made along the way.

The big point here is that *the needs of the philosopher and those of the empirical scientist diverge in important ways* (see Horst 1992). And this is reflected in the ways we tend to regard one another's projects. In my experience, scientists tend to be utterly mystified at what philosophers could desire beyond (*a*) a good model of mental processes, and (*b*) localization of the functional units of the model in the nervous system. To them, the issue of whether the connections found are reductions, or supervenience relations, or merely empirically adequate generalizations is virtually unintelligible, and surely of no interest for the practice of science. I think they are *right* to think this way *so long as the issue is one of empirical science*. Empirical science aims at finding the regularities and connections that are there to be found, and seeks as strong an explanatory relation as it can uncover. Science is blind to distinctions beyond empirical adequacy, as such distinctions cannot be decided by experiment. And from the empirical perspective, it counts as progress to assert what you have found, even if you think you should have found something more. But from the empirical standpoint, it is also true that theoretical ideologies (say, that we *must* have a *certain kind* of explanation, or a certain view of the world) stand in the dock against the evidence of data and successful theories, and not just the other way around. The Cartesian view of the essence of matter as extension implied that there *must* be a mechanical explanation of light and gravitation, and indeed of all material phenomena. But better experimentation and better theories forced us to abandon Cartesian physics. Likewise, field theories now stand alongside theories of contact interactions. And teleological categories in biology are being accepted against an older mechanistic ideology. A psychology with sufficient internal good-making qualities that was not a strong naturalization would itself call the need for strong naturalization into question.

I therefore think that philosophers have been wrong to yoke the *sci -*

entific importance of computational psychology to its potential *philosophical* benefits. (And hence the failure of CTM to produce those philosophical results need not impugn computational psychology as an empirical research programme.) My alternative suggestion is that we look at what computational psychology might provide in the way of good-making qualities *internal* to scientific practice, and that mathematical and connective maturity stand out in this regard. This alternative has implications for how we ought to go about the philosophical study of cognitive science. For example, if this approach is the right approach, the best way to study cognitive science as science would be through careful case studies and comparisons between different theories (much the way one studies the history of any other science). Such an endeavor lies outside the scope of this book. But it is possible briefly to assess some of the comparative merits of my alternative construal of the importance of computational psychology with approaches that bind science and metaphysics in a tighter yoke. Here are several advantages that I think my alternative approach enjoys.

10.6.1 A BETTER DESCRIPTION OF SCIENTIFIC PRACTICE

First, the alternative approach is in closer accordance with the facts of scientific practice. The kinds of metaphysical and epistemic constraints required for strong naturalization are often missing in paradigm examples of good theories in other sciences. Scientific progress has often come, for example, in the form of laws relating several variables in the absence of any metaphysical necessity or conceptual adequacy relating those variables, and also in the absence of any microexplanation of why the law should hold. Most laws, when they were discovered at least, expressed relationships that were metaphysically contingent and epistemically opaque. If this is good enough for, say, Newton's laws, why should we hold psychology to a more stringent standard?

It is also plainly the case that, say, experimental psychologists and psychophysicists do not seem to feel a need to vindicate the phenomena they study (at any rate, no more than do practitioners of any other sciences), and that phenomenological and intentional description often play important roles in the initial description of problems that it is the job of theoretical psychology to solve. And the kind of "explanation" of, say, "seeing a red square" that is sought by a vision theorist does not require anything like metaphysically sufficient conditions.

Finally, one can point to anecdotal evidence from joint meetings in which psychologists are frustrated and baffled by the problems that divide philosophers. My experience thus far has been that the characterization I have presented of where psychological concerns end and purely philosophical ones begin has been almost universally well received by psychologists, modelers, and neuroscientists, though often more controversial among philosophers.

10.6.2 IDEOLOGY AND THEORY—BAD PRECEDENTS

Second, one can hardly be optimistic about the precedents for holding scientific theories to the litmus of a particular metaphysical or methodological ideology. Cartesian mechanism, Humean views on induction and causation, logical positivism, behaviorism—these stand as just a few prominent examples of research programmes that were based on metaphysical or methodological views with heavy theoretical implications. All of them seemed very compelling in their time, as they canonized metaphysical views or forms of explanation that held a firm grip on the imaginations of their day. But each was eventually eroded by successful scientific work that belied their metatheoretic assumptions. It is one thing to study the mutual influence of scientific theory and metaphysics or theories of scientific method. It is quite another to take a view like strong naturalism and use it as a test for scientific legitimacy. This kind of move has a poor track record. Better to look at the kinds of explanations that psychological theory *does* provide and draw one's conclusions from there.

10.6.3 SCIENTIFIC PROGRESS WITHOUT NATURALIZATION

Third, as argued above, it is possible to have important kinds of scientific progress without also achieving interesting metaphysical results or solving traditional philosophical problems. A philosophy of cognitive science that is tied down to particular metaphysical goals is not free to assess the kind of good-making qualities that psychological theories may enjoy (and which perhaps some already do) that fall short of these goals. My approach, in contrast, emphasizes such achievements, while remaining open to the investigation of results with more metaphysical bite should they arise.

10.6.4 COMPARISONS WITH OTHER RESEARCH PROGRAMMES

Finally, my alternative approach yields a strategy for comparing "traditional" computational approaches to psychology that center on notions of "rules" and "representations" with rival approaches arising from other sources such as neuroscience and neural network theories. For most of the time that the computer metaphor has been exploited in psychology, other approaches have also been explored, even if they have only recently been brought to the awareness of a broad audience of philosophers. For example, the Mathematical Theory of Communication of Shannon and Weaver (1949) has been explored by Sayre (1969, 1976, 1986) as an alternative basis for characterizing mental processes from the early 1960s to the present, and "neural network" approaches based on attempts to provide a mathematical characterization of the interactions of large numbers of cells in the brain were pioneered by McCulloch and Pitts (1943) and have been developed over the space of more than three decades by researchers such as Grossberg (1982) and Anderson (1973), as well as more recent researchers better known to philosophers such as Rumelhart and McClelland (1986) and Smolensky (1988).

Each of these approaches has its own preferred mathematical tools (sometimes including new mathematical machinery developed to solve particular problems) and its characteristic formally exact models of processes underlying cognition. Much of the debate between proponents of different models centers upon the features gained and lost by different kinds of mathematical apparatus: for example, the use of differential versus difference equations, or additive versus multiplicative shunting. (A good survey of mathematical differences in neural modeling is found in Levine [1991].) A second area of difference, both within the neural network camp and between its members and traditional artificial intelligence comes in the relationships assumed between the project of mathematical modeling and the project of connecting the model upwards to the data supplied by psychophysics and downward to that supplied by neuroscience. Some models are designed only to fit particular data curves, while others are intended additionally to be neurologically plausible.

In short, there is a great deal to be understood about the major research programmes in this area by looking at their mathematics, looking at their commitments to forming ties to other domains, and looking at their strategies for doing so. This kind of approach has some hope of shining light on individual theories, and also of clarifying the real differences between them. By contrast, most of what has come out of the philosophy of psychology with respect to neural networks so far has been centered on one of two issues: (1) whether connectionist theories are really the same as (or compatible with, or reducible to, or implementable through) computational theories (and vice versa), and (2) how the availability of connectionist models vitiates Fodor's "only game in town" argument to the effect that we need intentional states because the only models we have to explain behavior require them as theoretical posits. Very little has been said of a philosophical nature about neural network theories in their own right, as opposed to how they compare to the kind of view espoused in CTM. Since I have argued that CTM does *not* in fact bear any philosophical fruit after all, I find comparisons with neural networks along *that* axis to be pretty much beside the point. Looking in detail at their mathematical repertoires and their commitments to kinds of interdomain connections in various directions, by contrast, gives us a concrete project in philosophy of science that we can sink our teeth into.

10.7 COMPUTATION AND ITS COMPETITION

This mention of competing research programmes provides a natural transition to a final point to be made in this chapter. What I have tried to provide here is an alternative approach to assessing the importance of the computational paradigm in psychology—a way of looking at the question, "If computational psychology is a successful research programme, what is it that it will have contributed to psychology?" My answer has been that computational psychology tries to endow psychology with two good-making qualities that have often been viewed as highly (even crucially) important to the maturation of sciences: namely, mathematical and connective maturity. But we should note that both the question and the answer are highly conditional: they concern what computational psychology *would* do if carried out successfully. Of course it is an open question whether it *can* or *will* be carried out successfully, so none of the preceding is meant as an endorsement of the computational approach to psychology as the *right* approach. It is *an* approach that is on the table, and as philosophers of science we are obliged to assess its promise.

At the same time, it is important to acknowledge that there are serious issues concerning the viability and prospects of the research programme. First, like any research programme, it may simply not succeed even on its own terms by failing to achieve any fundamental explanatory successes. Second, there have been serious arguments raised by writers like Dreyfus (1972) and Winograd and Flores (1986) to the effect that there are properties of the mind that symbol-manipulating systems cannot duplicate, and even more fundamental objections by writers like Ryle

(1949), Wittgenstein (1958), and the *Verstehen* tradition to the interpretation of psychological ascriptions now called the "theory-theory." I think these are all serious issues, and any one of them could turn out to have serious implications for the possible applications of rival models of the mind.

Finally, any successes of the computational approach to the mind in accordance with BCTM would also have to be assessed by comparison with the successes of rival research projects such as those arising out of neural network approaches or information theory. A brief list of issues might include but not be limited to:

The availability of exact mathematical descriptions for a wide variety of psychological phenomena.

The "naturalness" of these descriptions to their subject matter. (For example, the classical computational approach seems to have a natural way of approaching the attitude-content structure of intentional states. Do other approaches have an equally intuitive way of reflecting this feature in their models? Likewise, connectionist models seem naturally suited to modeling the behavior of fields of neurons, and information theory seems to have a natural way of talking about fidelity of intentional states.)

The comparative elegance of the models. (Can one approach supply straightforward descriptions and explanations where another requires a mass of ugly kludges?)

The tradeoffs between having a general framework (such as rule-conforming counter transformations or the technical notion of information) and having the freedom to employ an eclectic batch of mathematical tools.

The ways in which alternative research programmes are really competitors, and the extent to which they are ultimately compatible—because their formalisms turn out to be equivalent, for example, or because they are really engaged with different aspects of cognition or different questions about the mind.

The investigation of these questions will constitute a serious *philosophical* research programme in its own right, and will not be undertaken here. However, one important result of the discussion that has preceded in this book is the following: one might have thought that the approach to the mind found in CTM should enjoy pride of place over some of its competitors because it solves certain philosophical problems (explaining intentionality and vindicating intentional psychology) that its competitors have no strategies for solving. But I have tried to argue that it fails to solve these problems, and that its true benefits lie in how it might provide virtues wholly internal to the science of psychology. But without the philosophical claims to confer pride of place upon CTM, there is a level playing field. We may now assess the comparative merits of CTM, connectionism, and neuroscience on wholly scientific grounds as scientific research programmes. This, I believe, effectively separates a set of questions about the philosophy of mind (such as the mind-body problem and the question of the precise metaphysical relationship between mental states and the bodily states through which they are realized) from questions about the science of the mind (such as what the important goodmaking qualities are for such a science). And this, I believe, is progress.

CHAPTER ELEVEN

Intentionality Without Vindication, Psychology Without Naturalization

The argument thus far in the book may be summarized briefly as follows: (1) Advocates of CTM have claimed that viewing the mind as a computer allows us not only to make advances in empirical psychology, but also to satisfy the more specifically philosophical desiderata of supplying a (naturalistic) account of intentionality and vindicating intentional psychology by demonstrating its compatibility with causal-nomological psychological explanation, materialism, and the generality of physics. (2) CTM fails to make good on its claims to produce these philosophical results. But (3) a "bowdlerized" version of CTM can nonetheless provide a framework for an interesting empirical research programme in computational psychology, because what is needed for the "good-making" qualities internal to a science is much weaker than what is needed for strong naturalization or vindication. Psychology could attain a significant degree of internal mathematical maturity without any demonstrable connections between psychological categories and the categories of the physical sciences. And it could attain a great deal of connective maturity through localizations that were empirically adequate yet metaphysically contingent and epistemically opaque. Empirical science is largely blind to metaphysical modalities stronger than empirical adequacy, while questions about the metaphysical nature of mind-body relations are precisely the sorts of things that are of importance for strong naturalization and vindication. In short, science and metaphysics enjoy a substantial degree of mutual autonomy.

But if such issues are not the practicing scientist's concern, they certainly are the concern of philosophers. Or, more precisely, while there is an important kind of philosophy of science that examines precisely what particular sciences are about, there are also broader philosophical questions of a metaphysical bent as well. And neither the legitimacy nor the importance of the metaphysical questions is undercut by the fact that they are not questions for the empirical scientist. The reason that CTM was initially of interest was that it seemed to offer solutions to some problems that many philosophers in this century have regarded as difficult and important ones. Even if we can do good science without answers to the questions, it nonetheless behooves us *as philosophers* to see where we are left if our bowdlerized interpretation of computational psychology leaves them unresolved.

The reader will recall that there were two main issues that CTM sought to address. The first was to provide an account of the intentionality of mental states—and in particular, to do so in naturalistic, or nonintentional, terms. The second was to "vindicate" intentional psychology. And here there were three principal concerns: (1) that intentional explanation be (or at least point to) causal-nomological explanation, (2) that intentional realism be compatible with materialism, and (3) that intentional realism be compatible with the generality of physics. For a compatibility *proof*, however, you need *demonstrable* identities between mental and naturalistic states, and I have argued (*a*) that BCTM does not provide this and (*b*) that it is not to be expected. Luckily, empirical correlations of a weak metaphysical nature are good enough for science. But those who are concerned about the original motivating problems on *philosophical* grounds may still have reason to be worried about the resulting picture on those same grounds. It is thus my intention to address such concerns in this chapter.

11.1 THE CENTRAL PROBLEM OF MODERN PHILOSOPHY

It is not merely a matter of convenience that we divide the history of philosophy into "ancient," "medieval," and "modern" epochs. In each of these transitions, there was a distinctive new synthesis that arose to accommodate new foundational assumptions. Medieval philosophers needed to find a synthesis between the philosophy and science of the ancient world and the precepts of revealed religion. Modern philosophers needed to accommodate the changes in world view that accompanied the emergence of what has come to be called modern science in the sixteenth through eighteenth centuries. The medieval synthesis embraced a largely Aristotelian science whose teleological nature was applied indifferently to physics, biology, and psychology. There was no mind-body problem for the medieval mind, because the medieval notions of body and soul, while distinct, were not in conflict. Modern science, however, employed very different modes of description and explanation. On the Galilean view of science, embraced in different ways by philosophers as different as Hobbes and Descartes, the physical world is basically just a large collection of bodies in motion. These motions can be exactly described by mathematics, and the behavior of complex bodies is a consequence of (and can be derived from) the motions of their simple parts. Indeed, the guiding metaphor for many writers of this period was that of geometric proof: to understand a natural phenomenon, one first "resolves" it into its simple parts (corresponding to the definitions and axioms) and then "composes" the complex phenomenon from the simple (corresponding to geometric constructions and the proof of a theorem). The resolutive step requires hypothesis testing of some sort or other, but the compositive step is regarded as being necessary and epistemically transparent.

The problem comes in how to integrate this view of physical nature with a larger philosophical picture that includes mental, moral, and social phenomena as well. Hobbes sketched a strongly naturalistic programme in which politics is derived from psychology, complex psychological phenomena are derived from simpler ones, and simple ones are identified with motions of the body. But he never explained how this crucial identification is supposed to proceed, much less showed us how to *derive* phantasms from bodily motions in quasi-geometric fashion. And Descartes (who was, incidentally, the most important promoter of mechanistic physics, and even mechanistic psychology with respect to things like reflex action and perception [see *Treatise on Man,* AT XI.202]) points out at least three significant differences between the mental and the physical: the unmediated first-person *access* to mental states, the *metaphysical* "real distinction" between body and mind due to incommensurable essential properties, and the inability to give mechanistic *explanations* of the faculties of reason, language, and the will.

Both of these writers—Descartes by argument and Hobbes by exampie—provide early lessons in the fundamental problem facing modern metaphysics: that of how to find a single overarching philosophical framework that accommodates both our best way of talking about the physical world and the most natural ways of talking about the mind (i.e., in mentalistic terms). Now just how you *describe* this problem depends very heavily on your metaphysical assumptions. For example, if you buy into the kind of substantival metaphysics assumed by Descartes, you may be inclined to see the issue as one of how many kinds of substances there are, while this way of formulating the question may seem unintelligible if you reject the substantival metaphysics on which it depends. Perhaps the most neutral way of describing the problem is as follows: we have natural and successful ways of talking about the mind and about physical nature—different kinds of *discourse,* if you will—and they seem to be incommensurable. Neither seems to be reducible to the other, and there does not seem to be any common denominator that unites them. The problem is not that they are *inconsistent.* It is merely that they are not unified. Let us call this apparent chasm between mentalistic and physical discourse the "Cartesian Gap."

Now there are at least two kinds of problems that have traditionally led philosophers to feel uneasy about the Cartesian Gap. One is that the whole philosophical instinct is directed towards finding a world view that unifies all of our discourses. The philosophical impulse is an impulse towards unification. And hence there is something ugly and unsatisfying about the Cartesian Gap. The second problem is that we have clear intuitions to the effect that there *are* relationships between the physical and the mental—and important ones at that!—that we would like to be able to describe and explain. On the one hand, there is voluntary causation: volitions would seem to be causes of actions. On the other hand, there is perception, in which events in sensory nerves would seem to be causes of perceptual states. In addition, as was already clear to Descartes and his contemporaries, thoughts seem to bear some kind of special and intimate relationship to events in the brain. Action, perception, and the localization of mental functions are all things that we would like to be able to talk about. Incommensurably separate discourses about mind and matter leave out some of the phenomena that we would most like to explain. To remain content with the Cartesian Gap is to remain content with some amount of mystery, and indeed with a substantially greater helping of it than most philosophers are inclined to be content with.

11.2 THE "RECEIVED VIEW"

There are, of course, many kinds of philosophical theories directed at solving or *dis*solving this problem: various kinds of dualism, materialism, and idealism, as well as linguistic and social theories that rework the apparently fundamental metaphysical issues into epiphenomena of language, cognition, or sociality. CTM and many of its favorite adversaries (behaviorism, reductionism, eliminativism) share a grounding in a particular materialist approach to the Cartesian Gap: namely, a *normative* claim to the effect that particular kinds of connections to physicalistic discourse are *necessary conditions for the legitimation of mentalistic discourse*. One might call this view *normative naturalism*. CTM's advocates differ with behaviorists and reductionists on the *nature* of the needed connections. They differ with eliminativists on the issue of whether the grounding of the mental in the world of nature can in fact be accomplished. However, they generally share the assumption that if push comes to shove between naturalism and intentional realism, it is intentional realism that should be abandoned.[1]

Now one might well think that, since CTM's original appeal drew in large measure from its ability to address concerns that arose directly out of positivist, behaviorist, and reductionist projects, the perceived need to address those concerns would not long survive the demise of the projects that spawned them. But almost no one believes the verification theory of meaning anymore or embraces a reductionism of the form popular in the 1950s, and even methodological behaviorists are increasingly difficult to find. But concerns about vindicating intentional psychology live on. Indeed, the current orthodoxy in philosophy of mind—the "Received View," if you will—seems to treat as axiomatic the claims (1) that materialism is true, (2) that there are no mentalistic properties that are fundamental (as opposed to derived from more primitive physical properties), (3) that sciences must deal in causal-nomological explanations, (4) that the only legitimate entities are those that appear in the explanatory inventory of some natural science, and hence (5) that, however useful or well-confirmed mentalistic ascriptions and explanations may be, they are nonetheless in the position of needing to be justified on metaphysical grounds. According to the Received View, it is our discourses about the mind (whether scientific or commonsensical) that must answer to a materialist and naturalistic metaphysics and a causal-nomological view of science, and not the other way around. Fodor, for example, writes,

> The deepest motivation for intentional irrealism derives... from a certain ontological intuition: that there is no place for intentional categories in a physicalistic view of the world; that the intentional can't be *naturalized*....
>
> ... It's hard to see... how one can be a Realist about intentionality without also being, to some extent or other, a Reductionist. If the semantic and the intentional are real properties of things, it must be in virtue of their iden-

tity with (or maybe their supervenience on?) properties that are themselves *neither* intentional *nor* semantic. If aboutness is real, it must be really something else. (Fodor 1987: 97, 98)

Thus it becomes an intelligible move in the game to claim such otherwise outrageous things as that mental states are "explanatory posits" of "folk psychology" and might be "eliminated" if this "folk psychology" cannot be made into a rigorous causal-nomological science.

11.3 DIALECTICAL POSSIBILITIES

How does one arrive at the Received View? One story about the underlying train of thought goes as follows: You start by looking at a few physical sciences (particularly mechanics, thermodynamics, and chemistry) and a few notable episodes of scientific accomplishment involving intertheoretic connections (the atomic theory, the derivation of thermodynamic equations from statistical mechanics, etc.). On the basis of observations of these actual domains of scientific research and actual interdomain connections, you form *second* -order theories about the proper form of *all* scientific discourse (or even all metaphysically respectable discourse), and about the proper relations between discourses. The resulting view is one in which the objects of the special sciences are differentiated by the kinds of physical processes they study, and more particularly by the levels of *complexity* of the processes they study. The more basic sciences study simpler objects that form the proper parts of objects studied by the higher-level sciences, and ultimately you should be able to explain the higher-level properties as derivative from the lower-level properties. Let us call this picture the "Hierarchical Picture." The Hierarchical Picture is a second-order theory about the canonical form of discourses in the special sciences (and indeed for all discourses speaking about real objects) and about the connections between them. Because its paradigm examples all involve straightforwardly physical objects, a materialistic inventory seems implicit in the model. The Received View then applies the Hierarchical Picture as a *norm* for looking at actual discourse about the mind and actual attempts to form connections between mentalistic discourse and other kinds of discourse, such as that of neuroscience.

This is arguably a very charitable way of interpreting the emergence of the Received View. One might well point out, for example, that the essentials of the Hierarchical Picture were already present in Hobbes,

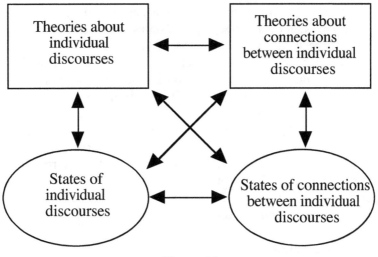

Figure 14

who tried to make the Galilean method of resolution and composition into the basis for a metaphysics long before most of the scientific accomplishments that might plausibly be thought to support the Hierarchical Picture. One might well take the view that it is the Hierarchical Picture that is driving the interpretation of science, and not vice versa. Or, one might point out the many advances in special sciences that would not have taken place had the scientists of the day placed a higher priority upon conformity with the Hierarchical Picture than with discoveries in their respective local discourses. But it is, I think, the kind of story that adherents of the Received View like to tell, and we can grant it for present purposes.

What I wish to point out about this story is the complex dialectic between four separate kinds of concerns (see fig. 14):

(1) the state internal to each of the separate discourses (e.g., the actual state of commonsense discourses about action, neuroscience, psychophysics, physics),

(2) one's *theories* about the forms individual discourse do or ought to take (e.g., the philosophy of psychology, the philosophy of physics),

(3) the state of connections between discourses (actual intertheoretic reductions, localizations, correlations), and

(4) one's theories about the forms connections between discourses do or ought to take (e.g., reductionism, supervenience, the Neutral Project).

Behind this prevailing naturalistic current in contemporary philosophy of mind is an assumption that the status of our discourse about the
mind is to be held up to the litmus of conformity with *metatheories*
about the nature of *particular* discourses and about the connections one
ought to be able to find *between* discourses. In particular, it is assumed
that *real* intentional states would play a role in a causal-nomological
science of the mind, and that they would ultimately be derivative from
nonintentional phenomena in much the way that, say, thermodynamics
is derivative from statistical mechanics. Thus this Received View can-
onizes certain metatheoretical assumptions about the nature of sciences
and the connections between science and ontology, and then applies
these assumptions as norms for judging the form of actual work in psychology and connections forged between psychology and other domains.

While the dialectic underlying the Received View may seem very
compelling, it is by no means the only serious approach that has been
taken. Recent philosophy of science, for example, has increasingly
moved towards an approach that looks first and foremost at the actual
dynamics of discourse within the special sciences—with the result, for
instance, that the teleological categories of evolutionary biology have
won increasing acceptance against older mechanistic objections. On
another front, writers like Ryle and Wittgenstein have insisted that intentional explanation is not scientific discourse, or even protoscientific
discourse, at all, and should not be forced into that model. And continental philosophy has decided that the *Geisteswissenschaften* or "sciences of mind and culture" are fundamentally different in their form
than the natural sciences, and have adjusted their metatheoretical views
in light of this observation.

This brings up two questions. First, does the Hierarchical Picture really capture the form of current work even in the natural sciences? And,
second, is the healthiest dialectic between (*a*) actual practice in the local discourses of the special sciences (and, for that matter, nonscientific
discourses) and (*b*) our metatheories of the same, one in which the metatheories are applied as a test for legitimacy of the assumptions of special discourses—or should things, perhaps, be the other way around?
And to these one might well add a third question: namely, do all of the
considerations relevant to the assessment of specifically scientific discourses apply more generally as well? In other words, must discourse
be *scientific* to be *respectable,* and must object kinds appear in scientific explanations in order to be *real?*

In the remainder of this chapter, I shall deal principally with the second question regarding the healthiest form of the dialectic between the successes of local discourses and our metatheoretical views about the forms of those discourses and the relations between them. I shall argue that the Received View has the proper dialectical relationship backwards, even by standards expressed in the pieties of many of its proponents. To wit, it is principally our metatheories that stand in the dock against successes of special discourses, and not vice versa. And, as a consequence, intentional categories stand in no need of vindication. Along the way, I shall point to a very few of the possible examples of important work in the physical sciences that have strayed from the course the Received View might have urged. I shall also make a case for separating science and ontology in such a fashion that the *metaphysical* legitimacy of mental categories in no way depends upon the possibility of a science that deals with them.

My plan, then, is to look at the assumptions underlying the perceived need for a vindication of intentional psychology, and to show how the dialectic that drives the perception of this need for vindication is wrongheaded. I shall first consider concerns about extending the causal-nomological model of explanation to psychology, and then treat of the desires for materialism and the generality of physics together.

11.4 PSYCHOLOGY, THE MENTAL, AND CAUSAL-NOMOLOGICAL EXPLANATION

First, consider the issue of whether intentional psychology can be made into a causal-nomological science. This issue concerns the dialectic between (*a*) the actual form of our understanding (both commonsensical and scientific) of mental states and processes and (*b*) the prevailing view that *scientific* discourse is causal and nomological in character. There are really two separate issues here, of course, though they tend to get conflated: (1) the issue of whether there can be a psychological *science* that is causal and nomological in character, and (2) the issue of whether entry into nomic causal relations is relevant to *ontology*. The relationship between science and ontology will be discussed later. Here I shall concentrate on the dialectic between the causal-nomological picture of science and the actual state of our discourse about the mental.

There are two basic ways of coming to the conclusion that intentional explanation cannot be causal explanation based on causal-nomological regularities. The first way is to think that intentional explanation is part and parcel of an enterprise called "folk psychology" which is an attempt at protoscientific explanation of behavior, and then conclude that it fails in its attempt to capture causal regularities in the intentional vocabulary. This view that commonsense intentional psychology is an attempt at causal-nomological explanation, sometimes called the "theory-theory," is fairly prevalent in cognitive science circles. It will be discussed in the following section. Before turning to it, it is important to see that there is also another way to come to the conclusion that intentional explanations cannot be causal-nomological explanations: namely, to believe that *they were never intended as such in the first place.*

11.4.1 THE VERSTEHEN ARGUMENT

The claim that mentalistic discourse does not even attempt to provide causal explanations has had important advocates in both the analytic and the continental traditions. It includes the *Verstehen* tradition in the *Geisteswissenschaften* on the continent, Ryle and Wittgenstein, and lately a related sort of view has attracted interest from analytic philosophers like Alvin Goldman (1992, 1993a, 1993b,). On this view, intentional explanation involves giving *reasons,* and giving reasons is different from giving causes. The kind of "explanation" we are involved in when we allude to people's beliefs and desires is not causal explanation, but something like *interpretation.* The goal of investigation framed in the intentional idiom is not knowledge of general laws of thought, but in *Verstehen,* or interpretive understanding, of human beings.

In a certain way, I am very sympathetic to this view, especially in the ways in which it stands in contrast to the theory-theory. It is no doubt true that "commonsense psychology" involves a "theory" in the weak sense that we make generalizations about how people are likely to think and act, and that these expectations, were we to express them, would be expressed in the intentional vocabulary. But it seems a bit dubious to equate this with an attempt to formulate causal laws, and very highly questionable to refer to intentional states as "theoretical entities" as that expression is sometimes used in the philosophy of science (i.e., to signify entities posited through a process of retroduction). Particularly when we refer to dispositional states such as beliefs, it seems quite reasonable to say that what we are doing is trying to paint a picture of the person's thoughts that makes sense of their words and deeds, and it would be somewhat strained to say that we are trying to provide the causes of their actions, much less that we are tacitly assuming universal laws formulable in the intentional idiom.

Nevertheless, it seems important here to separate two very different sorts of issues. We might, on the one hand, concentrate on issues relating to the nature of commonsense intentional psychology. On the other hand, we might undertake quite a different sort of inquiry into the prospects for a causal-nomological intentional psychology. It is useful here to distinguish four very different questions:

(1) Is the theory-theory an accurate representation of the nature of commonsense psychology? (That is, is commonsense psychology really an attempt at a scientific theory based on causal laws?)

(2) Is there a viable enterprise of interpretive psychology whose goal is *Verstehen?*

(3) If so, should we call it "science"?

(4) Is there a viable causal-nomological psychology that makes use of generalizations at the level of cognitive states and processes?

The reason that it is useful to separate these questions is that the answer to the fourth question—our main focus here—is really quite independent of the answers one gives to the other three. In point of fact, I think that one might have to give a somewhat mixed answer to the first question, since the characterization of commonsense views is likely to be a somewhat complicated undertaking; but I am inclined to side more with Ryle and Wittgenstein here than with the theory-theory. I take it that the second question can best be answered by looking at what is accomplished by attempts at interpretive psychology (e.g., psychoanalysis as interpreted by Ricoeur [1970]) in the long run. The third question is in large measure a matter of wrangling over words. In some ways, I think the German approach has some real advantages: if you have two categories of *Wissenschaften* you can then ask serious questions about how they differ; whereas if your distinction is between "science" and "nonscience," there is an unfortunate tendency to assume that everything that is nonscience is also non*serious* and non*rigorous,* and that all such enterprises are nonserious and nonrigorous in the same ways. All in all, though, I find this fight less interesting than I used to, and do not intend to belabor it here.

What seems crucially important here is that the viability of a causal-nomological intentional psychology is in no way threatened either by the possibility of interpretive psychology or by the possibility that commonsense psychology is not accurately conceived of as protoscience. One might demonstrate this point by way of a thought experiment. Suppose, for example, that someone were convinced by writers like Wittgenstein and Ryle that the "because" of intentional explanation (in ordinary usage) is not the "because" of causation. Suppose, furthermore, that she believed that ordinary intentional explanation was part of a project of interpretive understanding (*Verstehen*), and that there were disciplines—the *Geisteswissenschaften* —that attempted to approach such interpretative understanding in a systematic way. She might think, for example, that Ricoeur's Freud was an active and even an occasionally successful practitioner of such a discipline. Suppose now that one day she comes across something like Colby's (1975) attempts to implement models of Freudian theories in the form of computer programs. Upon examining these she might well feel that she has discovered a new possibility for explanation in the intentional idiom: in addition to the possibility of an interpretive psychology, it might be possible to systematically explain *causal* relationships among intentional states as well. It might even turn out that there are important connections between them—for example, that rules for deriving new cognitive state tokens are so formed as to preserve truth and maximize coherence and relevance. Our fictitious person could thus discover the idea of a nomological intentional psychology as a project quite orthogonal to commonsense psychology and interpretive psychology, but in no way in competition with them.

What I think this shows is that neither the Ryle-Wittgenstein view of intentional explanation nor the *Geisteswissenschaft* approach to psychology should be viewed as counting against the viability of a nomological intentional psychology. These views would, of course, undercut the claim that the success of commonsense psychology in allowing us to predict one another's behavior in day-to-day affairs provides direct evidence for the possibility of a nomological intentional psychology. But it need not rule out providing such evidence more indirectly: if we are successful at interpreting others, it might be in part because we have a kind of model in our heads of how people are likely to think and act given certain beliefs and desires, and this at least suggests the possibility of a nomological account of how different beliefs and desires in combination will produce other cognitive states and, ultimately, behaviors. In short, even if "folk psychology" is a dubious philosophical reconstruction, this does no fundamental damage to the project of nomological intentional psychology.

11.4.2 INTENTIONAL CATEGORIES AND CAUSAL RELEVANCE

Establishing compatibility between causal explanation and *Verstehen* (as two separate projects), however, is the easy part of the job. The hard part lies in making a case that intentional psychology is susceptible to being made into a "science" in the prevalent usage of the word in America, where it suggests a causal and nomological character. One important problem lies in the fact that intentional psychology individuates its objects semantically. But if it is to be a science, one would wish for the individuation of terms to take place in a fashion that captures causal regularities: if a belief that a cat is in the yard is used to explain different behaviors than a belief that a unicorn is in the yard, and we are talking about a kind of causal explanation, we should expect the difference in the causal powers of the two beliefs to be intimately connected to their difference in content. But it is hard to see how this might be the case, and it has generally seemed that there are only three basic possibilities: (1) there is some interpreter that is sensitive to semantic properties and is also the locus of the causal powers, (2) semantic properties are themselves causally efficacious, or (3) semantic properties are linked to some other kinds of properties that are causally efficacious. The problem with (1) is that it leads to a homuncular regress. The problem with (2) is that semantic properties just do not seem like things that have causal powers. The computer paradigm takes option (3) with respect to semiotic-semantic properties, which are unproblematically linked to nonsemantic properties through interpretive conventions, but it is not clear how mental-semantic or MR-semantic properties might similarly be coordinated with some other properties that are causally efficacious, or what those properties might be.

The issue here is one of trying to *see how something might be the case*—that is, how it might be that causal regularities of cognition could run parallel to semantic properties of cognitive states. And the issue of "seeing how *X might* be the case" shapes up very differently depending on what kind of evidence we have that *X* actually *is* the case. For example, many people in Newton's day (and perhaps many of us today) found the picture he presents of gravitation to be problematic because it involves action at a distance. The familiar paradigm of causal interaction had long been one of *contact* interaction, and it seemed—indeed, it still seems—hard to see how bodies could exert influence somewhere they are not. This difficulty in seeing how it *could* be so might have been seen as a compelling argument against causation that does not involve contact interactions, except for the fact that there was overwhelming evidence that such causal influence *did* take place. (Indeed, of the fundamental "forces" posited by contemporary physics, *none* of them aside from mechanical force involves contact interaction.)

I think that the same kind of benefit could, in principle, accrue to psychology: *if* there were to emerge an intentional psychology framed as a set of laws governing reasoning, and it had a sufficient degree of predictive accuracy, this would provide strong evidence that intentional states *do* have the kind of causal role assigned to them by such a theory. And it would provide such evidence *regardless of whether we can see a mechanism that could account for such causation.* The particular form of intentional causation *could be fundamental,* after all, like the particulars of gravitation or electromagnetism or of why particles behave in the precise way they behave when they collide. Or intentional causation could be opaque *to us* without being fundamental—it could be that we can have evidence that there are causal regularities at the level of intentional explanation, and it might also be the case that these are emergent out of some more basic kinds of regularities without our being able to know just what the relationship between levels is. (Even if computational formalisms are appropriate for psychology, it seems all too likely that we shall never know the details of interlevel relationships in detail.) In such an eventuality, its status would be not unlike that of Newtonian description of gravitational attraction, which supplied a nomological description (and hence conferred mathematical maturity) without explaining this behavior by positing an underlying mechanism (and hence did not supply additional connective maturity).

The first part of the answer to the second objection to intentional psychology, then, is this: computational psychology provides one of the first chances we have really had for making any realistic attempt at building nomological theories of cognition that treat intentional states as causally relevant in reasoning and behavior.[2] The programme is relatively young, and any possible model of the mind rich enough to test for predictive accuracy with respect to cognition would necessarily be orders of magnitude more complex than most of the fundamental laws in other sciences.[3] (Though many researchers seem to think that some of the initial results seem promising.) The obvious moral to draw is that we ought to let cognitive psychology mature as best it may, and see whether it *does* provide compelling evidence that semantic properties of intentional states seem to be at least correlated with causal regularities.

There is also a second response to this objection—one that draws more directly upon what is made explicit in BCTM. BCTM makes a fairly bald-faced assertion to the effect that intentional states are realized through naturalistic properties in a systematic way, such that (*a*) intentional modality is realized through functional relations and (*b*) mental-semantic properties are realized through some naturalistic properties (labeled "MR-semantic properties" until their true identity is discovered) possessed by cognitive counters. And it is important not to be fooled by the "semantic" in 'MR-semantic'. Remember that all 'MR-semantic properties' means is "those properties of cognitive counters, whatever they turn out to be, through which content is realized." In particular, the "semantic" in 'MR-semantic' is not so robust that our intuitions that "semantic properties" cannot be causally efficacious should transfer to MR-semantic properties. MR-semantic properties, if they are anything at all, are just naturalistic properties whose real identity we have not discovered, and so they seem perfectly respectable as possible explainers of causal regularities (though whether they capture the right regularities to make our intentional explanations into causal laws remains to be seen). If the research programme associated with BCTM can be carried out, then there *will* be a causal, nomological science to be carried out, *at least at the level of the realizing system*. That is, the system of cognitive counters will have lawlike causal regularities, and the MR-semantic properties through which mental-semantic properties are realized will be at least correlated with the causal properties of the cognitive counters of which they are the properties.

The difficulty here is this: the level of description for the realizing system is not a level of *psychological* description per se. What can the causal, nomological character of the reali*zing* system show us about the reali*zed* system, the system of intentional states and processes? Does the causal, nomological character of the realizing system automatically accrue to the system it realizes? Or does it, perhaps, suggest that the realized system also has such a character? There are, I think, two parts to the answer. First, even if we could not construe the system of intentional states and processes as a causal and nomological system, it would matter a great deal if we could show that this system is realized through a system that *did* have these virtues. For example, if there were overwhelming Rylean objections to interpreting intentional states as being even the sorts of things that can enter causal relations, the psychologist can at least take heart at the news that there is some *other* system of states and properties—those through which intentional states and processes are realized—that *can* enter into such relations, and moreover, that whenever one picks

out intentional states and processes, one picks out the states and processes through which they are realized as well. (If you go to the race-
track, you can bet on the horse or bet on the jockey on the horse; they
win or lose together!)

But I think one can make a case that at least some of the properties of
the realizing system can also accrue to the realized system as well.
Take our example of the Victoria Crown. The property of being the
Victoria Crown is realized through a particular bit of matter. Let us say
it weighs fifty pounds. Now there is nothing about the property of being the Victoria Crown that entails weighing fifty pounds. But the object that is the Victoria Crown does weigh fifty pounds, and does so
because the matter of which it is composed has that mass. This property
of the realizing matter accrues to the realized object as well. Similarly,
suppose Jones's generosity is realized through his giving $100 to the
Presiding Bishop's Fund for World Relief on Pentecost. The act takes
place at a particular time and has a particular beneficiary. Now we
should not say that Jones's generosity is an event taking place at a particular time, but we might say that it was *exercised* at a particular time.
And we should surely say that the Presiding Bishop's Fund was the
"beneficiary of Jones's generosity" and not just the recipient of $100.
So it seems that, at least in some limited ways, the realized system can
take on some of the properties of the realizing system. (It does not follow, of course, that the properties of the realizing system accrue to the
realized *property* —the property of generosity does not take on new
implications because of how it is realized.)

A detailed examination of the various possible relations between realizing and realized systems would probably require some very careful
metaphysical investigation. It seems quite reasonable, however, to suppose the following: if (1) a mental state X of type M is realized through
a natural state of type N, and (2) it is a law that N 's cause O 's under
condition C, and (3) C obtains, then (4) an O will come about, and (5)
X may be said to be a cause of O. (And this holds even if we do *not* say
"X is an N " but only "X is realized through an N -token.") Here the
causal powers of the instantiating type N accrue to the realized individual X, but not to the realized property M.

Now none of this precludes the possibility of saying that intentional
states have causal powers in their own right, and not just by virtue of
how they are realized. The point is merely that, if computational psychology as described by BCTM can be carried out at all, there will be
some naturalistic system through which intentional states and processes
are realized, and this system can be causal and nomological in ways that

are not at all problematic. If there *are* problems about intentional and semantic *properties* being of the right sort to ground causal regularities, perhaps this gap can be filled by way of appeal to the realizing system.

To summarize, there really are some prima facie difficulties with the attempt to construe intentional explanation as lawlike causal explanation, and hence to make intentional psychology into a "science" in the sense of the word that implies such causal lawlike explanation. The issue of the *nomological* character of intentional explanation, however, is best settled by letting the project of intentional psychology, supplemented by resources of the computer paradigm, flourish as best it may and seeing whether the project will pan out in the end. And if it does produce what look like nomological regularities, this in itself provides substantial warrant for suspecting that these regularities are *causal* in nature as well, *whether or not we can find or even imagine an underlying mechanism that could account for the causality.* In any case, if BCTM can be carried out, there will also be a system describable in wholly naturalistic terms through which the system of intentional states and processes is realized. And this system (*a*) can unproblematically be viewed as causal in nature, (*b*) is in one-to-one (or even many-to-one) correspondence with the system of intentional states and processes, and (*c*) may even confer its causal properties upon the things it realizes.

11.4.3 NEED MENTALISTIC DISCOURSE BE SCIENTIFIC TO BE LEGITIMATE?

To tell the awful truth, though, I have my doubts about whether there can be a causal-nomological science of the intentional. Apart from all the problems in formulating actual theories with so many hard-to-isolate, mutually dependent variables, I share the Wittgensteinian suspicion that ordinary-language belief ascriptions are not causal explanations; and hence, whatever a computational psychology *might* do, it would not render whatever ordinary language is doing in such cases scientific, but add a new kind of discourse, perhaps only loosely inspired by the original. I also believe in free will, which seems in tension with a thoroughgoing nomological psychology. There are also other reasons that I find harder to articulate. *But none of this makes me doubt the existence or legitimacy of mental states.* And this is because I do not think that the considerations that exclude an object from a specifically *scientific* ontology (i.e., the domain of a science) exclude it from ontology generally. I will discuss some parts of this issue here and others later in the chapter.

First, different kinds of discourse have different purposes, and the conditions for legitimation for a given kind of discourse vary with its purpose. The natural sciences aim at describing and explaining the regularities of nature. The good-making qualities of the natural sciences are thus conditioned by the practical constraints governing what counts as a good enough explanation, the need for truth-conditional evaluation, and the availability of real regularities to be found in nature. Most of our human discourses, however, have different sets of constraints. Many speech acts do not have truth conditions at all, but other kinds of felicity conditions (see Austin 1962; Searle 1969). Indeed some entire language games may lack truth conditions, though of course they have felicity conditions of some kind. For example, if two people play mental chess, saying "King to Queen one" constitutes a move, not an assertion. Now some people would go so far as to think that mental state ascriptions are similarly nonassertoric. And perhaps *some* mental state ascriptions are not assertions and do not have truth conditions. But I think many of them are assertions and are subject to truth-conditional evaluation.

What I think they might lack is a nomic character. Categories of nonscientific discourses can be legitimate and can pick out real objects even if those categories do not pick out things that are the subjects of natural laws. Dollars can be bills or coins, and there are no physical laws regarding dollars; yet I do not conclude that I am broke.[4] I suspect that there are no laws regarding Toyotas, but I do not feel trepidation when I look out in the driveway every morning in fear that my car was an illusion. I suspect that there are no natural laws applying only to jigs, but I am not dissuaded from buying recordings of Irish music as a result. There are no physical laws regarding numbers, but that does not excuse me from balancing my check book properly or tallying my taxes. Nor do my doubts about natural laws governing dollars (or Toyotas or jigs) give me reason to accuse my employer (or my car dealer or the Bothy Band) of fraud. In general, the failure of a category to appear in a science does not cause me to doubt the reality of things to which that category is supposed to apply, or to doubt the legitimacy of the category. Indeed, the only cases in which I am inclined to make inferences in anything like this way are those in which an object-kind is introduced by hypothesis within the context of a scientific theory (e.g., phlogiston). I do, of course, doubt the existence of trolls and unicorns. But the reason I doubt them is that there seems to be no reliable evidence of their existence. It is not because there are no theories in which the kind "troll" or

"unicorn" plays a nomic role. If I were inclined to doubt trolls and unicorns on *that* basis, I would have to doubt dogs and Toyotas as well. In
short, what you need to qualify a category for legitimacy in a scientific
theory is something far stronger than what you need to qualify it for
mere ontological legitimacy. Scientific kinds are reasonably orderly.
Ontology is a motley.

11.5 INTENTIONALITY, MATERIALISM, AND
THE GENERALITY OF PHYSICS

A separate set of issues lies upon a second axis: the relationship between
the current state of discourse about the mental and metatheories about the
relationships one ought to find between separate discourses. CTM was
supposed to "vindicate" intentional psychology by demonstrating the
compatibility of intentional realism with materialism and the generality
of physics. Such a vindication is only necessary, however, on the assumption that compatibility with materialism and the generality of physics is a condition for the legitimacy of psychology as a science or of
mental states as real entities. This position takes a particular picture of
how various discourses do or ought to fit together—a picture in which all
special discourses are in some sense a special case of physics—and applies it as a norm for evaluating the current state of discourse in psychology and commonsense discourse about the mental. (Indeed, if you reverse the dialectical relation, so that the commitment to mental states is
held constant and the claims of materialism are weighed in the balance,
computers might be thought to provide a "vindication of materialism"
instead of a vindication of intentional psychology.)

In order to better assess the strength of arguments against intentional
realism based upon concerns about materialism and the generality of
physics, it may be helpful to make explicit some of the prevalent views
about ontology and its relation to science in the context in which the
apparent need for vindication arises. In particular, I shall discuss three
views whose popularity is to no small extent traceable to Quine's influence.[5] First, Quine is, of course, well known for the "desert landscape"
plea for the impoverishment of the ontological inventory and the association of this view with a presumption in favor of some form of materialistic monism. (I personally have always found it curious that people
found this view appealing, and even more curious that they regarded an
expression of taste as an argument. I like lush landscapes and seashores;
and, more to the point, would be disinclined to try to describe a rain-

forest as a desert just because I liked deserts.) Second, Quine was influential in the dissemination of the now widespread view that all ontological questions are essentially questions to be answered by science. And, finally, Quine seems to bear much responsibility for the currency of the view that the business—indeed, the *only* business—of ontology is to *provide an inventory of the basic kinds of things.* I think that most philosophers who do not have one foot (or at least a few toes) outside of late-twentieth-century analytic philosophy simply take this last view for granted. For an eye-opening discussion of how ontology has been viewed historically, I would point them to the entry on ontology in the *Encyclopedia of Philosophy.*[6] In brief, ontology has traditionally involved not only the question "What (kinds of) things are there?" but a number of questions about the *nature of being* —for example, "What is it to be a thing?" "What is it for *X* and *Y* to be the *same* thing?" "Is 'being' said in the same sense about different kinds of things?"

For the record, I think that all of these Quinean views are very deeply wrong. However, I shall not undertake a full-scale assault upon Quinean ontology here. Instead, I shall confine myself to making two much more modest points. First, there are certain ways that ontology can, in principle, extend beyond what science can talk about, with the consequence that intentional realism can be ontologically respectable even if intentional psychology cannot become a mature science. Second, if one takes seriously the notion that science ought to guide ontology, one ought to take the attitude that a successful psychology committed to intentional realism would give us warrant for believing in intentional states and processes, whether they be compatible with materialism and the generality of physics or not—and indeed that if push comes to shove between intentional realism and materialism and the generality of physics, it is the latter that stand in the dock. Hence intentional psychology does not stand in need of vindication on this score.

11.5.1 WHY ONTOLOGY EXTENDS BEYOND SCIENCE

There are several reasons that ontology extends beyond science. First, ontology has traditionally been concerned with questions above and beyond those concerned purely with inventory. It has been concerned with questions about the nature of being and unity, and with such issues as whether there are different kinds of principles for "being" and for "being one thing" for different kinds of "objects." Granted that there is a kind of "being" that applies to material simples, the question remains of whether there are other kinds of "being" that apply to other things. Peter van Inwagen (1990, 1993), for instance, has recently defended the position that there is also a principle of individuation to be found in something like the "transcendental unity of apperception"—namely, that subjects capable of a certain kind of thought are thereby *things* in a sense denied to mere physical aggregates. Aristotle (*Metaphysics,* IV-VIII) thought that "being" was said in a primary sense only of members of what we today might call "natural kinds," that it was said in derivative senses of properties and relations, and that it was somewhat of a stretch to say it of mere matter at all. Charles Sanders Peirce (and perhaps the late Plato) saw three realms of being: matter, abstract objects such as mathematical objects, and minds.

It is important to see that such thinkers were not simply engaged in projects of empirical science that somehow went awry. They are engaged, in part, in linguistic and conceptual analysis; but insofar as our *ideas* of being and unity reflect how things really are, they are engaged in attempts to clarify the nature of being and unity as well. Such a project may have results that stand outside of empirical science: for example, if there is a sense of "being" that applies to abstract objects, those objects stand outside of the realm of empirical science. If van Inwagen is right that there is a principle of being and unity that attaches to a thing by virtue of being a thinker, then there is a reason to include thinkers in your basic ontology even if wherever there is a thinker there is also a body of a specific kind (physical or functional) as well. That is, if there is a principled reason for allowing things other than physical simples to count as basic kinds, then one cannot simply assume in advance that the simplest inventory needed for science (by that meaning the inventory of simplest *parts* employed in our scientific theories) is all we need for ontology.

A second reason for seeing ontology as extending beyond science is that some things that might be thought of as objects and unities seem to stand outside of the domain of empirical investigation—most notably, abstract objects such as numbers and sets. I take it that there are serious questions about the ontological status of abstract entities. It may or may not be possible to settle such questions at all; but it seems clear that empirical science cannot settle them, nor can such answers as might emerge from the analysis of empirical science. Perhaps there are also other sorts of objects (or possible objects) that are similarly unsuitable for empirical investigation of the sort conducted in the sciences—for example, God, angels, objective values—in which case there are other ontological issues that fall outside of the domain of science.

A third reason (already touched upon) that ontology can extend beyond science attaches to the *systematic* character of science. Scientific description applies to the universe only insofar as it is describable in ways involving systematic features of structure and causation. Many of our object categories, however, are not implicated in such systematic relations: the category "lamb chop" has little to do with how to carve nature at the joints and much to do with how we carve meat at the joints. Yet there are surely sensible questions to be asked about the senses in which it is right to say a lamb chop "exists" and is "one thing," how these differ from the senses in which "unity" and "being" may be predicated of living animals, simple particles, or numbers, and why a lamb chop is not just a "heap" of atoms. Such questions fall within the scope of what has historically been called "ontology," they seem like sensible questions, and they may shed some light on the questions that do have something to with science. Moreover, the inventory view of ontology accepts as a premise one possible answer to questions of the sort raised here—that is, it is one contender in a field of possible answers about how we ought best to talk about various sorts of things that common sense treats as "objects." Hence there are ontological questions (i.e., questions about what ontological approach to adopt) that cannot be addressed within the inventory approach.

This last consideration brings up the possibility that there are ontological questions to be asked about intentional states that are not questions about the analysis of intentional psychology as a science. If one takes it that one has reason to believe there are either dispositional beliefs or inner episodes such as occurrent judgments, one might sensibly ask ontological questions about them: for example, is "being" applied to such states as it is applied to objects such as living beings, or as it is applied to properties or states, or in some other fashion? And these questions are no less proper if there is no science of intentional states. In short, even if intentional psychology fails as science, this does not have the implication that intentional states are any less ontologically respectable than any number of other things that do not fall into the categories used for explanation in an ideal science—that is, they are no worse off than dogs or lamb chops or numbers. (Likewise, the consciously accessible occurrent states are no more "theoretical" in nature than are dogs, lamb chops, or numbers.)

11.5.2 INTENTIONAL PSYCHOLOGY, ONTOLOGY, GENERALITY

If this last sort of consideration is helpful with respect to intentional *realism* generally, it does nothing for intentional *psychology* as an attempt at a rigorous science. Here we need to explore the relationship between the project of intentional psychology and commitments many have to materialism and to the generality of physics. Now while Quineans officially state that what we should allow into our ontological inventory is the simplest inventory needed for science, in practice their emphasis tends to be on the simplicity rather than upon what is needed for science. That is, there are really three claims to be distinguished here:

(1) Ontology ought to include *all* entities that are required for scientific explanation.

(2) Ontology ought to include *only* entities that are required for scientific explanation.

(3) Given a choice between two different scientific pictures, one should opt for the one which posits the fewest basic entities.

I wish to argue, first, that anyone who accepts (1) ought to take the success of a special science as evidence that the entities it describes and posits actually exist and, second, that if push comes to shove, getting good explanations at the right level of description is a more important value than is having a very simple inventory.

Different sciences are distinguished from one another in large measure by the proprietary vocabularies they use and the descriptive and explanatory categories they employ. Categories such as "fault line," "high pressure system," "predator," and "desire" are employed by geology, meteorology, biology, and psychology, respectively. For these special sciences to give the kinds of explanations they need to give, they in some sense *need* such categories, and in many cases need to posit "entities" corresponding to them. But what ontological conclusions ought we to draw from the success of a science? Perhaps we can draw only prima facie conclusions, but it seems that we ought to count the success of a theory as evidence for the existence of the objects to which the theory is committed. Such commitments can, of course, be undermined by competitor theories; and theories can come into conflict with one another. But if we are to take the enterprise of the special sciences seriously at all, we have to be willing to entertain a prima facie commitment to both the things they explain (earthquakes, storms, wolves eating sheep, decision making) and the things invoked to explain them (fault lines, high pressure fronts, predation, desires, etc.).

Now to all of this the Quinean has a perfectly straightforward response: namely, that he is quite willing to admit that fault lines and predators "exist" in some sense, but that *ontology,* in his sense, is not interested in the rich abundance of being that admittedly is to be found in the universe, but in the basic entities, properties, and relations out of which that abundance is generated—the things out of which everything else is made and which are themselves reducible no further. And thus it is important to separate two different issues about ontological status: (a) the distinction between *legitimate* entities and *pseudo* -entities, and (b) the distinction between *basic* entities and *compound* entities. For it is clear that there are all sorts of entities—rabbits, rabbit forelegs, buttonholes, Corinthian columns, Wagner operas—that are unproblematically real (as opposed to fictitious or unreal), yet are neither included in nor reducible to the explanatory vocabulary of the sciences. The "desert landscape" approach does not work like the replacement of phlogiston with oxygen, but like the explanation that water is H_2O. The point is not that, in some intuitive sense, there "aren't any" rabbits or buttonholes—or, for that matter, mental states and processes. The point, rather, is that simplicity of *basic* ontological inventory is to be viewed as a virtue for a theory, and that rabbits and buttonholes (and perhaps even mental states) are complex phenomena whose ultimate parts are all of a very few kinds—namely, the kinds of basic particles recognized by an ideally completed physics.

There are, of course, several ways one could interpret this kind of principle of simplicity. The extreme position is that of thinking there is an a priori case for monism. However, a more sensible way of looking at the principle of simplicity is to see it as a kind of *maxim* or *guiding principle* for doing science. Ontological parsimony might be seen as one of the "good-making" qualities of science, a part of the "elegance" that has apparently proven a good guide to finding viable theories in physics in this century. Such a principle must, however, be played off against other principles: a theory with a larger basic inventory might well be preferable to a more frugal theory if it also has greater explanatory power or more elegant laws. There is a point at which a "taste for desert landscapes" would cease to be a reasonable inclination towards elegance and begin to degenerate into a mania for monism. In particular, if one truly believes that entities that are needed for science are thereby ontologically warranted as well, it is important not to dictate to science in advance what entities it is allowed to need.

Now if one takes this point seriously, the whole rationale behind criticizing intentional psychology on the basis of a possible incompatibility with materialism seems wrongheaded. If science were really to dictate ontology, the proper strategy would seem to be to wait and see if we could get a good explanatory psychology that could capture the relevant generalities about thought and behavior, then see what entities it was ultimately committed to. Writers like Fodor (1975, 1987) and Pylyshyn (1984) have argued (quite persuasively to my mind) that there is a broad range of psychological phenomena for which the only kinds of explanations we have that seem to capture the right generalizations are cast in the intentional idiom. The explanatory success of such theories, and the lack of competitor theories, they rightly argue, provides significant warrant for assuming, provisionally but with confidence, the existence of the entities posited in the explanations.

This, however, makes a case only that intentional states are *legitimate* entities rather than pseudo-entities, not that they are ontologically basic. The point to be made there, however, seems perfectly straightforward: if you are methodologically committed to letting what is needed by science into your ontological inventory, and you have a psychology that provides warrant for the existence of intentional states, then you have at least a prima facie commitment to whatever kinds of things intentional states turn out to be. If they can be accommodated within a materialist inventory, hooray for simplicity. *But if they cannot be so accommodated, so much the worse for materialism.* It is one thing to be committed to letting science determine what falls within the basic inventory; it is quite another to let science do so, *but only so long as the results are consistent with materialism.* On the one approach, the "vindication" of intentional psychology will stand or fall with its explanatory success, and the question of whether intentional states are *basic* in the inventory will be answered by analysis of the relationship between the resulting psychology and other sciences. On the other approach, the "vindication" of intentional psychology would consist in its being held to a standard of ontological orthodoxy. Unless some compelling a priori argument for materialism can be marshaled, it is hard to see why either science or ontology ought to be held to such a standard.

I confess that I have never found anything attractive about materialism in any case, but it seems to me that even those who *do* find it attractive ought to consider the following scenario very carefully: suppose that computational psychology (or some other research programme) were to bring intentional explanation to a stage of considerable mathematical and connective maturity and to supply general explanations that displayed a good measure of predictive accuracy. This is, I think, the scenario that most advocates of CTM think is suggested by current research.

Suppose, further, that an analysis of the resulting psychology revealed a commitment to something over and above what we were committed to by physics. Do we then (a) throw out our psychology even though it is respectable in relation to the values internal to science, or do we (b) decide that we now have good scientific grounds for rejecting materialism? I believe that (b) would be the more reasonable course to take in such an eventuality. But perhaps more to the point, if (b) is the more reasonable option to choose if push comes to shove between intentional psychology and materialism and the generality of physics, then it is likewise wrong to hold intentional psychology to proving its compatibility with materialism in advance. If a successful intentional psychology could call materialism into question, it is quite wrongheaded to expect intentional psychology to justify itself in advance by demonstrating compatibility with materialism.

11.6 THE COMMITMENTS OF THE SPECIAL SCIENCES

There is also another and more fundamental way of opposing the Quinean view of the relationship of ontology and the sciences. The Quinean view seems to equate "the entities to which science is committed" with the basic components out of which the objects described by the sciences are composed. One can, of course, equivocate on words such as 'object' and 'being' so that this view is necessarily true. But there is also a sense in which the "objects" to which the special sciences are "committed" are not the basic particles of the physicist, but things like *high pressure fronts, fault lines,* and *paranoid delusions.* For the generalizations of the special sciences are cast in vocabularies that are distinctive of those sciences, and unless generalizations can be made *at those particular levels,* one ceases to have explanations that are specifically meteorological, geological, or psychological. Now if one thinks that one is ontologically committed to the things one quantifies over in the statement of the laws of ultimately completed sciences, and the special sciences need to be formulated in ways that quantify over things other than the simple objects of physics, it would seem that the level of *objecthood* to which science commits us is not restricted to that of fundamental particles, but to objects of any of the types that are required for scientific explanation.

It is tempting to think that the distinction between type and token physicalism will take the bite out of this problem posed by the special sciences, but I think this is not fully true. We have seen in recent years that some kinds of explanation that seem useful in such fields as computer science, biology, and psychology—notably functional explanation and explanation involving such categories as "selection" and "adaptation"—cannot be reduced to statements in physics. And thus any laws formulable in terms of such explanations cannot be translated into laws cast wholly in the vocabulary of physics, with the result that the kinds of explanation that are now flourishing in several of the special sciences are incompatible with *type* physicalism. But *token* physicalism is seen as avoiding this problem, since what it claims is not that all object and event types correspond to physical types, but merely that every individual object has a physical description and every individual event is predictable under a physical description by way of laws cast in the vocabulary of physics.

It does seem correct to say that the kinds of explanation we see in the special sciences do not have the prima facie incompatibility with token physicalism that they have with type physicalism. And token physicalism may or may not be true—it is not really my concern to argue that question here. What I do wish to argue is two things: First, the special sciences need to quantify over things that fall into the kinds picked out by their proprietary vocabularies, and when their categories have a one-to-indefinitely-many realization relation (as in the case of functional kinds), there is no way of specifying the same classes of objects in the vocabulary of physics. The second thing I wish to argue is that the success of a special science should urge upon us a commitment to its theoretical posits that is at least as strong as is our commitment to the propositions (1) that all of the things it names have descriptions in the vocabulary of physics or (2) that the events it describes could be predicted by laws cast in the vocabulary of physics. Hence the success of intentional psychology would commit us to the existence of intentional states regardless of whether this result was inconsistent with materialism. That is, *it is token physicalism and the generality of physics that need to be tested against the special sciences, and not vice versa.*

To repeat an earlier point, claims made for materialism and the generality of physics may be taken in at least two ways: they can be taken as metaphysical theses or as maxims guiding science. If they are taken as theses, they can be supported either on the basis of a priori claims or on the basis of considerations involving empirical theories. Now it is true that someone who was convinced she had an a priori argument for materialism and the generality of physics might have strong reasons to hold the special sciences to the task of proving themselves compatible with these positions. (Likewise, it seems true that someone convinced of an a priori argument *against* materialism might not take any scientific evidence as warrant for materialist conclusions.) But it is not my impression that most modern materialists embrace these theses on the basis of a priori considerations. (Which is just as well, as the denial of materialism does not appear to involve one in contradiction.) Instead, there are two basic lines of argumentation that one tends to hear in support of such views: (1) the argument from simplicity (the "desert landscape" approach), and (2) the argument from the "collective evidence of modern science." In order to argue for materialism and the generality of physics on grounds such as these, it is necessary to adopt the premise that the results of the sciences can authoritatively determine the answers to ontological questions. But if the results of the sciences can be used to argue for materialism, they can, in principle, be used to argue against materialism as well, and likewise for the generality of physics. Considerations of simplicity might favor a materialist theory, but only if none of the explanatory force of the special sciences is bought at the price of incompatibility with materialism. Then one has to choose between a simpler theory that does not explain well and a more profligate theory that has greater explanatory scope. The criterion of simplicity does not tell us how to choose between theories that differ with respect to explanatory power. Likewise, if a proof that the entities required for a special science are just physical composites counts as evidence for materialism, a proof that such entities are not physical simples or composites ought to count against materialism, at least if one really thinks that ontology ought to be accommodating to successful science. And this would seem to commit us to pursuing promising candidates in the special sciences *first* and drawing our ontological conclusions *afterwards,* rather than the other way around.

Nonetheless, one might have some reason to view materialism and the generality of physics as having more of a normative status than some other kinds of claims. More specifically, one might wish to regard them not so much as *claims at all,* but as something on the order of *maxims of scientific theory construction.* That is, we have a kind of picture of what we think the overall story about the world ought to look like, and we quite reasonably try very hard to make the scientific and metaphysical stories we tell about the world exhibit the virtues of this picture. For example, we think the universe is orderly and rational (i.e., its order is of a sort comprehensible to our rational faculties), and hence we tend to look for ways to replace lots of piecemeal generalizations with a single overarching principle. We try to explain new phenomena in terms of laws with which we are already familiar. And, more to the point, we think it would be very elegant if we had a model of the world in which there are a few basic simple entities that interact in a few well-defined ways, such that we can describe all other objects as combinations of these simples and predict all events in terms of the laws governing interactions of the simple units. We thus approach the actual scientific theories we have with an eye towards goals such as subsumption of laws under more general laws, microexplanation of phenomena at one level in terms of the interactions of their components, simplification of the ontological inventory by analyzing objects into their constituents, and so on. The *limiting case* of this sort of procedure is a science in which basic entities are all of one basic sort (e.g., material bodies) or a few basic sorts (leptons, mesons—substitute this year's list of basic particles) and all events that take place can be given an explanation at the level of interactions between the simple units—that is, the case in which an atomistic materialism and the generality of physics hold true.

But it is important to see that what we have here is a set of maxims for scientific theorizing that is *guided by a particular view* of what a picture of the world *should* look like. It would be a grave error to pass subtly from the view that we ought to try very hard *to see whether* the entities posited by psychology are physical entities, to the claims (*a*) that we have *shown* that they *are* physical entities, or (*b*) that they *must* be physical entities. It is simply false to say that any such thing has been shown. There is a serious and long-standing discussion about the question of the unity of the sciences, and it is unresolved. Deeply held pictures of what explanations ought ultimately to look like are easily mistaken for necessary truths, or truths that have been demonstrated satisfactorily. But these assumptions about what an ultimate theory would look like—a picture that looks a lot like Tractarian metaphysics minus the Tractarian account of language—might well prove incorrect. Our assumptions about what scientific explanation ought to look like have very often been wrong in the past, and one ought not bank on them too heavily.

In particular, one ought to find a way of applying methodological maxims in a way that does not prevent seeing what is really out there. For example, the maxims should be directed towards taking the special sciences that do in fact develop and attempting to unify and simplify them to such extent as proves possible. They should not be oriented towards assuming in advance what degree of simplicity and unification one wants and then discounting theories inconsistent with these somewhat arbitrary desiderata. But this requires that we allow the special sciences to flourish and that we take descriptive and explanatory success seriously, even when it conflicts with our ontological views. In other words, it means that mature sciences should shape our views about materialism and the generality of physics rather than the other way around.[7]

If this is the case, however, *there is no need to vindicate intentional psychology* against charges of incompatibility with materialism and the generality of physics. It may or may not prove to be the case that intentional psychology is thus incompatible. But it is not in need of vindication for two reasons. First, compatibility could only definitively be established for a psychology in a form far more mature than its present form. Second, commitments to materialism and the generality of physics are not things that have already been established as true against which scientific theories need to be tested; rather, it is the results of successful science that will determine whether materialism and the generality of physics are in fact correct. Of course, psychology and other special sciences may never reach a stage of maturity at which such claims can properly be assessed; but let us judge the success or failure of this maturation by standards internal to science, and not by tests of metaphysical orthodoxy.

11.7 FINAL WORDS

While the final picture I have presented does not do all of the things that CTM was touted as doing, I hope that I have made a case (1) that computational psychology could, in principle, do some very important things for *empirical science* and (2) that the specifically *philosophical* desiderata of strongly naturalizing intentionality and vindicating intentional psychology should never have been viewed as imperatives in the first place.

On the one hand, computation may provide the mathematical resources for a successful psychology of cognition. On purely scientific grounds, this is a good thing even if it does not bring a solution to the mind-body problem in its wake. And on metaphysical grounds, those inclined to strong naturalism on the basis of what they have seen happening in the sciences can only be consistent by holding their strong naturalism accountable to the ultimate state of psychology, and not the other way around. As for the vindication of the mental, I am inclined to view the situation in the following way: If Smith accuses Jones of trespassing on Smith's property, Jones may very well need to vindicate himself by showing that he did not break the law. But if Smith accuses Jones of trespassing on *Jones's own* property, there is no need for vindication, because walking on your own property is not a crime. Similarly with intentional states: there is no need to vindicate them, because failure to conform with materialism and the generality of physics are not philosophical crimes. Strong naturalizers may not *like* unreduced mental kinds, much as Smith may not like Jones's domestic perambulations. But you need more than a violation of taste or ideology to call for a vindication. (Of course, one might well have concerns in the opposite direction. That is, one might view successes in intentional psychology, and the inability of nearly four centuries of modern philosophers to reduce the mental to the physical, as casting substantial doubt upon materialism and strong naturalism. If that is the real concern of naturalizers, it seems to me to be a concern that is well founded.)

Where does one go from here? I think that the results of this inquiry point in two directions. First, there are questions that are purely about the philosophy of psychology: For example, how do rival research programmes in psychology today confer good-making qualities upon the psychological enterprise? To what are they committed? What are their underlying methodological assumptions? In short, it is important to do careful case studies in contemporary psychology just as it is to do other case studies in the history and philosophy of science. Second, the present discussion of the relationship between psychology and the metaphysics of the mind has only scratched the surface. If "naturalism" has become a kind of shibboleth in recent times, this disguises the fact that there is enormous variety in the kinds of projects that are called "naturalistic." I have given some reasons here for skepticism about naturalistic platitudes, and suggested that the Neutral Project is all that one really needs for science and all one is likely to get in metaphysics. But really this is not the end of that topic but only the beginning. We need a much more serious examination of the roots of contemporary naturalism and the assumptions it encodes. We need a more thorough examination of how well its normative models of explanation and of intertheoretic relations are supported by actual examples of scientific theories even in the natural sciences. We need a more systematic categorization of kinds of "explanation," and an application of these to actual work done, both in psychology and the other sciences.

My hope is that this book will have separated what is truly useful about the computer paradigm from false hopes based upon incautious uses of language. If I am right, both science and metaphysics will be better off if they see that beyond a certain point they must follow separate paths. I do not claim to know whether the computational path in psychology will end up leading to Oz or leading nowhere. But I think we will be better off if we turn around the map.

Symbols and Machine Computation

Chapter 5 of this book presented a brief discussion of the tokening of symbols in computers. More specifically, it did two things: (1) it gave a brief examination of how the semiotic categories described in the Semiotic Analysis of chapter 4 can apply to things in computers, and (2) it distinguished these categories from an implicitly functionalist usage of the words 'symbol' and 'semantics' found in the works of some writers dealing with the computer paradigm. Chapter 5 offered only a cursory discussion of symbols in computers, as the main point there was simply to convince the reader of the facts that (*a*) symbols (in the usual semiotic sense of that term) are indeed present in production-model computers, and (*b*) most of our talk about "symbols in computers" is best cashed out precisely in terms of these semiotic categories. This appendix, by contrast, offers more of a full-dress examination of how the Semiotic Analysis might be applied to symbols in computers. In light of the distinctions made in chapter 4, this analysis can be broken down into the following four questions:

(1) *The Marker Question:* In what sense(s) and under what conditions can computers be said to store marker tokens?

(2) *The Signifier Question:* In what sense(s) and under what conditions can the marker tokens in computers be said to be signifiers (i.e., to have semantic properties)?

(3) *The Counter Question:* In what sense(s) and under what conditions can arrangements of marker tokens in computers be said to be *syntactic* arrangements?

(4) *The System Question:* In what sense(s) and under what conditions can the regularities of computer state changes be said to be governed by *syntactic* or *formal rules?*

The basic forms that answers to these questions may take have already been laid out in the technical locutions developed in chapter 4. These provide schemata for talking about how *anything* can be a marker, a signifier, or a counter. What remains is to fill in the details of how *computers* can contain such entities.

An object is said to be a marker, a signifier, or a counter by virtue of its relationships with *conventions* and *intentions*. It is the existence of marker-, signifier-, and counter-establishing conventions that render an object *interpretable* as a marker, signifier, or counter, and the *possibility* of such conventions that render it *interpretable-in-principle* as a marker, signifier, or counter. It is by virtue of *authoring intentions* that an object is said to be *intended as* a marker, signifier, or counter; and an object is said to be *interpreted as* a marker, signifier, or counter by virtue of the way it is *construed* or *interpreted* by someone who apprehends it. These modalities, moreover, are applicable to objects of any conventionally determined type. Notably, larger units of language games, such as proofs of theorems, are subject to the same distinctions: a series of marker tokens can only count as a token proof if there is a language game which involves proofs, and whose criteria for being a proof are met by the series of marker tokens. If computer state changes are to be viewed in terms of formal rules, one must therefore also ask how their having a description that involves formal rules is dependent upon conventions and intentions.

This appendix will separate the questions that deal with *convention* from those that deal with *intention*. The discussion of convention will examine conventions by which the contents of storage devices in computers count as marker tokens, how those marker tokens are conventionally associated with semantic interpretations, and how their arrangements count (by convention) as *syntactic* arrangements. The section on intention will discuss the ways in which the states of the computer are further involved in networks of human intentions—notably, those of computer designers, programmers, and users.

A.1 THE DESIGN PROCESS AND SEMIOTICS

Computers come into relationship with conventions and intentions at two junctures: in the *design* of hardware and software by engineers and programmers, and in the *use* of computers by end-users. Both the design process and the user's understanding of the computer must be articulated in terms of tasks the computer is to perform—tasks such as the evaluation of mathematical functions, the storage and editing of text, the statistical analysis of data, the proof of logical theorems, etc. Such an understanding of what computers do *presumes* that computers are in some sense "symbol manipulators." But neither designers nor endusers are usually concerned with questions about just *how* the things they are dealing with can be *said to be* symbols, or how the processes can be said to *count as* analyses or proofs. From the design standpoint, one begins with some set of tasks one wishes to automate—the evaluation of some set of mathematical functions, for example—and then sets out to design a system that is appropriate to that task. This involves (*a*) determining what sorts of *things* need to be represented, (*b*) deciding upon a way of representing those things, (*c*) determining what *processes and relationships between those things* need to be captured, and then (*d*) designing hardware or software that will manipulate the representations in a way that mirrors the relationships between the things represented and tracks the processes which they undergo.

The design process thus proceeds "top-down" from an informally specified task, and from there to a more rigorous description of the task, thence to a system of representations and functions, down to the "functional architecture" of the program or the machine, which is in turn realized through the hardware. The semiotic questions to which we need answers, on the other hand, require us to proceed "bottom up" from the hardware through functional architecture to the levels at which the machine may be said to be storing markers, counters, and signifiers and performing operations such as evaluating functions. The designer is concerned with questions about *what* system of representations and functions will allow the computer to perform certain tasks. But we are concerned with the more basic question of *how* and *in what senses* what the computer does can be *said to involve* representations and formal operations in the first place.

The discussion here will thus proceed in very much the opposite direction from the design process. Whereas the design process proceeds from the assumption that one can unproblematically speak of computer representations and operations and seeks the right functional architecture, we shall begin by assuming that it is unproblematic that computer hardware and software are functionally describable and that this description forms a level of analysis distinct from the description of the computer in physical terms (since the same functional architecture can be realized through different components). We shall then ask how and in what senses the objects picked out by the functional description can further serve as the basis for the tokening of markers, signifiers, and counters, and how the functionally characterized state changes can be deemed to be rule-governed processes.

A.2 THE FUNCTIONAL LEVEL

Digital computers are *functionally specifiable devices*. That is, the workings of a digital computer can be seen wholly in terms of the inter-relationships of its components and relations to inputs and outputs. Each of the components, moreover, is itself a digital device—i.e., one that is capable of being in some integral number of mutually exclusive states, and is always in one of those states whenever the computer system is in operation. (Most production model computers are composed of *binary* digital devices—i.e., devices capable of exactly *two* stable states.) The state of the entire computer at any time t is a function of the states of its many components. We shall follow Turing (1936) in refer-ring to this overall state of the machine as its "complete configuration." The complete configuration of a machine may be viewed as an ordered *n*-tuple of the states of its *n* components.

It should perhaps be noted at least in passing that viewing the com-puter as a digital device requires a certain amount both of abstraction and of idealization from its physical description. One *abstracts* away from properties of the machine that are irrelevant to its functional de-scription (e.g., its weight and color); but one also performs a more im-portant abstraction in treating as equivalent phenomena that may be different for the physicist's purposes (slightly different voltage levels that do not affect the behavior of a circuit, differences in timing that are fine-grained compared to the clock speed of the machine, etc.). One *idealizes* the behavior of the computer by making certain background assumptions that may not always be true in vivo—for example, that electric current of the proper voltage is running through the machine. Change the voltage or cut off the power supply, and of course the func-tional description no longer describes the actual behavior of the ma-chine. When one treats the "proper" background conditions as given, one is making an idealization, albeit an innocent one. Similar idealiza-tions are necessary for most if not all nomic descriptions in the scienc-es.

Many of the components of a digital computer are devices with in-puts and outputs. To take an example, an electronic AND-gate circuit has two or more input leads and one output lead. Each lead has an ac-tive or "on" state (characterized by some physical property, such as a high voltage level) and an inactive or "off" state. The circuit, moreover, is so designed that the output lead will be active just in case every input lead is active. The AND-gate is *functionally specifiable* in the sense that the state of the output lead may be viewed as a *function* of the states of the input leads. (This is a use of the term 'function' in the strict mathematical sense.) One may represent the functional configuration of the AND-gate by way of a table showing the state the output lead will be in for each configuration of input leads:

Input A

AND	ON	OFF
ON	ON	OFF
OFF	OFF	OFF

Input B is the label for the rows (ON, OFF).

Figure A.1

In a similar fashion, the entire computer may be characterized by a function table—called the *machine table* for that computer—which specifies, for each complete configuration, what the next states of the various components will be. In a relatively few cases, the computer will be completely deterministic, and its function table will specify, for each complete configuration, the next state of *every* component. (This is the case, for example, with the machine Turing describes.) In most cases, however, the computer will have input devices, and these will include transducers whose states are partially determined by environmental conditions to which they are designed to respond. In these cases, the machine table maps from complete configurations to equivalence classes of complete configurations, where complete configurations in the same equivalence class are identical except for the states of input transducers. (Alternatively, it maps from ordered pairs [complete configuration, input configuration] to complete configurations.)

It is possible for computers and other functionally specifiable devices to have isomorphic machine tables and yet be built from different components. To use the simple example of the AND-gate, such a circuit can be realized in many ways. It can be built from vacuum tubes, for example, or from transistors. The resulting circuits are different *physically* because they are made from different components, but are *math-functionally equivalent,* because the mappings from inputs to outputs are isomorphic. In similar fashion, two computers can be math-functionally equivalent even if they differ in physical structure, so long as they share a machine table. Descriptions at the math-functional level are thus not reducible to physical descriptions, since there are many ways that a given math-functional description can be realized.

The functional level of description thus involves a significant abstraction from the various levels of physical description, since it picks out equivalence classes of objects by virtue of the interrelationships of their component parts while abstracting from the physical nature of those parts. Yet functional description picks out real relationships between components which are physical particulars, and what it picks out does not depend on convention. To get from physical description to functional description, one need only *abstract;* no *interpretation* is necessary.[1]

An individual object may be subject to more than one functional description, and thus may be describable as being a computer of more than one type. (Similarly, if one is creatively inclined, it is possible to find a way of describing any collection of objects as a system of digital devices—with the consequence that any collection of objects has a description as a computer.) Computers can, moreover, be subject to multiple functional descriptions *in ways which are connected to the intentions of the designers and programmers.* A program running on a computer causes it to function in particular ways, and the way a computer works by virtue of running a particular program may itself be described by a function table. Such a table will not be inconsistent with the machine table for that computer, but will not involve some of the complete configurations that appear in the machine table at all and will treat others as equivalent. The resulting table, moreover, may very well be isomorphic to the machine table for some other computer, in which case the program run on the first computer may be said to *emulate* the second. The functionally describable system of relationships set up by such a program is sometimes called a *virtual machine* because the first computer functions like the second while running the program.

The term *functional architecture* will here be used to denote any functionally describable system of interrelationships in a computer, be they those realized through a program or those of the hardware of the system. (This is probably the most prevalent use of the expression 'functional architecture', but it is worth noting that some writers [notably Pylyshyn 1980, 1984] reserve the expression 'functional architecture' for the functionally described *hardware.* Here the broader use will be adopted, since from a functional standpoint there is nothing of unique interest about the hardware used in computers. If a distinction need be made, one may simply qualify the expression with the words 'hardware' or 'program'.)

A.3 FUNCTIONAL ARCHITECTURE AND SEMIOTICS

The components picked out by the functional description also play roles in the way the computer is used. Some of them are used for *storage* of symbolic representations. Others are used to manipulate such representations in a way that is useful for the evaluation of mathematical and logical functions and the manipulation of symbol strings. Those used for storage are relevant to the discussion of computer markers, signifiers, and counters. Those used in programs that govern the state changes the computer is to undergo are relevant to the System Question.

It is important to note, however, that while the functional architecture of the computer may render it *suitable* for the storage of markers, signifiers, and counters, there is nothing about functional architecture in and of itself that *makes* what is in computer storage a marker, signifier, or counter. The functional architecture is indeed designed so as to accommodate representations, and the operations the computer is built to perform are designed so that the changes in representations that they induce will be interpretable as derivations in accordance with syntactically based rules. But it is only by virtue of *conventions* and *intentions* that the storage locations picked out by functional description and realized through physical parts of the machine can *count as* storing marker tokens. And the reason for this is perfectly straightforward: an object is only said to be a marker by virtue of its relationships to conventions and intentions.

It is useful for purposes of analysis to treat each level of convention as characteristic of a particular level of analysis. At the *marker level,* marker conventions allow things picked out by functional description to count as marker tokens. Once one has a set of marker conventions, one can then adopt syntactic conventions and construe arrangements of them as syntactic arrangements (at the *counter level*) and adopt semantic conventions linking markers with interpretations (the *signifier level*). Finally, once one has adopted syntactic conventions for expressions, one can apply conventions that allow one to interpret the state changes induced by a program or a hardware function as governed by syntactically based rules (the *system level*). The resulting hierarchy may be represented by a diagram (see fig. A2).

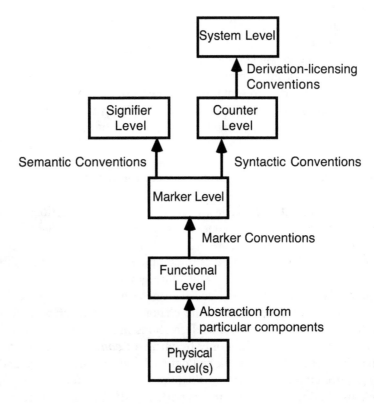

Figure A.2

Several features of this model are worthy of emphasis. First, there is a fundamental difference between the physical and functional levels, on the one hand, and all of the higher levels on the other: namely, the former do not involve convention, and the latter do. Indeed, each level above the functional level is reached by way of an additional set of conventions. A second important feature of this analysis is its treatment of the relationship between the signifier, counter, and system levels. Here the signifier and counter levels are treated as being *parallel* and *independent*. The reason for this is simple: there are semantic conventions that are not dependent upon syntax and syntactic conventions that are not dependent upon semantics. This does not mean that *no* semantic conventions presume any notion of syntax, or vice versa. (The meanings of complex words are sometimes a function of the syntactic arrangements of bound and unbound morphemes, for example, and the meanings of sentences a function of the meanings of the words.) What it does mean is that there is no absolute priority between syntactic and semantic conventions. There is, however, a priority relationship between the counter and system levels: in order for a succession of computer states to be interpretable as in accordance with a formal rule, it must be the case that the individual states be interpretable as counter tokens. (Otherwise there is no syntactic structure to them on the basis of which the rules can be applied.) But the reverse does not hold: one can very well interpret a series of computer states as a series of syntactically structured entities without interpreting the *series* as something licensed by formal rules.

The structure of this model seems clearly licensed by the semiotics developed in chapter 4. The relationships and priorities it picks out are those relevant to computer semiotics. For other purposes—such as those of the system designer—they admittedly might prove irrelevant or even confusing. To answer the questions of *how* and *in what senses* computers can store and manipulate symbols, however, one must proceed upwards from the level of functional architecture and ask how conventions make storage locations interpretable as bearing marker tokens, and how these in turn are interpretable as signifiers and counters. One may then ask how computer processes involving these conventionally determined entities are interpretable as involving formal rules. Having answered these questions about the role of convention in the semiotics of computer storage, we may then ask how the things that are *interpretable* in these ways are related to actual human intentions.

A.4 MARKERS IN COMPUTERS

How, then, are the storage locations that are picked out by functional description and realized in actual computers through particular hardware components to count as markers? The question is best approached through examining a paradigm example. Consider the devices employed for circuit storage in production-model computers. The circuit storage of production-model computers generally consists of a series of *bistable circuits or flip-flops.* These circuits have two output leads (generally designated by the numerals 0 and 1), and have an internal configuration such that exactly one of the output leads will be at some particular higher voltage level (e.g., +10V relative to ground) and the other at some other, lower level (e.g., -10V). The *state* of the circuit is determined by which output is at the high voltage level: if the 0-output is at +10V, the circuit may be said to be in its 0-state, and if the 1-output is at +10V, it may be said to be in its 1-state.[2]

It is important to note, however, that describing a circuit in digital terms—namely, as having a "1-state" and a "0-state"—is *not* tantamount to describing it as *storing numerals* or *numbers*. The digital description of a circuit is an abstraction from its many possible physical states, an abstraction that picks out two equivalence classes of physical states. In the case of the bistable circuit, the equivalence classes are picked out both *architecturally* (by the behavior of the circuit in and of itself) and *functionally* (by its relationship to the rest of the machine). Architecturally, the structure of the circuit is such that there are some states that are stable and some that are not. The stable states fall into two clusters: those that involve voltages very close to +10V at the 0 output and those that involve voltages very close to +10V at the 1 output. The states within each of these clusters may be treated as equivalent, and any state within the one cluster may be called a "0-state" and any state within the other cluster a "1-state." If the computer system has been properly engineered, these same equivalence classes will also be those picked out by the functional description of the computer—i.e., it will be differences in the digital state of the circuit, and only such differences, that will have an effect on how it influences other components of the system. There will be cases in which the state of that circuit will help determine the next state of the overall system, but *only* those differences picked out by the digital description (and not, for example, minor differences in voltage level at the outputs) will make a difference in the behavior of the overall machine. The labels '0-state' and '1-state', however, are just convenient labeling conventions we have chosen to provide, and signify no special relationship to particular numerals or numbers. (We might as well have called them the "cat-state" and the "dog-state" or the "Isaac-state" and the "Ishmael-state.")

Other digital storage media function in an analogous fashion. In each case there are atomic storage locations that are ordered and capable of some integral number of discrete states. In the case of magnetic disks and tape, for example, the storage locations are regions of the disk or tape, and the physical property that determines the state of a location is the magnetic flux density at that location. In the case of paper cards and tape, the storage locations are again regions of the card or the tape, but the physical property that determines the state of the location is the presence or absence of a perforation at that location. In each of these cases, one may call one of the two possible states of each location the "0-state" and the other the "1-state." Similar descriptions can be given for other media, such as magnetic cores and holographic disks.

While neither the internal structure of the storage location nor its functional role makes it count as *being* or *storing* a marker token, there are conventions whereby either the *states* of single atomic storage locations or *patterns of states* found across series of such locations can serve as the criteria for conventional types. One could, for example, adopt a convention for the tokening of markers in individual flip-flops. The convention would set up two marker types: the criterion for one would be the circuit's being in its 1-state, and the criterion for the other would be the circuit's being in its 0-state.

Such a convention would be of very limited use, however, since it would limit one to two marker types. One can obtain a more flexible convention by typifying markers according to the *pattern* of digital states *across a series* of atomic storage locations. A string of n storage locations, each of which is capable of i different states, can hold any of i^n different *digital patterns*. In production-model computers, bistable circuits in memory generally function in groups—most often in groups of eight, sixteen, or thirty-two—and the most elegant way of understanding the coding schemes used with computers involves treating groups or sequences of atomic storage units as storage locations for markers. The criteria for marker types are *patterns* or *sequences* of 0-states and 1-states present across the series of circuits making up the group. The further conventions by which computer states can count as representations of numbers, text, etc., involve assignments of interpretations to states typified by such *digital patterns*.

This way of describing the patterns in storage has several important advantages. First, it provides a way of seeing that the *same pattern* can *literally be present* in two very different storage media. Second, as a result of this, if digital patterns are used as criteria for marker types, the typification of markers can be independent of the nature of the storage medium—i.e., the same marker types can be used whether one is dealing with bistable circuits, magnetic tape or disk, paper tape or cards, holographic disks, etc. So long as a storage medium is composed of atomic units that have a digital description, series of atomic units can hold a digital pattern, and a convention may be employed whereby objects possessing digital pattern P_i are interpretable as tokens of marker type M_j. One may, moreover, adopt a canonical notation for digital patterns which can be used to represent them regardless of what medium they are present in. In the case of binary patterns, the numerals 0 and 1 may be used, and the pattern represented by a concatenation of these numerals: for example, 00001111, 101, etc. These sequences of token numerals are representations of patterns. The individual numerals do not represent anything in particular, and the pattern itself is not composed of numerals or numbers, though a sequence of numerals is itself an object in which such a pattern is present.

While any binary pattern whatsoever can serve as the criterion for a marker type (and hence any object possessing a binary pattern is *inter-*

pretable-in-principle as a marker token whose type is typified by that pattern), only a very small subset of such patterns is actually employed in the design, programming, and use of a computer. In general, computers are designed with several basic kinds of operations in mind—notably, logical operations, mathematical computations, and operations upon text and other symbol strings—and the designers usually decide upon efficient ways of storing text and representing mathematical and boolean values. Efficiency may require that different computer operations work upon strings of different lengths—e.g., one might use a sixteen-bit storage location to represent an integer, but employ a larger location to represent a floating-point number. The design process will thus characteristically involve developing conventions for several sets of marker types. Some of these will be fixed-length types—e.g., sixteen-bit patterns for integers, thirty-two-bit patterns for floating points, sixty-four-bit patterns for machine language instructions—while other types may be defined by a rule that allows for strings of variable length.[3] (A LISP machine, for example, would be designed to work with LISP files which are conceived of as *lists,* which are among other things concatenations of markers. A list, moreover, can be of any length.) Additional conventions may be supplied by the programmer, whose program may require data structures that use fixed-length strings of lengths other than those used for general machine functions or variable-length strings that are governed by rules other than those directly accommodated by the design of the machine. (A LISP interpreter, for example, may be run on a computer that is not itself a LISP machine.) It is in virtue of such conventions that a pattern across a storage location may be said to be *interpretable as* a marker token.

A.5 COMPUTER SIGNIFIERS

Like other markers, those in computers can be used to bear semantic values if there are conventions linking marker types to semantic interpretations. There are two basic kinds of conventions that link marker tokens in computers with semantic interpretations. Conventions of the first kind, which will here be called *representation schemes,* associate marker types with semantic interpretations: for example, with the boolean values *true* and *false,* with integers, or with floating point numbers. The second kind, which will here be called *coding schemes,* do not associate marker types directly with semantic interpretations, but rather associate marker types with other marker types. The ASCII code, for example, associates the alphanumerics and other graphemic characters with computer marker types typified by binary strings. Although a coding scheme does not itself involve any direct association of marker types with semantic interpretations,[4] it indirectly allows for the representation in the code alphabet (e.g., in the set of marker types used for ASCII coding) of anything that may be represented in the source alphabet (e.g., the set of graphemic characters used in written English).[5]

Consider the kind of representation scheme often employed for the representation of integers in computer storage. Such schemes generally employ fixed-length storage locations to store representations of integers. Contemporary computers tend to use sixteen-bit or thirty-two-bit locations, but for ease of notation let us discuss a convention which employs an eight-bit location. Each eight-bit storage location is a series of eight binary storage units, each of which is either in its 0-state or its 1-state. The eight-place series of binary locations carries a binary pattern of length eight. There are 2^8 or 256 such patterns, and this set of patterns can provide the criteria for 256 marker types. A number of notational conventions are employed to indicate such patterns, and generally are treated as equivalent. The most perspicuous way to note a marker type is to use the string of 0s and 1s which serves as the canonical representation of the binary pattern characteristic of that type. Other notations, however, are possible, and may be advantageous for reasons of brevity. A representation of a binary pattern may also be read as a representation, in base-2 notation, of an integer, and the pattern can be more briefly noted by the decimal or hexadecimal (base-16) notation for that same integer. (Thus the binary pattern whose canonical notation is 11110000 might also be noted by the decimal string 240 or by the hexadecimal string $F0.)[6]

The most commonly used representation schemes for the integers also exploit the relationship between the notation for the pattern present across a series of binary storage locations and the base-2 notation for integers: since the canonical representations of binary strings can also serve as representations of integers in base-2 notation, it is convenient for an interpretation scheme to assign to a string of binary digits the integer that string would represent if interpreted under base-2 conventions. However, since there are both positive and negative integers, one of the digits of the string is used to indicate the sign of the number. Here, then, is a sample convention for the interpretation of markers typified by eight-digit binary strings as representations of integers:

(1) If the first digit of the canonical representation of the string is a 0, take the remaining seven digits and interpret them as a representation of an integer in base-2 notation; this is the number the marker represents under this convention.

(2) If the first digit of the canonical representation of the string is a 1, take the remaining seven digits and interpret them as a representation of an integer in base-2 notation; multiply this number by -1; the resulting number is the number the marker represents under this convention.

While computers are often thought of primarily as "number crunchers," and while the bulk of computation done by production-model computers may well be numerical, representation schemes can be devised which will link computer marker types with any interpretations one might like. A series of binary storage locations n bits long can hold any of 2^n binary patterns. If these binary patterns are used as the criteria for a set of marker types, these marker types can then be associated, via signifier conventions, with as many as 2^n interpretations. For example, the boolean values *true* and *false* can be represented in a single binary storage location, with a 1-state indicating one value and a 0-state the other. Similarly, if one wished to be able to represent the twelve apostles, the set of sixteen four-unit binary patterns would provide criteria for enough marker types to ground an unambiguous representation scheme, even if one included Judas and Paul. One would simply employ a convention, for example, whereby the marker type typified by the binary pattern 0000 would stand for Peter, 0001 for Andrew, etc.

Computer markers can also be used to store meaningful data in a way that does not depend upon conventions that directly associate computer marker types with semantic interpretations. A second sort of convention—a *coding scheme* —takes marker types that are already employed in meaningful language games (e.g., the graphemic characters used for inscriptions of natural languages) and associates them with marker types in a "code alphabet" such as a set of marker types typified by binary patterns. The ASCII coding scheme is probably the most familiar example of a scheme that associates computer markers with markers of other types. The scheme takes marker types characterized by seven-bit binary patterns and associates them with a set of graphemic symbols which includes the upper- and lower-case letters, the numerals, punctuation symbols, and a number of additional frequently used graphemes, plus several less familiar types used to represent the backspace, spacebar, and return keys found on a computer keyboard. Unlike the mapping involved in the representational scheme for the integers, the mapping involved in the ASCII convention is most easily expressed not by a rule that maps marker types onto interpretations, but by a table that associates binary patterns with the graphemes with which they are paired under the convention.

The encoding of text under the ASCII convention is fairly straightforward. A file containing text is simply a sequence of storage locations, each of which bears a pattern that renders it interpretable as a marker suitable to the ASCII convention.[7] But the convention for ASCII coding is a convention for the coded representation of *text* —a convention whereby *sequences* of binary strings can encode *sequences* of graphemes. The purpose of the "encoding," moreover, is not to make the message *unreadable,* but rather to make it suitable for storage in a computer, and hence to make it *readable* through the mediating operations of the machine. It is thus reasonable to view an ASCII file produced by a word processing program as containing text in a natural language, albeit in a notational form that differs from written language. It is not natural language text *by virtue of consisting of seven-digit binary patterns,* of course; rather, it is natural language text because there are conventions for the graphemic representation of linguistic items and further conventions for the translation of these graphemic representations into ASCII notation. A properly encoded message retains its imputed semantic value, and does not lose it just by virtue of the encoding. The explanation of how the marker type may be said to be associated with an interpretation, however, is slightly more complex in the case of symbols in a code, because there are two levels of convention involved: (1) the *coding* conventions associating items in the source alphabet with items in the code alphabet (e.g., the ASCII code), and (2) the *semantic* conventions associating items in the source alphabet (e.g., strings of graphemes) with interpretations. (One might, of course, view phonemically based written language in a similar fashion—i.e., as involving coding conventions associating phonemes with graphemes and semantic conventions associating strings of phonemes with interpretations.)

A.6 THE COUNTER QUESTION

The syntactic properties of computer markers depend in similar fashion upon the web of conventions and intentions of designers, programmers, and users. Since computer storage locations are arranged in series, there can be sequences of markers in storage. Some language games involve syntactic patterns whose criteria pick out equivalence classes of marker sequences. Insofar as a marker token fits into one of the slots of such a pattern, it may be called a *counter* in that language game. The kinds of "fit," moreover, are the four modalities of chapter 4. In the case of computer storage, there can be several different sources for the relevant conventions and intentions. In pure cases of coding, such as ASCII files containing text, the coded sequence inherits the syntactic properties of the uncoded sequence. In the case of more structured representations such as records containing multiple fields, the conventions are those of the programmer who designed the representation structures. Finally, in some cases, such as parsing programs and theorem provers, the syntactic structures are picked out by more flexible rules. The computer can be made to be "sensitive" to syntactic structure—i.e., can be so designed that its operations *covary with* the presence and absence of a particular structure—but the computer cannot literally be said to *recognize* syntactic structure.

A.6.1 CODING AND INHERITED SYNTAX

The storage of text in computer storage media is often accomplished through the use of a coding scheme. When this is the case, the text file is stored in a series of storage locations, each of which holds a marker of a type appropriate to the coding scheme that is being employed. The file is thus a series of tokens from the "code alphabet," each of which is associated by the coding scheme with some type in the "source alphabet." In many cases, there is an almost exact correspondence between the series of markers in computer storage and the series of graphemes that would appear in a printed representation of the same text.[8]

Graphemic characters are used in language games that have syntactic structures. Notably, they are used for writing text in a natural language. This means that there are conventions whereby strings of graphemes can count as written tokens of words, sentences, assertions, etc. Token sentences in natural languages have syntactic structures, and this is true independently of whether the sentences are spoken or written. And so a token string of graphemes can (by virtue of conventions) count as having a syntactic structure. Encoding the string of graphemes by substituting for each the binary string it is associated with by the ASCII convention preserves both the ordering of the marker tokens and the syntactic properties that they have by virtue of being written language. If a coding scheme is used for encoding text, the coded message preserves the syntactic properties of the original.

Here, then, is one way that markers in computers can be said to have syntactic properties: if (*a*) there are conventions setting up a one-to-one coding scheme whereby a set C of computer markers are used to encode some other set M of markers, and (*b*) there is a language game G which uses markers from M, and (*c*) G is syntactically structured. In such cases, strings of markers in the computer preserve the syntactic structures of G.[9] In this sense, all of the syntactic features of natural language can be preserved in computer storage in precisely the way that they can be preserved in printed text.

To most of these features, however, the computer is likely to be little more sensitive than is a printed book. Research in artificial intelligence has made some inroads into sentence parsing, but the average computer does not have a sentence parser. Nor does it need one in order to store coded natural language text, any more than a book needs a parser in order to contain written sentences. The computer's insensitivity to grammatical features of text it stores places limits on what it can *do* with that text—e.g., its ability to respond to questions or requests for deductions—but does not impair its ability to *store* syntactically structured text.

A.6.2 STRUCTURED REPRESENTATIONS

Computer design and programming can and does make use of syntactic structures to which the computer is sensitive as well. In programming, for example, it is common to create complex representational structures by combining simpler ones. Suppose, for example, that a researcher in the social sciences is engaged in an experiment in which he uses a questionnaire with fifty true-false questions. He wishes to store the results of the questionnaire in a data base and then run several statistical analysis programs on his data, and wishes to index the answers from each questionnaire by the Social Security number of the participant. An efficient and intuitively appealing way of organizing the data is to think of the information from each questionnaire as one *record,* and the entire data base as a series of records. Each record holds an encoded social security number and fifty representations of answers to questions.

Such a record could be stored by means of a complex marker structured in the following way: (1) a series of nine seven-bit locations holding the ASCII encodings of the digits of the Social Security number, followed by (2) a series of fifty one-bit locations holding representations of the boolean values *true* (represented by a 1-state) and *false* (represented by a 0-state). What has just been articulated is a convention for a complex data type. The convention specifies not only the coding and representational schemes to be employed, but the syntactic structure of the record as well. And *this* sort of syntactic structure the programmer *would* assure that his program was sensitive to, since he wants to be able to have access to different kinds of information and wants to be able to perform different operations on different kinds of information.

A.6.3 RULE-GOVERNED SYNTACTIC STRUCTURES

A slightly more complex example along similar lines could be provided by a program such as a parser or a theorem prover. Here, however, the syntactically structured types are not set up in terms of fixed-length representational structures, but are articulated in terms of rules governing classes of concatenations of markers. A program designed to generate derivations in the sentential calculus, for example, might check a representation of a proposition to see if it fit any of a number of syntactically structured patterns such as the following:

[negation sign],[wff]

[wff],[implication sign],[wff]

[wff],[disjunction sign],[wff]

[wff],[conjunction sign],[wff]

Checking a string of markers against such templates might involve a fairly complicated test procedure, especially since the operation might have to be recursive to determine whether a given substring is a *wff*. The aim of such a procedure, however, is simply to provide a means of determining when a representation is of one of the syntactic types relevant to derivations in the propositional calculus. And it is important to distinguish the ability to *determine* whether the syntactic structure is present from the *fact* that it is present. A computer marker string can *have* a syntactic structure just by virtue of being associated with some syntactically structured *human* language game, as in the case of stored text, without the computer being sensitive to the syntax. It is, however, possible in some cases to make the computer sensitive to the syntax—namely, in those cases where the syntactic class can be picked out by a rule that operates upon marker concatenations.

A.6.4 THE NATURE OF THE COMPUTER'S
SYNTACTIC "SENSITIVITY"

But just what does the computer's "sensitivity" to syntactic structure amount to? In this sort of case, a computer may be said to be "sensitive to" a syntactic pattern P just in case (1) there is a functionally describable operation provided by the hardware or the programming which (2) takes marker strings as input and (3) whose output depends precisely upon whether the input string can be construed as having pattern P. What this most assuredly does *not* involve is any *understanding* of syntax on the part of the computer. The computer's "sensitivity to syntax" is a matter of computer operations on marker strings tracking conventionally sanctioned syntactic construals of those strings. The computer does not *recognize* syntactic patterns *as* syntactic patterns. To *recognize* something, it would have to be a cognizer; and to *construe* patterns as *syntactic,* it would have to be privy to the conventions of a language game. It may well be that it is logically possible for there to be computers which were also cognizers and could share our conventions; but it is not necessary to posit such things about actual computers in order to explain what they already do. For this all one need see is that the functional architecture of the computer and the programs it runs can be so designed that computer operations will be in a relationship of *causal covariance* with syntactic patterns licensed by (human) conventions.

A.7 FORMAL RULES AND THE SYSTEM QUESTION

Thus far the discussion of computer semiotics has been confined almost exclusively to questions about symbol *tokens.* And what has been said about the way these are related to conventions is very similar to what might have been said about printed text on a blackboard or a page of a book. Yet computers differ from blackboards and books in one very important respect: whereas blackboards, books, and computers can all serve as media for storing symbols, computers can also in some sense *manipulate* symbols, while blackboards and books cannot. Moreover, the manipulations may be construed as corresponding to formal rules. To see how this is so, it is useful to separate two issues: first, how individual *functions* performed by either circuitry or software can be viewed as corresponding to formal rules; second, how the *machine table* of the entire computer may be viewed in such a fashion.

First, consider a program that is designed to evaluate mathematical functions such as integrals. The designer of the program presumably will know a number of techniques for integrating different sorts of expressions—indeed, the familiar methods for integration involve a piecemeal set of rules for expressions with different syntactic forms. What the designer does is to find an efficient way of encoding a wide range of expressions that will preserve the syntactic features relevant to the integration methods, and then write a set of procedures corresponding to the different integration techniques. Each procedure will contain two parts: a section which *tests* a marker string supplied to the program to see if it has the right syntactic form for that integration technique, and a section which *generates* a new marker string if the result of the test is positive. The new marker string should correspond, under the designer's coding conventions, to the result of integrating the expression encoded by the string sent to the program.

How does what the procedure does end up counting as a derivation in accordance with a formal rule? It does so because (1) the procedure generates a new marker string only if the input string is within an equivalence class whose members are interpretable[10] as being of a syntactic type T (i.e., only if it is of the right counter type), (2) the output string it generates is interpretable as being of a syntactic type U, and (3) there is a language game L containing a syntactically based rule whereby expressions of type T license the derivation of expressions of type U. There is thus an analogy between aspects of the language game on the one hand—namely, (*a*) the derivation-licensing rules, (*b*) the expressions already derived, and (*c*) those expressions whose derivation is licensed on the basis of (*a*) and (*b*)—and aspects of computer procedures: namely, (*a′*) the hardware or software that performs the operation, (*b′*) input strings that cause the hardware or software to perform its operation, and (*c′*) the output strings produced as a result.

A.7.1 CAUSALITY, FUNCTIONAL ARCHITECTURE, AND FORMAL RULES

The computer operation, however, has two aspects: *causal* and *symbolic. And it is essential that the two aspects be distinguished.* For the causal propensities of any computer operation can be described in purely functional terms (i.e., as a function to and from equivalence classes of digital patterns) without any mention of symbols or syntax. Understanding the operations of the computer function in *syntactic* terms involves the *imputation* of symbolic status to what is in the computer, and this involves convention. It is, of course, the task of the designers of hardware and software to make the functional architecture that is to perform manipulations of symbols correspond to formal rules. It may additionally be the case that all functions performable by a computer can be interpreted as rule-based counter manipulations. But the fact that a physical process has a functional description no more *makes* it the application of a syntactic rule than the fact that a storage location is picked out by a functional description of a computer makes the state of that location the tokening of a marker. Functional description simply does not suffice for syntax, because syntax requires convention as well.

A computer function is thus interpretable as a derivation in accordance with a formal rule if three conditions hold. First, there must be a functionally describable, causal covariation between inputs and outputs for that function. Second, those inputs and outputs must be interpretable in terms of syntactic categories. And third, there must be formal rules available under which counters of the types that cause the function to produce an output would also license the derivation of counters of the types produced by the function in response to such input.

It is absolutely essential to note that there is *no* mention here of the computer *dealing with* syntax *as syntax*. Rather, the functional architecture is designed in such a fashion that the causal covariations will track syntactic relationships. It is thus misleading at best to say that a computer "knows" or "deals with" syntax but not semantics. Functional architecture can support state changes that track the formal relationships used in syntactic derivation techniques. But it can in similar fashion support both static data structures and state changes that track semantic relationships. There may be a large difference in *degree* between what syntactic features can be supported and what semantic features can, but the computer's relationships to syntax and semantics are quite parallel: it is said to involve each only by virtue of conventions.

We should thus be very careful to make the distinction between saying that state changes happen *in accordance with* a rule (i.e., in a fashion interpretable as licensed by such a rule) and saying that they involve the *application* of a rule. Computers do not *apply* rules. Or even if there were some computer that was part of our linguistic community and could make semiotic conventions, this would be something *more* than what we are talking about when we say, in general, that computer state changes are "rule-governed." Here we merely mean that computer storage locations are interpretable as counters, and that the computer operations that change the states of computer components are interpretable as functions to and from equivalence classes of counters, and that such functions may be expressed in the form of syntactically based rules.

A.7.2 THE MACHINE TABLE AND FORMAL RULES

It is not only small, local computer functions that may be viewed as rule-governed in the sense articulated above. The entire machine may also be described in such a fashion. The entire computer, after all, is characterized by its machine table, which is a function table from complete configurations to equivalence classes of complete configurations.

Seldom are there explicit syntactic conventions for interpreting the complete configuration of a machine as a syntactically structured marker string, but one could create such conventions. (Turing [1936], for example, gives such conventions for describing the machine state of the machine he describes.) The complete configuration, after all, is an ordered n-tuple of the machine's n constituent components. If one employs a marker convention that would link each possible component state with a marker type, each complete configuration is interpretable as a string of markers. Once this has been established, it is simple to prove that the machine may be regarded as "rule-governed." Simply take each complete configuration to be the sole member of a syntactic class (i.e., a counter type). For each complete configuration, there is a set of complete configurations which can be the next state. (The members of a class differ only with respect to the states of input transducers.) Each of these next states has an interpretation as a counter. To view the machine as rule-governed, one need only view each mapping from a complete configuration C to its set of possible successors as the function table for a rule which licenses the derivation of any of the counters corresponding to the successors whenever the counter corresponding to C is tokened by the machine.

A.8 COMPUTERS AND INTENTIONS

This brings to a close the discussion of *conventions* associating the states of computer storage conventions with semantic interpretations. There is little difficulty in seeing that marker tokens in computers are often *interpretable* under widely employed conventions as having semantic values. Given the representation scheme for integers outlined above, for example, any series of storage units possessing an eight-unit binary pattern is straightforwardly interpretable as a representation of some integer under that convention in much the same way that a sequence of numerals is interpretable as a representation of some integer. The role of *intentions* and *actual interpretations* in the semantics of computer markers, however, is not so straightforward as it is in the case of inscriptions. The reason for this is that someone who sets out to inscribe a meaningful message has some sort of awareness of the marker token he is authoring and has at least implicit knowledge of the semantic convention under which the marker token is to count as a signifier of a given sort, while the computer user may be quite unaware of the coding and representation conventions through which he stores his data, and almost certainly lacks any direct access to the storage media and the marker tokens found therein. With the possible exceptions of perforated tapes and cards, computer storage media store markers in forms that human perceptual faculties are ill suited to perceive. Circuits and disks are generally hidden from the user, and even when they are not, his eyes and ears and fingers cannot perceive voltage levels or flux densities. In addition, the user is quite unlikely to have any explicit knowledge of the conventions for marker tokening. The average user of a word processor has a dim understanding that what he has typed at the keyboard ends up "in the computer" or "on the disk," but probably has no knowledge of the coding conventions employed, the marker types they use, or the physical properties through which the marker tokens are realized in circuit memory and disk segments.

There is, in effect, a kind of division of duties between the designers and programmers of the computer, on the one hand, and the users, on the other. The design of hardware and software is generally pursued with the understanding that the computer will be used in tasks that involve certain kinds of representation, and the functions that are built into the hardware and software are designed so that they will manipulate marker tokens in ways that track meaningful relationships between the things that the markers are to be used to represent. If, for example, one wishes to build a circuit or write a program which performs computer addition (i.e., which takes pairs of representations of numbers and returns a representation of their sum), one must know how the numbers are to be represented, since the choice of a representation scheme will have consequences for how the circuit or the program must be designed. Indeed, one needs a clear idea of what representational scheme one is going to employ in order to write any program that is to induce state changes that track meaningful relationships, be it a theorem prover, a language simulator, a numerical analysis program, or a chess program.

To these conventions the average computer user is usually oblivious. He has direct access only to input devices, such as the keyboard on which he types, and output devices such as monitors and printers, through which he has access to graphic and graphemic representations of data stored in the computer. If, for example, he is entering and editing text through a word processing program, he may very well think of himself as typing words of written English on a roll of "virtual paper" that scrolls past on his monitor. He may be completely ignorant of how text storage is accomplished in the computer; indeed, the question may never occur to him. He does, however, perform actions which he intends to result in there being some kind of linguistic representations— *text,* in effect—"in the computer," however vague or misguided his beliefs about how that is achieved.

These two difficulties—the lack of direct access to marker tokens and the user's ignorance of marker conventions—complicate the question of how things in computers can be *intended as* and *interpreted as* marker tokens. What may be said of things that are interpretable as marker tokens and caused by a computer user will depend on what kind of access the user has to the markers and what kind of understanding he has of the conventions, and these factors are subject to a great deal of variability. At one extreme, there are some marker tokens which are interpretable under conventions as markers, counters, and signifiers, but really are never authored or apprehended at all. At the opposite extreme, a computer engineer using a program that allows him to sample and alter the states of specific storage locations has very *reliable* if not *direct* access and knows the conventions he is working with. In his case, the fact that his access is mediated by software and hardware does not seem all that significant. But what about the user of a word processing program, for example, who is untutored in programming and computer science? Are the things that get into storage as a result of his actions intended as marker tokens? Are the things in storage that result in text appearing on his monitor interpreted as marker tokens?

The answer to this question can be *yes* so long as one allows that the user's relatively vague intention that his actions at the keyboard count as "typing something into the computer" can connect his actions with marker conventions of which he does not know the specifics. This suggestion is in some sense parallel to Burge's insistence that the use of a word in a natural language ties one's utterances and even one's beliefs to the standard meaning of the word, even when one does not have a full grasp of that meaning. The suggestion here is similar: the intentions involved in "typing text into a computer" involve assumptions to the effect that (1) there is marker tokening going on, (2) this corresponds in a fairly straightforward way to the kind of tokening that goes on when one types on paper, and (3) the marker tokening is connected in regular ways to keystrokes on the input side and output in graphemes on paper or a monitor.

One may make a similar case to the effect that under certain conditions the apprehension of graphic or graphemic representations on paper or monitors may also amount to the apprehension (with varying degrees of directness) of marker tokens in storage. The example of the electrical engineer using a program that displays representations of binary strings found in storage locations is the clearest case of someone interpreting the contents of a storage location as a marker of a certain sort, but some case can be made for anyone who thinks of the output as "coming from the computer" and believes that it was "in the computer" in the form of marker tokens.

If this analysis is right, then in cases like word processing, the computer user's intentions alone are sufficient for determining what the markers created as a result of his actions (e.g., his typing on a keyboard) are *intended as*. Recall the definition of authoring intentions for signifiers:

(S2): An object X may be said to be *intended (by S) to signify (mean, refer to) Y* iff

(1) X was produced by some language-user S,

(2) S intended X to be a marker of some type T,

(3) S believed that there are conventions whereby T-tokens may be used to signify Y, and

(4) S intended X to signify Y by virtue of being a T-token.

The definition allows some degree of latitude in the extent to which S must be cognizant of either the marker or signifier conventions that are relevant to his inscription or utterance. People can intend to use words meaningfully, for example, even if they are unsure, confused, or even mistaken about how they are spelled or pronounced. They can also intend to use a word meaningfully even if they are unsure, confused, or mistaken about the semantic value(s) the word carries under standard semantic conventions. (A foreigner who mixed up the English words 'dog' and 'cat' might *intend* to express the belief that the cat is on the mat by saying, "The dog is on the mat." His utterance would not, however, be *interpretable* under English conventions as an expression of the belief that the cat is on the mat.) Similarly, a user of a word processor has some understanding that he is "typing text into the computer" even though he may be unaware of just what that amounts to, or even have erroneous beliefs about what is involved. (He might believe, for example, that tiny letters are being inscribed in ink somewhere inside the machine.)

In general, explicating the exact ways in which particular marker tokens in computer storage may be said to be meaningful representations will involve tracing out the web of human conventions and intentions that have played a role in the etiology of those particular tokens. In particular, one must attend to the representation and coding schemes intended by the programmers and engineers on the one hand and the authoring intentions of the user on the other. Carrying out such an explication will generally prove significantly more involved than carrying out an explication of the meaningfulness of an inscription or utterance—in part because there are more people's intentions to take into account, and in part because there are more processes mediating the symbol user's access to the symbols. In principle, however, there is no problem in saying of markers in computer storage that they have semantic properties under any of the four modalities outlined in chapter 4.

Notes

INTRODUCTION

1. Notable to my mind are Cummins (1989), which addresses the use of the word 'representation' in cognitive science in ways that are similar to those suggested here, and von Eckardt (1992), which takes a more comprehensive look at the philosophy of cognitive science as philosophy of science. While I think that von Eckardt is inadequately sensitive to subtle problems in the notion of 'representation', this kind of book is a very welcome development in the field.

CHAPTER 1

1. Fodor (1987, page 6) says that "the predictive adequacy of commonsense psychology is beyond rational dispute."
2. Pylyshyn, for example, writes that there are "regularities and generalizations which can be captured using cognitive terms that could not be captured in descriptions using behavioral or physical (neurophysiological) terms." (1984, page 7) And Fodor writes in a similar vein that "[w]e have no idea of how to explain ourselves to ourselves except in a vocabulary which is *saturated* with belief/desire psychology." (1987, page 9)
3. Fodor thus writes that "[a] lot of what common sense believes about the attitudes must surely be false," (1987, page 15) and Pylyshyn opines that it may be that folk psychology's "set of terms needs to be augmented or pruned, that many of the beliefs expressed in folk psychology are either false or empirically empty, and that many of its ex-

planations are either incomplete or circular." (1980, page 113)

4. Fodor (1981, pages 18-19), for example, writes that there are "lots of interesting generalizations to state about how the propositional attitudes that an organism has are affected by aspects of its experience, and its genetic endowment, and by the other propositional attitudes that it entertains. One of the goals of a (cognitive) psychology is to state the generalizations and, whenever possible, to systematize and explain them."

5. Pylyshyn (1980, page 112) writes that "its most serious shortcoming, from the point of view of the scientific enterprise, is that the collection of loose generalizations that makes up this informal body of knowledge is not tied together into an explicit system."

6. Fodor's "cruder but more intelligible" gloss would also seem to indicate that beliefs with different contents are typified by the same functional relationship R, but differ with respect to the type of representation to which the organism is related. So if Jones' believing that two is a prime number consists in Jones standing in relation R to a representation of a type MP that means that two is a prime number, then Jones' believing that it is raining must consist in Jones being in the same relation R to a representation of a different type, MP^*, such that tokens of MP^* mean that it is raining. Conversely, when Jones has different cognitive states that share a common content, the different states are characterized by different relations to tokens of the same symbol type. For example, if Jones both believes and hopes that it is raining, his believing that it is raining consists in his being in relation R to a token of type MP^*, and his hoping that it is raining consists in his being in some other relation R^* to a token of type MP^*.

7. First, it is not clear whether the *same* relation R is characteristic of belief for *different cognizers*, or whether the functional relations characteristic of propositional attitudes such as believing or desiring may differ from organism to organism. That is, it is unclear whether *Smith's* believing that two is a prime number (as opposed to *Jones'* believing it) would involve Smith being in the same relation R to a mental representation meaning that two is a prime number that Jones is in when *he* believes that two is a prime number, or whether it would involve Smith being in some *other* relation S to such a representation. This question is highly significant in its bearing upon how CTM might guide empirical research, but will not have an effect upon the results of this inquiry.

Fodor's articulation of the nature of cognitive states is also unclear on the question of whether bearing a particular functional relation to a mental representation is to count as a *sufficient* condition for being in a particular cognitive state, or whether it is merely a *necessary* condition.

What Fodor says is that *"to believe that* such and such *is to have* a mental symbol that means that such and such tokened in your head in a certain way." But it is not fully clear whether Fodor means to say (a) "to believe that such and such is *just* to have a mental symbol that means that such and such tokened in your head in a certain way," or (b) "to believe that such and such is *to be in a state that includes* having a mental symbol that means such and such tokened in your head in a certain way." On interpretation (a), Fodor's account offers a sufficient condition for an organism O's being in a particular cognitive state; on interpretation (b), it offers only a necessary condition. Interpretation (a) seems the more natural reading of the text in question, but the reader is naturally guarded about drawing too strong an inference from a characterization that is described by the author as "crude."

8. Fodor reserves the spelling of intentionality with a 't' for "contexts connected with intent." (See note 2 on page 318 of *Representations*.)

9. There is, of course, a significant history of controversy about whether intentional explanation is a form of causal explanation, and the issues that stand between, say, Vienna Circle Positivism on the one hand and Ryle, Wittgenstein and the *Verstehen* tradition on the other. In particular, there has been controversy about the nature of intentional explanation, notably about whether such explanations provide only *reasons for* action, only *causes* of action, or both. Much of the Continental tradition in philosophy and the human sciences has held that explanation in the human sciences gives reasons for action and provides interpretive understanding (*Verstehen*), whereas explanation in the natural sciences gives causes of events and provides explanations. In the English-speaking world, Wittgenstein, Ryle and other ordinary-language philosophers have advanced similar views to the effect that intentional explanations provide reasons for action and do not name causes. Much of the English-speaking philosophical tradition, on the other hand, has followed the Logical Positivists in assuming that the causal, nomological mode of explanation employed in the natural sciences is normative for the application of the word "science", with the implication that psychological explanation must either be causal and nomological or else be unscientific. (The contribution of CTM to this discussion will make up the greater part of Chapter 2 of this book.)

10. While computational procedures (such as integration techniques and column addition) and formalized mathematical systems (like Hilbert's geometry) may both be said to involve "formal rules," the "rules" they involve are of different sorts. Formalized mathematical systems such as formal logic and formal geometry contain *derivation-licensing*

rules, but no *rules of procedure*. That is, they involve rules that govern the *validity* of derivations, but give no sort of guidance as to *what to derive*. The rule of *modus ponens*, for example, licenses as valid derivations of an expression of the form q whenever one has already derived expressions of the form $p \supset q$ and p, but does not tell a logician that he *should try* deriving q as a part of a particular proof. Computational algorithms, on the other hand, contain rules of procedure: rules specifying what is to be done under particular circumstances. The algorithm for column addition does not say that one *may* put at 7 in the next column of the sum when one finds a 3 and a 4 in the next column of the two addends, it specifies that this is what *must* be done if one is to apply the algorithm. Yet both the procedural rules involved in algorithmic procedures and the licensing rules involved in formal logic, geometry, etc., may be said to be "formal" in the sense that *the applicability of the rules depend entirely upon the syntactic forms of the expressions* and not upon their semantic values.

11. There are practical problems with this. For example, if C is a storage register of the same size as A and B (e.g., 32 bits), then a substantial portion of the possible summations will result in an "overflow" of register C—i.e., the absolute value of the sum will be too large to be represented in a register of the size of C. One may avoid this problem by making C one bit longer than A and B, but this has the effect of making the domain of C different from the domain of A and B, making it hard to find a tidy way of characterizing just what mathematical structure is implemented through the circuitry. One cannot say, for example, that one has a subgroup of the integers under addition, since the domain and range of the function are different subsets of the integers.

CHAPTER 2

1. Related in conversation by Professor Sellars.

2. Brentano's distinction between "mental" and "physical" phenomena seems to have been meant as a distinction between intentional states and qualia, not (*pace* Professor Chisholm) as a distinction between what English-speaking philosophers would normally call mental and physical *objects*. (For a convincing argument to this conclusion, cf. McAlister, Linda (1974).) The directedness of cognitive states sets them apart from both sorts of "physical" things, but the specific nature of Brentano"s distinction is a consequence of the fact that his "empiricism" was of the philosophical rather than the experimental sort.

3. Chisholm's criteria for the "intentionality" of sentences are formu-

lated in terms of features of the subordinate clauses of the verbs used to denote propositional attitudes, which do not admit of existential generalization or substitutivity of identicals. Uses of referring terms that do not admit of existential generalization or substitutivity of identicals are also called "opaque." Thus a considerable number of writers use the expression "intentional" to denote a class of sentences marked by opacity of reference. (Some writers, notably Fodor and Searle, employ a separate spelling for this Chisholmian usage of the word: namely 'intens*ionality*'. This usage is not to be confused with the "intension" of the Port Royal logicians, which is defined in contrast with extension.)

4. See, for example, Rudolph Carnap (1938). See pp. 144-146 for the reduction of biology to physics, 146 ff. for the reduction of psychology.

5. Hempel (1949), in Block (1980), page 18. It is worth mentioning that in a preface to the reprinting of this article in Block (1980), Hempel indicates that he has long since rejected logical behaviorism and that the article was "far from representing [his] present views." (p. 14) Logical behaviorists, however, were often at pains to distance themselves from claims about (a) the ontology of mind and (b) the proper methodology for psychology. Hempel, for example, clarifies the relation of logical behaviorism to existence claims involving the mental by saying that

> Logical behaviorism claims neither that minds, feelings, inferiority complexes, voluntary actions, etc., do not exist, nor that their existence is in the least doubtful. It insists that the very question as to whether these psychological constructs really exist is already a pseudo-problem, since these notions in their "legitimate use" appear only as abbreviations in physicalistic statements. (Hempel (1949), in Block (1980), page 20.)

As for psychological methodology, Hempel writes that his thesis, though

> ...related in certain ways to the fundamental idea of behaviorism, does not demand, as does the latter, that psychological research restrict itself methodologically to the study of the responses organisms make to certain stimuli. (Hempel (1949), in Block (1980), page 20.)

6. Cf. the discussion of this development in Herbert Feigl, "The "Mental" and the "Physical"" (1958) especially pages 371–2.

7. Putnam and Oppenheim write, for example:

> It is not absurd to suppose that psychological laws may eventually be explained in terms of the behavior of individual neurons in the brain; that the behavior of individual cells—including neurons—may eventually be explained in terms of their biochemical constitution; and that the behavior of molecules—including the macro-molecules that make up living cells—

may eventually be explained in terms of atomic physics. If this is achieved, then psychological laws will have, in *principle*, been reduced to laws of atomic physics. (Putnam and Oppenheim (1958), page 7.)

8. The precise form of logical behaviorism is best understood as a result of the epistemological concerns of early Vienna Circle Positivism. Skinner's operationalism is a result of his concern about methodological rigor. The reductive physicalist's views are best understood, by contrast, in terms an atomistic view embracing (a) an ontological commitment to materialistic monism, (b) a commitment to the "completeness" or "generality of physics"–i.e., the view that every token *event* may be explained wholly in physical terms, and (c) a metatheoretical assumption that the *properties* relevant to high-level sciences like psychology can, in principle, be identified with (co-extensive) complex physical properties, even if the *predicates* of physics are not *synonymous* with those of the special sciences.

9. One also finds phenomenalism in the early positivist writings and neutral monisms.

10. The scare quotes are Turing's. Cf. Turing (1936): page 231.

11. Hilary Putnam, "Minds and Machines," (1960) is the classic source in philosophy of mind for the Turing machine analogy. Fodor (1965): "Explanations in Psychology", and (1968) "The Appeal to Tacit Knowledge in Psychological Explanation" espoused a similar functionalist view, but did not make specific use of the Turing machine analogy.

12. Fodor alludes to both problems when he writes, "Functionalism tends to vindicate Realistic construals of mentalistic etiologies; but it does not, per se, vindicate a Realistic reading of etiologies in which appeals to propositional attitudes figure." (1981: page 24)

CHAPTER 3

1. In the Introduction to *Representations*, for example, Fodor writes:

...computers are symbol-driven symbol-manipulators: their programs are sets of semantically interpreted formulae and their typical operations consist in the transformation of sets of semantically interpreted formulae. Only because the first is true can we think of the machine's operations as *rule-governed* (*un* interpreted strings of formulae aren't *rules*), and only because the second is true can we think of the machine's operations as *eventuating in proofs* (sequences of strings of uninterpreted formulae aren't *proofs*). (Fodor (1981), page 23, author's emphasis.)

2. Cf. his discussion of levels of analysis on page 34.

3. Searle defines "Intentional causation" in the following fashion:

[I]f x causes y, then x and y stand in a relation of Intentional causation iff
1. Either (a) x is an Intentional state or event and y is (or is part of) the conditions of satisfaction of x
2. or (b) y is an Intentional state or event and x is (or is part of) the conditions of satisfaction of y
3. if (a), the Intentional content of x is a causally relevant aspect under which it causes y
   if (b), the Intentional content of y is a causally relevant aspect under which it is caused by x. (Searle (1983), page 122–123.)

The bestowal of intentionality upon an utterance is a case of "Intentional causation" because the performance of an illocutionary act involves the "meaning intention" that the utterance should take on the conditions of satisfaction of the state it is to express, and one of the conditions of satisfaction of the meaning intention is that the utterance should take on the conditions of satisfaction of the state expressed. In terms of Searle's definition of "Intentional causation", the meaning intention plays the role of x and the bestowal of meaning upon an utterance plays the role of y. The relationship between the speaker's intentions and the intentional properties of the illocutionary act is thus (in part) a causal one.

4. Cf. Fodor (1987, 1990), Dretske (1983, 1986), Millikan (1983), Field (1978), Loar (1982, 1985), Putnam (1983), Cummins (1989). Cummins (1989) provides a useful survey of theories of content for mental representations.

5. Cf. Aristotle, Chapters 1 and 2 of *Categories*, *Topics* 106b35–38, *Metaphysics* IV.2, 1003a32–1003b1.

6. Searle's discussion of the relationship between illocutionary acts and intentional states clearly involves a kind of logical dependency, since the "essential condition" of an illocutionary act is that it express a particular intentional state. Moreover, Searle's account of linguistic meaning and the intentionality of illocutionary acts is not an empirical account, but an *analysis* of what it is for something to be an illocutionary act. On Searle's view, it would be quite incoherent to say that an illocutionary act was (e.g.) an assertion that two is a prime number, but not an expression of a belief that two is a prime number. Searle would, of course, recognize that one could insincerely express such a belief if one were lying, or joking or acting. But Searle's use of the word 'express' allows that such acts would still express a belief, albeit insincerely. To use the terminology of *Speech Acts*, such acts would meet the "essential condition" but not the "sincerity condition" for assertions.

CHAPTER 4

1. On this basis he can also make guesses about such matters as whether the script he is dealing with is phonetic (in which case one can expect a high frequency of long repeated character strings) or ideographic (in which case the frequency of repeated strings will characteristically be much lower), and whether Tanganjikan makes use of prefixes, suffixes or other bound morphemes which might indicate that the language is inflected.

2. The written form of Old English employed several graphemes that have since been dropped. One of these, called Thorn, was used to represent sounds represented in modern English orthography by the letters 'th'. And so the word 'the' was represented by a Thorn and an 'e'. Since Thorn came to resemble a 'Y', writers of Modern English have sometimes come to the mistaken conclusion that 'ye' was a variant form of the definite article, with the consequence that one finds establishments with such linguistically confused names as "Ye Olde Englishe Bed and Breakfast".

3. I confine myself for the moment to words with semantic values. Other sorts of words, such as articles and prepositions, would require slightly different treatment. These words, however are not signifiers, and the technical term 'signifier' is *not* intended to correspond exactly to the term 'word'.

4. In the case of spoken language, for example, different features of sounds are significant in differentiating one phoneme from another. In Indo-European languages, for instance, there are usually variants on consonants which differ only with respect to whether a vowel sound is produced. In English, each "voiced" stop has an "unvoiced" variant which differs only in that one does not sound a vowel while making it: [b]/[p], [d]/[t], [g]/[k], and similarly for the fricatives: [z]/[s], [v]/[f], [j]/[ch]. In some languages, however, voicing is not a significant feature, and the sounds [b] and [p] would be treated as equivalent. In similar fashion, it is notorious that East Asians who learn English as a second language have difficulty distinguishing our two "liquids", [l] and [r]

5. This is quite within the ordinary use(s) of the term "intend". If, for example, someone believes that 'melody' is spelled with two l's, writes it thus, and is then asked "Did you intend to write down two l's here?", there is a sense in which the answer should be "Yes." That is, he wrote it just the way he thought it should be written - it was not a flaw in execution. He could express this by saying "That's exactly what I intended to write", even though he had no "intention" to do so in the stronger

sense of having a conscious plan to do so. It is worthy of note that this sense of "intention" falls *between* the broader sense of "intention" and "intentionality" which refers to all cognitive states with content and the narrower sense which has to do with purposeful action.

6. These points are not meant to be extendable to all human actions, or even to all actions involving the creation of symbols. There is certainly abstract art, for example, that is susceptible to many interpretations, none of which was intended by the artist. And in written text, both the author's level of awareness of and his responsibility for the interpretation of what he produces seems in general to decrease with each higher level of interpretation. Writers seldom inscribe characters without meaning them to be specific letters and are probably more clear about what they mean by a given word than what the "meaning" of an entire poem or novel might be. The point of the distinctions made above, however, is not that there is always a privileged intended interpretation of any act whatsoever involving symbols, but that, *when* there is such an intent, it has a special status.

7. The variable x must denote an actual object if we are to say that it *is* interpretable-in-principle as signifying Y. If we wished to speak about what properties an object would have, if it existed, we should say that "x *would be* interpretable-in-principle as signifying Y."

CHAPTER 5

1. The point is not, of course, that a two-place predicate cannot express a property that is closely related to that expressed by a three-place predicate, but merely that they do not express the same property.

2. I am quite nervous about what is involved in thematizing or reifying thought in this way. It seems to me that thoughts are primarily acts, and hence the most accurate way of describing them is in the form of a verb rather than a noun. Thematizing thought makes it all too tempting to assume that there is some *thing* called the "thought" that is separable from the thinker. If we mean it is separable in the way that a shape is separable from an object or a dance from a dancer, that is fine. But it is not separable in the sense of being an isolable *part* of the act of thinking.

3. It is important to realize that this is an idealization. Change the voltage coming from your wall socket *significantly* and your computer will behave differently. Its behavior will seem like gibberish to you, but it is exhibiting a different functional architecture. The digital description of the machine treats things like voltage level as constant, and

hence is an idealization, the way gravitational laws abstract away from influence of mechanical force and electromagnetism.

4. I think it is fairly clear that these will not do as analyses of what is orindarily meant (even by professional philosophers) by 'designation' and 'interpretation'. We can designate things with which we do not enter into causal relationships. (Perhaps the clearest examples are abstract objects such as numbers.) Likewise, there are things that are paradigm cases of objects of interpretation that are not processes at all, much less processes the interpreter can carry out. Symbols in a language are an embarrassingly obvious example.

5. Although Turing's methodology is applicable to all computable functions, his examples are taken largely from problems involving computable numbers because they involve "the least cumbrous technique." (page 230) Presumably base-2 notation is also employed for reasons of simplicity.

6. Cf. the footnote on page 249, where Turing writes, "If we regard a symbol as literally printed on a square we may suppose that the square is $0 \leq x \leq 1$, $0 \leq y \leq 1$. The symbol is defined as a set of points in this square, viz. the set occupied by printer's ink."

7. Turing (1936), page 251. The insertion of the word 'human' is added here to avoid possible misunderstanding, but is clearly justified by Turing's usage throughout the article.

8. This may explain the bizarre equation of the syntactic with the non-semantic in Fodor.

CHAPTER 6

1. It is true, for example of the views expressed in Grice (1967, 1969), Austin (1962), Lewis (1969), Strawson (1964), Schiffer (1972) or Bach and Harnish (1979).

2. One could offer a theory of mind that employed another notion of "representation," (e.g., pictures, maps) but it would be a distinctly separate theory from CTM. It would have different strengths and weaknesses (e.g., it would not yield a syntactically-based account of productivity and systemmaticity), and would require a separate analysis. Even more fundamentally, however, it is not clear at the outset that there *can be* a general account of representation, because it is not clear that there is one property called "being a representation" that is common to symbols, pictures, maps, schematic diagrams, flow charts, and the other things to which the word 'representation' is applied. Any attempt to supply a "general account of representation" would have to wait for a

careful analysis of several *specific* kinds of "representation" (i.e., analysis of the uses of the word 'representation' as it is applied to apparantly distinct paradigm examples) before one could decide whether there was some feature they had in common, or whether they were called by the same name in virtue of family resemblance, or simply homonymously. This would be a worthy investigation, but stands outside the scope of this book. It is worth noting, however, that only one basic kind of general account has historically been offered. On this account, put forward by writers as diverse as Thomas Reid, Edmund Husserl, A.J. Ayer and Daniel Dennett, R is a representation of X just in case some P uses R to stand for X. One might note that this "general" account carries the same dangers of interpretive regress as the semiotic account presented in Chapter 4.

3. This criticism was raised most forcefully by Rob Cummins, who read a draft of the manuscript for this book. It was also raised by one anonymous referee of an article developing the same view.

4. Richard DeWitt (1993) makes a similar point in his "Vagueness, Semantics, and the Language of Thought."

5. One might additionally observe that there is no such thing as "the pattern" associated with, say, the letter Eta. As Hofstadter (1985) has argued, there are infinitely many patterns that can count as Etas. Moreover, what can count as an Eta *in situ* is highly context-dependent, so it will not do to simply take the whole set of patterns that can *ever* count as Etas and treat that set as constitutive of Eta-hood. This is arguably even more true with phonemes than with graphemes. (Opera-goers are quite familiar with this: on the high notes, all of the vowels tend to gravitate towards [a].)

6. It was Rob Cummins who initially made me see this point during his NEH Summer Seminar on Mental Representation in 1991. Rob was kind enough to show me that the point was already pretty clearly implicit in my analysis of symbols and syntax. But, however clear the implication might have been, it had been entirely lost upon me until pointed out. To the best of my knowledge, neither he nor anyone else has really explored the point in print. But as far as I know, the original insight was his and not mine.

7. Indeed, Davidson takes this view to the logical conclusion that differences in usage between speakers amount to a difference in language, since a language is determined by a unique mapping from expressions to interpretations. Thus it is ideolects (at particular times) that are languages in Davidson's sense. There is no such "language" corresponding to the public language *English*, since there are many variations upon English in individual idiolects.

8. The word 'model' here is, of course, ambiguous. In the terminology of set theoretic modeling, it is the interpretation of the set theoretic construction—i.e., the mathematical domain—that is called a "model". Here I am using terminology in precisely the opposite way, taking the set theoretic entity to be a "model" in the sense that one speaks of "models" in the sciences. Thanks to Sanford Shieh for alterting me to possible confusions on this point.

9. It is the problems with semantically closed languages that lead Tarski to another important conclusion: namely, that the T-equivalences for a language L and the truth theory $T(L)$ may not be articulated in L (else L would be semantically closed, hence inconsistent, hence unsusceptible to a truth definition), but must instead be articulated in a metalanguage M, which contains L as a proper subpart, but also is "essentially richer" in that it contains variables of a higher logical type. (p. 55) To this Tarski adds the following crucial point, articulated not so much as a logical necessity as a desideratum:

> It is desirable for the metalanguage not to contain any undefined terms except such as are involved explicitly or implicitly in the remarks above, i.e., terms of the object language; terms referring to the form of the expressions of the object language, and used in building names for the expressions; and terms of logic. In particular, we desire *semantic terms* (referring to the object language) *to be introduced into the metalanguage only by definition.* For, if this postulate is satisfied, the definition of truth, or any other semantic concept, will fulfill what we intuitively expect from every definition; that is, it will explain the meaning of the term being defined in terms whose meaning appears to be completely clear and unequivocal. (pp. 54-55)

CHAPTER 7

1. See Haugeland (1981), pages 28-31; Dennett, "Intentional Systems" in Dennett (1978).

2. The following account would, in fact, have to be modified to account for differences in register size or for addition overflows. This might present some problems about how the function instantiated might literally be said to be addition, but those issues will not be addressed here.

CHAPTER 8

1. Cummins (1989) makes a similar distinction between "the problem of meanings" (i.e., meaning-assignment) and the "problem of

meaningfulness."

2. I think it may be useful to treat abstract objects as a separate category from natural objects, and properties that are purely formal in nature as a separate class from natural properties. For one thing, natural and mentalistic properties can enter into systems sharing a formal description. If the abstract characterization is to be shared by two systems, one natural and one non-natural, it seems wise to treat the purely abstract characteristics as neutral between nature and the mental if one is to avoid the begging of ontological questions.

3. At least they purport to do this. It looks as though this requires a qualitative notion of "information," though, and neither writer seems to have supplied this. Information in the technical sense is a purely quantitative notion (a scalar one, at that). Unless one can cash out some notion of "patterns of information" (patterns characterizable in information-theoretic terms?), it is hard to see how an information-theoretic account can provide more than an account of fidelity of perception.

4. Here, by the way, is one of several places where one might find substantial sympathy between certain strains of cognitive science and the transcendental philosophies of Kant and Husserl. This is a matter that has not been done adequate justice by discussions of these philosophers and their relationship to cognitive science.

5. In brief the problem as it arises for causal covariation is that it stems from the force of Cartesian-style thought-experiments. It seems logically possible that there be beings who have experiences just like ours, but whose experiences are caused by malicious demons, or wicked Martian neuroscientists, or whatever. Intuitively, one should say that the content of such beings' mental states would be just what ours are, even though the content of their thoughts never corresponds to its cause. Here causal covariation does not provide even a successful demarcation criterion for meaning-assignments.

6. When I have discussed this topic with Sayre, this has been the position he has articulated on this matter.

CHAPTER 9

1. I am inclined sometimes to weaken this to the statement that "*A* is a conceptually adequate explanation of *B* just in case the conceptual content of *A*—*augmented by nothing more than purely formal (i.e., mathematical) resources*—is enough to derive the conceptual content of *B* without the addition of contingent bridge laws." To derive thermodynamic equations from statistical mechanics, for example, one

needs computational techniques that are not, strictly speaking, present in the mechanical equations themselves.

2. This notion of strong naturalization has its roots in a view of science that emerged around the beginning of the seventeenth century in the "method of resolution and composition" espoused by Galileo and adopted from him by Hobbes, and the "mechanical philosophy" championed by Descartes. Key to this approach to science is the idea that explaining something involves breaking it down into its constituent parts (the resolutive step), examining the properties of those, and then *deriving* the properties of the whole from the properties of the parts (the compositive step). Significantly, the notion of "derivation" here seems to be *geometric* rather than *logical* in origin: complex processes are "derived" from simpler ones in a fashion analogous to geometric construction rather than logical deduction. Thus Hobbes' simple objects in *De Corpore* are not simple material solids, but points in motion, from which it is possible to "derive" (i.e., to construct) first planar and then solid figures.

3. I am, however, in sympathy with Searle's (1993) argument to the effect that these states are only called *mental* by virtue of their relationship to the conscious states, which are "mental" in the first instance.

4. The reader may note that verbs of perception are systemmatically ambiguous between reports of veridical intentional states ("No, I wasn't hallucinating, I really *saw* her,") and reports of intentional character alone. ("And in my dream, I saw my dead grandmother sitting there looking at me.")

5. Some readers will perhaps note at this point I am flouting Sellars' points about the "Myth of the Given." I was never convinced by Sellars' on this issue, however, and a response to Sellars seemed to wide a detour to include it in this book.

6. Phenomenological content may *determine* broad content partially, if the phenomenological content involves a rule like "'P' means the stuff, whatever it is, that causes experiences like *this*."

7. Richard DeWitt suggested in response to a draft of this chapter that not all presentations have a phenomenology, but rather only the conscious ones do. There is surely an important distinction here between the elements (and episodes) of visual experience that are consciously accentuated and those that are not. I think there can probably even be episodes of *vision* that are truly non-conscious. But I have stipulatively reserved my intentional modalities such as VISUAL-PRESENTATION for states that are conscious. I should rather express DeWitt's point by saying that within conscious perceptions of, say, a room full of objects, some aspects are given greater *attention* than oth-

ers, and this "turns up the gain", as it were, on their phenomenology. I should say that features of a scene that are not focal but are nonetheless truly perceived do have a phenomenology, but that it is very unobtrusive when not attended to.

8. In fairness, Descartes does point out elsewhere the extent to which our knowledge of our own minds is fallible. (*Cf. Principles* I.67 (AT VIIIA.32-33)) Descartes does not think that all introspection is incorrible; merely that we can *sometimes* have clear and distinct knowledge of our own mental states, and in these instances we cannot be mistaken. My point here will once again ride roughshod over the objections of writers like Sellars (1956) and Garfield (1988), who dispute this kind of incorrigibility.

9. Cf Husserl, *Ideas*, §§137, 145. Kant says things that can be interpreted in a similar manner in the A version of the Transcendental Deduction of the Categories: *KRV*, A 108.

10. On this view it is in fact likely that our "discourse about the mental" will itself be a distortion of its subject-matter. For discourse is linguistic, and language "thematizes" its subject-matter—it treats it as an object. Thus Husserl points out that thematizing the self and thematizing thoughts distorts them and Kant points out that we know the transcendental ego only through the transcendental unity of apperception and cannot know it as noumenon except through the postulates of practical reasoning.

11. It is, of course, possible to simply *ignore* intentional content or deny the phenomenological side of intentionality. I take it the Millikan's analysis, for example, which has the conclusion that the intentionality of mental states is radically dependent upon history, could be seen simply as an account of something other than intentional character, and an account which has nothing to say about intentional character.

12. I think this line of argument finds kindred spirits in Baker (1987) and Garfield (1988).

13. Again, a somewhat parallel argument is to be found in Garfield (1988).

CHAPTER 10

1. I am also inclined to believe that one important role played by metaphor is to lead towards mathematization. Often, what is crucial about a successful metaphor is that the source domain of the metaphor has a formal description that the target domain shares.

2. Thomas Kuhn writes in a similar vein that "unlike Kepler's Laws, which are the astronomical culmination of the Copernican Revolution, the Newtonian universe is a product of more than Copernicus' innovation....Our problem now becomes larger than the Copernican Revolution proper." *The Copernican Revolution* (Harvard U. Press, 1957), page 231.

3. It is also worth noting the influence of Neoplatonism in Copernicus reasoning. A great portion of his argumentation, and also Kepler's, draws heavily on the Neoplatonic tradition.

4. Johannes Kepler, *On the Motions of Mars*. (Prague, 1609).

5. This result is first achieved in Kepler's Third Law, which relates the orbital velocities of the planets in different orbits. This law was published ten years later than the first two, in *Harmonies of the World* (1619).

6. A similar point can be made about other theorists of this era. Kepler, for example, attempted to develop an explanation of planetary motion in terms of a combination of magnetism and another force called the *anima motrix*. As Kuhn writes, "Few of Kepler's successors took his physical theory...as seriously as they took his mathematical description of the planetary orbits." (1957: 246) Kuhn also explores the mechanisms discussed by Borelli and Hooke, which likewise contribute little to the understanding of planetary orbits.

7. Einstein's picture of space/time *might* be viewed as explaining gravitation in terms of curvature of space/time, but now it is that curvature that is fundamental and unexplained.

8. One sees this perhaps most clearly in the notable gap between the expressed desire of writers like Hobbes and Leibniz to give a precise "calculus" of thought and the somewhat loose characterization of relationships between kinds of mental states that Hobbes gives in *Leviathan*.

9. Descartes quite explicitly treats the body as a machine in *Dioptrics*, *Treatise on Man*, *On the Human Body*, *The Passions of the Soul* and Book V of the *Discourse on Method*. In this last work, he argues that humans are distinguished from animals by the fact that they have two capacities that cannot be duplicated by mechanical means: namely, language and general reasoning. (AT VI.56—57)

10. Much of subsequent Continental philosophy has emerged precisely from disagreements that writers like Merleau-Ponty, Sartre and Derrida had with Husserl's account of intentionality, and so Husserl's work on that subject is perhaps the most important background reading for studying Continental philosophy, in addition to its intrinsic interest for the student of intentionality.

11. Note that the issue here is one framed wholly in terms of the relations between the intentional characters of different intentional states, and not their veridicality. The nature of the modality RECOLLECTION also implies additional felicity conditions that regard veridicality as well: to be a completely felicitous recollection of a perceptual experience of Y, it is not enough that the experience be founded on a previous perceptual gestalt; it mus must also be the case that that perceptual gestalt was in the right relationship to Y to be a successful seeing of Y.

CHAPTER 11

1. This view is probably not universal, but it is widely held. I asked Fodor about this explicitly at a conference at SUNY Buffalo in 1992, and he said that he would give up intentional realism if it were incompatible with materialism. On the other hand, Dretske said at a conference at VPI in 1994 that he would give up materialism under the same conditions.

2. Part of the qualification that it provides only "one of" the first chances stems from the fact that the mathematical machinery involved in neural network approaches was growing apace with "orthodox" computation.

3. The main reason I think such a model would have to be extremely complex is that in cognition there seem almost always to be many mutually dependent variables. If we are dealing at the level of intentional psychology, one's decisions are made against an enormous background of tacit assumptions and beliefs and desires that may never emerge as occurrent judgements and wishes. If we are dealing at the level of, say, perceptual modeling, the various parts of the brain that are implicated in perceptual processing are highly-interconnected, and seem to exhibit significant feed-forward *and* feed-back relations. To get even a minimally decent model that duplicates, say, visual performance with respect to subjective contour features, you need models of several individual modules with very particular architectures and also need a model of how they inter-relate. There seem to be more levels of complexity and interaction here than in, say, the way thermodynamic phenomena are related to statistical calculations over mechanical interactions of gas particles.

4. Indeed, it occurs to me that I can rightly be said to "have $10,000" even if I possess no currency. Much of our "possession" of money is realized through representations in bank computers. My having some sum of money "in the bank" is realized through a particular binary pat-

tern being instantiated in my bank's computer!

5. While these views are loosely inspired by Quine, I should be careful in how I attributed any of them to him personally. Quine is a subtle man.

6. Macintyre, Alisdair: "Ontology." *The Encyclopaedia of Philosophy*.

7. In actual practice, of course, there will be trade-offs in both directions. Even if the Ptolemaic system had perfect predictive success and a viable model of how the celestial speheres behaved, the inelegance of the hodge-podge of epicycles might lead one to doubt the truth of the theory.

APPENDIX A

1. This is not to say, of course, that no interpretation is involved in any particular act of understanding. In some sense some kind of interpretation is always involved in cognitive processes. The distinction here, though, is between objects of types which depend essentially upon conventions and objects of types which do not depend upon conventions. *Dog*, for example, is a natural kind. Any thing that is a dog would be a dog even if there were no conventions or languages or language users. Similarly, the freezing point of water would still be thirty-two degrees Farenheit even if that scale of measurement had never been adopted, in the sense that *our* expression "thirty-two degrees Farenheit" picks out a temperature, and the properties water has at that temperature are not dependent upon conventions, even though there are necessarily conventions involved in the use of that expression or any other expression. By contrast, an object can only be a US dollar bill if there there are conventions to the effect that there is a political entity called the United States, that it is governed in a certain fashion, etc. Functional descriptions are free of convention in the sense that they pick out properties that a system would have regardless of what conventions might be in force.

2. Of course the circuit's state changes are not really instantaneous, and there are bound to be minor deviations from the canonical description of a given circuit, such as minor variations in voltage. Physical descriptions usually do involve some abstraction and some idealization. But while the description of a bistable circuit in digital terms may involve abstraction and idealization, it is neither fictitious nor convention-dependent. On the one hand, the description does pick out properties the circuit really has, even if there are purposes for which the de-

scription is not adequately fine-grained. On the other hand, the description of the bistable circuit in digital terms is precisely the right description if one wishes to discuss how the circuit interacts with the rest of a computer system whose state changes can depend on which output lead is at the higher voltage level, but never depend upon the transient intermediate states the circuit may undergo or upon minor variations in voltage. For the designers of computer systems, of course, it may be quite a significant practical problem to get the CPU of the machine to sample storage units only when they are in stable states and to adjust the tolerances of various components so that the deviations from true digitalness really do not matter. And of course a defective bistable circuit may fail to meet the digital description for circuits of its type and fail to interact properly with the rest of the machine. Descriptions of the sort given above are given under the presumptions that the components are not defective and that the system is properly designed.

3. In some cases, the same set of patterns may be used in multiple coding and representational schemes in a single machine. One might, for example, have a machine which uses sixteen bit locations both for the representation of integer values and for the encoding of graphemic characters. As far as the semiotics of the situation goes, the definitions in Chapter 4 provide for two ways this could be dealt with. One way is to have two sets of markers which have the same criteria but are employed in different language games. The other way is to have one set of markers to which different reprsentation and coding schemes can be applied. Which way the conventions actually work is a question of *fact*, since conventions are historical entities.

4. One could choose to regard markers stored in ASCII code as *representations* of the graphemic characters with which they are associated, but there seems little reason for doing this.

5. The expression 'coding scheme' is generally employed by computer scientists for both types of convention. Thus one will find computer texts speaking of a "coding scheme" for representing the integers, where here the same convention will be called a representation scheme. In coding theory, however, the expression 'code' is used quite explicitly to denote a mapping between sets of *symbols*. (See, e.g., Chapter 1 Abramson (1963)) The differences between representation and coding may not be important for computer scientists or coding theorists, but this investigation is concerned with symbols and representation, and thus the distinction between conventions which associate markers with interpretations and those which associate them with other markers is quite relevant.

6. Hexadecimal notation employs the letters 'A' through 'F' for units

of ten through fifteen, respectively, and hexadecimal strings are conventionally flagged as such by the prefix of the dollar sign.

7. Most computers use storage locations with more than seven binary storage locations to store characters. In such cases, only seven locations are relevant to the marker typing which serves underlies the application of the ASCII convention.

8 Some of the code markers in conventions like ASCII, however, do not correspond to graphemes, but serve as markers for the ends of lines, for tabulation, etc. The documents produced by your word processor as files on disk do not contain just items corresponding to the letters you type, but also symbols (not normally displayed on your screen) that encode font and style information in formats such as the RTF format.

9. It would be possible to think of the use of graphemes and computer markers as notational variants for G, so far as the formal aspects of the language game are concerned. It is less clear that this would be permissible in talking about the pragmatic aspects of the language game and the encoding process.

10. Or at least interpretable-in-principle. There may well be computer functions whose operations were never considered in syntactic terms by the designers.

Bibliography

Abramson, Norman. 1963. *Information Theory and Coding.* New York: McGraw-Hill.

Anderson, James. 1973. "A Theory for the Recognition of Items from Short Memorized Lists." *Psychological Review* 80: 417-438.

Austin, J. L. 1962. *How to Do Things with Words.* Cambridge, Mass.: Harvard University Press.

Bach, Kent, and Robert Harnish. 1979. *Linguistic Communication and Speech Acts.* Cambridge, Mass.: MIT Press.

Baker, Lynne Rudder. 1987. *Saving Belief: A Critique of Physicalism.* Princeton: Princeton University Press.

Blackburn, Simon. 1984. *Spreading the Word: Groundings in the Philosophy of Language.* New York: Oxford University Press.

Block, Ned. 1978. "Troubles with Functionalism." In *Perception and Cognition: Issues in the Foundations of Psychology,* ed. C. W. Savage. Minnesota Studies in the Philosophy of Science, vol. 9. Minneapolis: University of Minnesota Press.

————, ed. 1980. *Readings in the Philosophy of Psychology.* 2 vols. Cambridge, Mass.: Harvard University Press.

Block, Ned, and Jerry A. Fodor. 1972. "What Psychological States Are Not." *Philosophical Review* 81(2): 159-181. Reprinted in Block 1980.

Brentano, Franz. 1874. *Psychologie yore empirischen Standpunkt.* Leipzig: Duncker and Humblot. Published in English as *Psychology from an Empirical Standpoint,* ed. Linda L. McAlister, trans. D. B.

Terrell, Antos C. Rancurello, and Linda L. McAlister (London: Routledge and Kegan Paul, 1973).

Burge, Tyler. 1979. "Individualism and the Mental." In *Studies in Epistemology,* ed. P. French, T. Uehling, and H. Wettstein. Midwest Studies in Philosophy, vol. 5. Minneapolis: University of Minnesota Press.

———. 1986. "Individualism and Psychology." *Philosophical Review* 95: 1, 3-45.

Carnap, Rudolph. 1928. *Der Logische Aufbau der Welt.* Berlin: Weltkreis.

———. 1936-1937. "Testability and Meaning." *Philosophy of Science* 3: 419-471; 4: 1-14. Reprinted in Feigl and Brodbeck 1953.

———. 1938. "Logical Foundations of the Unity of Science." In *International Encyclopedia of Unified Science,* vol. 1, no. 1, ed. Otto Neurath. Chicago: University of Chicago Press. Reprinted in Marras 1972.

Chisholm, Roderick. 1957. *Perceiving: A Philosophical Study.* Ithaca: Cornell University Press.

———. 1968. "Brentano's Descriptive Psychology." In *Proceedings of the Fourteenth International Congress of Philosophy, 1-9 September 1968, Vienna,* vol. 2. Reprinted in revised form in McAlister 1976.

———. 1983. "Belief as an Intentional Concept." In *On Believing: Epistemological and Semiotic Approaches,* ed. Herman Parret. Berlin: W. de Gruyter.

———. 1984a. "What Is the Problem of Objective Reference?" *Dialectica* 38: 131-142.

———. 1984b. "The Primacy of Intentionality." *Synthese* 61: 89-110.

Colby, K. 1975. *Artificial Paranoia.* New York: Pergamon.

Cummins, Robert. 1975. "Functional Analysis." *Journal of Philosophy* 72: 741-760.

———. 1983. *The Nature of Psychological Explanation.* Cambridge, Mass.: MIT Press.

———. 1989. *Meaning and Mental Representation.* Cambridge, Mass.: MIT Press.

Dennett, Daniel. 1976. "A Cure for the Common Code?" In Dennett, *Brainstorms.* Cambridge, Mass.: MIT Press.

———. 1978. *Brainstorms.* Cambridge, Mass.: MIT Press.

———. 1987. *The Intentional Stance.* Cambridge, Mass.: MIT Press.

Descartes, René 1985. *The Philosophical Writings of Descartes.* Trans. J. Cottingham, R. Stoothoff, and D. Murdoch. 3 vols. Cambridge: Cambridge University Press.

DeWitt, Richard. 1993. "Vagueness, Semantics, and the Language of Thought." *Psyche* 1(1): 1-19.

Dretske, Fred. 1981. *Knowledge and the Flow of Information.* Cambridge, Mass.: MIT Press.

———. 1988. *Explaining Behavior: Reasons in a World of Causes.* Cambridge, Mass.: MIT Press.

Dreyfus, Hubert. 1972. *What Computers Can't Do.* New York: Harper and Row. (Rev. ed., 1979.)

———. 1992. *What Computers Still Can't Do.* Cambridge, Mass.: MIT Press.

Dreyfus, Hubert, and Harrison Hall, eds6 1984. *Husserl, Intentionality and Cognitive Science.* Cambridge, Mass.: MIT Press.

Edwards, Paul, ed. 1967. *The Encyclopedia of Philosophy.* New York: Macmillan.

Fechner, G. T. 1877. *In Sachen der Psychophysik.* Leipzig: Breitkopf and Härtel.

———. 1882. *Revision der Hauptpunkte der Psychophysik.* Leipzig: Breitkopf and Härtel.

Feigl, Herbert. 1958. "The 'Mental' and the 'Physical'." In *Concepts, Theories and the Mind-Body Problem,* ed. Herbert Feigl, Michael Scriven, and Grover Maxwell. Minnesota Studies in the Philosophy of Science, vol. 2. Minneapolis: University of Minnesota Press.

Feigl, Herbert, and M. Brodbeck, eds. 1953. *Readings in the Philosophy of Science.* New York: Appleton-Century-Crofts.

Feigl, Herbert, Michael Scriven, and Grover Maxwell, eds. 1958. *Concepts, Theories and the Mind-Body Problem.* Minnesota Studies in the Philosophy of Science, vol. 2. Minneapolis: University of Minnesota Press.

Feyerabend, Paul. 1958. "Explanation, Reduction, and Empiricism." In *Concepts, Theories and the Mind-Body Problem,* ed. Herbert Feigl, Michael Scriven, and Grover Maxwell. Minnesota Studies in the Philosophy of Science, vol. 2. Minneapolis: University of Minnesota Press.

Field, Hartry. 1972. "Tarski's Theory of Truth." *Journal of Philosophy* 59: 347-375.

Fodor, Jerrold. 1965. "Explanations in Psychology." In *Philosophy in America,* ed. M. Black. London: Routledge and Kegan Paul.

———. 1968. "The Appeal to Tacit Knowledge in Psychological Explanation." *Journal of Philosophy* 65(20): 627-640.

———.1975. *The Language of Thought.* New York: Thomas Crowell.

———. 1980a. "Methodological Solipsism Considered as a Research Strategy in Cognitive Science." *Behavioral and Brain Sciences* 3:

63-73.

————. 1980b. "On What Only Brains Can Do." *Behavioral and Brain Sciences* 3: 4-31.

————. 1981. *RePresentations*. Cambridge, Mass.: Bradford Books/MIT Press.

————. 1982. "The Myth of the Computer." *New York Review of Books* 29(7): 3-6.

————. 1987. *Psychosemantics*. Cambridge, Mass.: Bradford Books.

————. 1990. *A Theory of Content and Other Essays*. Cambridge, Mass.: Bradford Books.

Fraser, Bruce. 1981. "Hedged Performatives." In *Syntax and Semantics III: Speech Acts,* ed. P. Cole and J. Morgan. White Plains, N.Y.: Longman.

Garfield, Jay. 1988. *Belief in Psychology: A Study in the Ontology of Mind.* Cambridge, Mass.: MIT Press.

Geach, Peter Thomas. 1957. *Mental Acts*. London: Routledge and Kegan Paul; New York: Humanities Press.

Goldman, Alvin I. 1989. "Interpretation Psychologized." *Mind and Language* 4: 161-185.

————. 1992. "In Defense of the Simulation Theory." *Mind and Language 7(1-2): 104-119.*

————. 1993a. "Consciousness, Folk Psychology, and Cognitive Science." Consciousness and Cognition: An International Journal *2(4): 364-382.*

————. 1993b. "The Psychology of Folk Psychology." *Behavioral and Brain Sciences* 16(1): 15-28, 29-113.

Grice, Paul. 1957. "Meaning." *Philosophical Review* 66: 377-388.

————. 1969. "Utterer's Meaning and Intentions." *Philosophical Review* 78: 147-177.

Grossberg, Stephen. 1980. "How Does the Brain Build a Cognitive Code?" *Psychological Review* 87: 1-51.

————. 1982. *Studies of Mind and Brain*. Boston Studies in the Philosophy of Science, vol. 70. Dordrecht, The Netherlands: D. Reidel.

Haugeland, John. 1978. "The Nature and Plausibility of Cognitivism." *Behavioral and Brain Sciences* 2: 215-226.

————. 1985. *Artificial Intelligence: The Very Idea*. Cambridge, Mass.: MIT Press.

————, ed. 1981. *Mind Design*. Cambridge, Mass.: MIT Press/Bradford Books.

Hempel, Carl. 1949. "The Logical Analysis of Psychology." In *Readings in Philosophical Analysis,* ed. Herbert Feigl and Wilfred Sellars. New York: Appleton-Century-Crofts, 1949. Reprinted in

Block 1980.

Hilbert, David. 1899. "Grundlagen der Geometrie." In *Festschrift zur Feier der Enthüllung des Gauss-Weber-Dekmals in Göttingen.* Leipzig.

Hofstadter, Douglas R. 1985. *Metamagical Themas.* New York: Basic Books.

Horst, Steven. 1990. "Symbols, Computation, and Intentionality." Ph.D. dissertation, University of Notre Dame.

———. 1992. "Notions of 'Representation' and the Diverging Interests of Philosophy and Empirical Research." Proceedings of the Conference on Cognition and Representation, State University of New York, Buffalo, April 3-5. Spiral-bound photocopy.

———. 1995. "Eliminativism and the Ambiguity of 'Belief'." *Synthese* 104(1). Hull, Clark. 1943. *Principles of Behavior.* New York: Appleton-Century-Crofts.

Husserl, Edmund. 1900. *Logische Untersuchungen.* Halle a.d. S.: M. Niemeyer. Published in English as *Logical Investigations,* trans. J. N. Findlay (New York: Humanities Press, 1970).

———.1913. *Ideen zu einer reinen Phänomenologie und phänomenologischen Philosophie.* The Hague: Martinus Nijhoff. Published in English as *Ideas: General Introduction to Pure Phenomenology,* trans. W. R. Boyce Gibson (New York: Humanities Press, 1931).

———.1950. *Cartesianische Meditationen und Pariser Vorträge.* Ed. S. Strasser. The Hague: Martinus Nijhoff. Published in English as *Cartesian Meditations,* trans. Dorian Cairns (The Hague: Martinus Nijhoff, 1960).

———. 1954. *Die Krisis der europäischen Wissenschaften und die transzendentale Phänomenologie.* Ed. Walter Biemel. The Hague: Martinus Nijhoff. Published in English as *The Crisis of European Sciences and Transcendental Phenomenology,* trans. David Carr (Evanston, Ill.: Northwestern University Press, 1970).

Jackson, Frank. 1982. "Epiphenomenal Qualia." *Philosophical Quarterly* 32: 127-136.

Kanizsa, Gaetano. 1976. "Subjective Contours." *Scientific American* 234(4): 48-52.

———. 1979. *Organization in Vision.* New York: Praeger.

Kant, Immanuel. 1968. *Critique of Pure Reason.* Trans. Norman Kemp Smith. London: Macmillan; New York: St. Martin's Press.

Katz, Jerrold J. 1977. *Propositional Structure and Illocutionary Force.* New York: Crowell.

Kripke, Saul. 1971. "Naming and Necessity." In *Semantics of Natural Language,* ed. D. Davidson and G. Harman. Dordrecht: D. Reidel.

Kuhn, Thomas. 1957. *The Copernican Revolution*. Cambridge, Mass.: Harvard University Press.

Lashley, K. S. 1929. *Brain Mechanisms and Intelligence*. New York: Dover.

Lehrer, Keith. 1989. "Conception without Representation— Justification without Inference: Reid's Theory." *Noûs* 23(2): 145-154.

Levine, Daniel. 1991. *Introduction to Neural and Cognitive Modeling*. Hillsdale, N.J.: Lawrence Erlbaum.

Lewis, David. 1969. *Convention: A Philosophical Study*. Cambridge, Mass.: Harvard University Press.

Loar, Brian. 1982. "Conceptual Role and Truth Conditions." *Notre Dame Journal of Formal Logic* 23: 272-283.

————. 1985. "Social Content and Psychological Content." In *Contents of Thought: Proceedings of the 1985 Oberlin Colloquium in Philosophy*. Tucson: University of Arizona Press.

Macintyre, Alisdair. 1967. "Ontology." In *The Encyclopedia of Philosophy*, ed. Paul Edwards, vol. 5. New York: Macmillan.

Marr, David. 1982. *Vision: A Computational Investigation into the Human Representation and Processing of Visual Information*. New York: W. H. Freeman.

Marris, Ausonio, ed. 1972. *Intentionality, Mind and Language*. Urbana: University of Illinois Press.

Martinich, A. P., ed. 1985. *The Philosophy of Language*. New York: Oxford University Press. (2d ed., 1990.)

McAlister, Linda. 1974. "Chisholm and Brentano on Intentionality." *Review of Metaphysics* 28(2). Reprinted in McAlister 1976.

————. 1976. *The Philosophy of Brentano*. London: Gerald Duckworth.

McCawley, J. 1973. "Remarks on the Lexicography of Performative Verbs." Proceedings of the Texas Performadillo Conference, University of Texas at Austin.

McCulloch, Warren, and Walter Pitts. 1943. "A Logical Calculus of the Ideas Immanent in Nervous Activity." *Bulletin of Mathematical Biophysics* 5: 115-133.

Millikan, Ruth Garrett. 1984. *Language, Thought and Other Biological Categories*. Cambridge, Mass.: MIT Press.

Nagel, Thomas. 1974. "What Is It Like to Be a Bat?" *Philosophical Review* 83: 435-450.

————. 1986. *The View from Nowhere*. New York: Oxford University Press.

Newell, Allen. 1986. "The Symbol Level and the Knowledge Level."

In *Meaning and Cognitive Structure,* ed. Zenon Pylyshyn and William Demopoulos. Norwood, N.J.: Ablex.

Newell, Allen, and Herbert Simon. 1975. "Computer Science as Empirical Inquiry." Presented as their 1975 Turing Lecture. Reprinted in Haugeland 1981.

Newton-Smith, W. H. 1981. *The Rationality of Science.* Boston: Routledge and Kegan Paul.

Oppenheim, Paul, and Hilary Putnam. 1958. "The Unity of Science as a Working Hypothesis." In *Concepts, Theories and the Mind-Body Problem,* ed. Herbert Feigl, Michael Scriven, and Grover Maxwell, 3-36. Minnesota Studies in the Philosophy of Science, vol. 2. Minneapolis: University of Minnesota Press.

Popper, Karl. 1959. *The Logic of Scientific Discovery.* New York: Basic Books.

Putnam, Hilary. 1960. "Minds and Machines." In *Dimensions of Mind,* ed. S. Hook. New York: New York University Press.

———. 1961. "Brains and Behavior." Read as part of the program of the American Association for the Advancement of Science, Section L (History and Philosophy of Science), 27 December. Reprinted in Block 1980.

———. 1967. "The Nature of Mental States." In *Art, Mind and Religion,* ed. W. H. Capitan and D. D. Merrill. Pittsburgh: University of Pittsburgh Press. Reprinted in Block 1980.

———. 1975. "The Meaning of 'Meaning'." In *Language, Mind and Knowledge,* ed. Keith Gunderson. Minnesota Studies in the Philosophy of Science, vol. 7. Minneapolis: University of Minnesota Press.

———. 1983. "Computational Psychology and Interpretation Theory." In *Philosophical Papers: Vol. 3, Realism and Reason.* London: Cambridge University Press.

Pylyshyn, Zenon. 1980. "Computation and Cognition: Issues in the Foundation of Cognitive Science." *Behavioral and Brain Sciences* 3: 111-132.

———. 1984. *Computation and Cognition: Toward a Foundation for Cognitive Science.* Cambridge, Mass.: Bradford Books/MIT Press.

Pylyshyn, Zenon, and William Demopoulos, eds. 1986. *Meaning and Cognitive Structure.* Norwood, N.J.: Ablex.

Reid, Thomas. 1983. *Thomas Reid's Inquiry and Essays.* Ed. Ronald E. Beanblossom and Keith Lehrer. Indianapolis: Bobbs-Merrill.

Ricoeur, Paul. 1970. *Freud and Philosophy: An Essay on Interpretation.* Trans. Denis Savage. New Haven: Yale University Press.

Rumelhart, David E., James McClelland, and the PDP Research Group. 1986. *Parallel Distributed Processing: Explorations in the Micro-*

structure of Cognition. Cambridge, Mass.: MIT Press.

Ryle, Gilbert. 1949. *The Concept of Mind*. New York: Barnes and Noble.

Sayre, Kenneth. 1969. *Consciousness: A Philosophic Study of Minds and Machines*. New York: Random House.

———. 1976. *Cybernetics and the Philosophy of Mind*. Atlantic Highlands, N.J.: Routledge and Kegan Paul.

———.1986. "Intentionality and Information Processing: An Alternative Model for Cognitive Science." *Behavioral and Brain Sciences* 9(1): 121-138.

———. 1987. "Cognitive Science and the Problem of Semantic Content." *Synthese* 70: 247-269.

Schiffer, Stephen R. 1972. *Meaning*. Oxford: Clarendon Press.

Searle, John. 1969. *Speech Acts*. London: Cambridge University Press.

———. 1971. "A Classification of Illocutionary Acts." *Language in Society* 5: 1-23.

———. 1980. "Minds, Brains and Programs." *Behavioral and Brain Sciences* 3: 417-424.

———. 1983. *Intentionality. An Essay in the Philosophy of Mind*. London: Cambridge University Press.

———.1984. *Minds, Brains and Science*. Cambridge, Mass.: Harvard University Press.

———. 1990. "Is the Brain a Digital Computer?" (APA Presidential Address). *Proceedings and Addresses of the American Philosophical Association* 64(3): 21-37.

———. 1992. *The Rediscovery of the Mind*. Cambridge, Mass.: MIT Press.

Sellars, Wilfrid. 1956. *The Foundations of Science and the Concepts of Psychology and Psychoanalysis,* ed. H. Feigl and M. Scriven. "Empiricism and the Philosophy of Mind." In Minnesota Studies in the Philosophy of Science, vol. 1. Minneapolis: University of Minnesota Press.

Shannon, Claude, and W. Weaver. 1949. *The Mathematical Theory of Communication*. Urbana: University of Illinois Press.

Skinner, B. F, 1938. *The Behavior of Organisms*. New York: D. Appleton-Century.

———. 1953. *Science and Human Behavior*. New York: Macmillan.

Smolensky, Paul. 1988. "The Proper Treatment of Connectionism." *Behavioral and Brain Sciences* 11(1): 1-74.

Stevens, Stanley Smith. 1951. *Handbook of Experimental Psychology*. New York: Wiley.

———. 1975. *Psychophysics: Introduction to Its Perceptual, Neural and Social Prospects.* Ed. G. Stevens. New York: Wiley.

Stich, Stephen. 1983. *From Folk Psychology to Cognitive Science.* Cambridge, Mass.: MIT Press/Bradford Books.

Strawson, P. F. 1964. "Intention and Convention in Speech Acts." *Philosophical Review* 73: 439-460.

Sur, Mriganka, Preston E. Garraghty, and Anna W. Roe. 1988. "Experimentally Induced Visual Projections into Auditory Thalmus and Cortex. (Ferrets)." *Science,* no. 242: 1437-1441.

Tarski, Alfred. 1944. "The Semantic Conception of Truth and the Foundations of Semantics." *Philosophy and Phenomenological Research* 4: 341-375. Reprinted in Martinich 1985.

―――. 1956a. *Logic, Semantics, Metamathematics.* London: Oxford University Press. (2d ed.: Hackett Publishing, 1983.)

―――. 1956b. "The Concept of Truth in Formalized Languages." In Tarski, *Logic, Semantics, Metamathematics.* London: Oxford University Press. Reprinted in Martinich 1985.

―――. 1956c. "The Establishment of Scientific Semantics." In Tarski, *Logic, Semantics, Metamathematics.* London: Oxford University Press.

Todorovic, D. 1987. "The Craik-O'Brien-Cornsweet Effect: New Varieties and Their Theoretical Implications." *Perception and Psychophysics* 42: 545-560.

Tolman, Edward Chase. 1932. *Purposive Behavior in Animals and Men.* New York: Century Company.

Turing, Alan. 1936. "On Computable Numbers." *Proceedings of the London Mathematical Society* 24: 230-265.

Van Inwagen, Peter. 1990. *Material Beings.* Ithaca: Cornell University Press.

―――. 1993. *Metaphysics.* Boulder: Westview Press.

Vendler, Zeno. 1972. *Res Cogitans.* Ithaca: Cornell University Press.

Von Eckardt, Barbara. 1993. *What Is Cognitive Science?* Cambridge, Mass.: MIT Press.

Wang, Hao. 1963. *A Survey of Mathematical Logic.* Peking: Science Press.

Watson, John B. 1913a. "Image and Affection in Behavior." *Journal of Philosophical Psychology and Scientific Methods* 10: 421-428.

―――. 1913b. "Psychology as the Behaviorist Views It." *Psychological Review* 20: 158-177.

―――. 1914. *Behavior: An Introduction to Comparative Psychology.* New York: Holt.

―――. 1924. *Behaviorism.* New York: Norton.

Weber, Max. 1959. *The Methodology of the Social Sciences.* New

York: Free Press.

―――. 1964. *The Theory of Social and Economic Organization*. New York: Free Press.

Weizenbaum, Joseph. 1976. *Computer Power and Human Reason: From Judgement to Calculation*. New York: W. H. Freeman.

Winograd, Terry, and Fernando Flores. 1986. *Understanding Computers and Cognition*. Norwood, N.J.: Ablex.

Wittgenstein, Ludwig. 1958. *Philosophical Investigations*. Trans. G. E. M. Anscombe. 3d ed. New York: Macmillan.

―――. 1961. *Tractatus Logico-Philosophicus*. Trans. D. F. Pears and B. F. McGuinness. New York: Humanities Press.

Index

www.ingramcontent.com/pod-product-compliance
Lightning Source LLC
Chambersburg PA
CBHW071356050326
40689CB00010B/1662